Y0-DCC-120

A Primer on Environmental Decision-Making

A Primer on Environmental Decision-Making

An Integrative Quantitative Approach

by

Knut Lehre Seip
Department of Engineering,
Oslo University College,
Oslo, Norway

and

Fred Wenstøp
Department of Strategy and Logistics,
BI Norwegian School of Management,
Oslo, Norway

A C.I.P. Catalogue record for this book is available from the Library of Congress.

ISBN-10 1-4020-4073-3 (HB)
ISBN-13 978-1-4020-4073-3 (HB)
ISBN-10 1-4020-5067-4 (e-book)
ISBN-13 978-1-4020-5067-1 (e-book)

Published by Springer,
P.O. Box 17, 3300 AA Dordrecht, The Netherlands.

www.springer.com

Printed on acid-free paper

TABLE OF CONTENTS

PREFACE

Inspiration: We became inspired to write a book like this way back in the 1980s when we worked together in an oil pollution prevention program, charged with recommending a national optimal level of preparedness against oil spills. One of us was a natural scientist working in biology and chemistry, the other a managerial scientist working in decision analysis. We realized quickly that the combination could be fruitful. We saw how a structured approach can make a project move forward smoothly and efficiently. We realized that the key is to first clarify values, and then to determine how they should be measured in terms of decision criteria, the field of the decision analyst. Then follows the consequence analysis, which is the arena of natural scientists and economists. At the end, however, somebody has to actually make decisions, and that implies weighting the criteria. We realised that this is intrinsically a question of subjective preferences, where regard for the wishes of stakeholders is important. This book is an attempt to bring these insights together, in a comprehensive framework that draws on knowledge from the natural sciences, economics as well as management.

The book is based on experience from environmental and business projects, as well as studies in preparation for lectures on decision analyses. One of us gives lectures mainly with a business orientation (FW), and the other with an environmental orientation (KLS). The lectures have been given mostly at the master's level, most often with a strong emphasis on project work. We have also given guest lectures in European and Asian countries. It has always been an inspiration to discover new environmental challenges and other preferences that mirror the living conditions in the countries we visit. In particular, having students from different cultures in the same project groups has been enchanting for us, as we believe it has been for the students.

The book's personality: We present first a framework for decision-making, and then elaborate on the different subjects in subsequent chapters. The intention is always to provide readers with tools for approaching real problems. The challenge is to learn how to screen environmental questions for validity, solvability and methodological approach. This requires a certain level of superficiality at the outset, but within constraints. At its best, it separates what is important from what is peripheral and subject to budgetary restraints.

Quantitative models: Environmental sciences have developed an abundance of dose-response relations, and we have presented a selection of particularly simple models that have proven useful. Usually there are more extensive models that could do a better job, but they do not have the same pedagogical merit.

Thank you: Among the people that have read the whole document, or bits and pieces and who gave us valuable advice and criticism, we would like to thank: Rune Baklien, Mounir Banoub, Marek Biesiada, Birgit Cold, Edward J.W. Chikhwenda, Sylvain Doledec, Jadwiga Gzyl, Chris Johnson, Gordon H. Orians, Hans Martin Seip, Val Smith, Pierre Sagnes. Bernard Stazner, Elizabeth Skjelsvik, Kouadio Tano and last, but not least, our students who learned that two things are important for good learning: enthusiasm and the feeling of contributing new knowledge to an important part of society.

Web Page: On the web pages of the book we have included additional basic equations that have been developed in ecological and environmental studies. The authors welcome any suggestions for new equations that can replace or supplement the ones that we have selected. We also welcome new applications to supplement our own selection. Please contact Knut Lehre seip via email: knut.lehre.seip@iu.hio.no or Fred Wenstøp via e-mail Fred.Wenstop@bi.no

Part 1

ENVIRONMENTAL DECISION MAKING

Overview

Chapters

1. INTRODUCTION.
2. DECISION-MAKING CONCEPTS: Introduction of basic concepts in environmental decision-making.
3. GETTING STARTED: A step by step outline of the rational decision analysis process.
4. INTEGRATED ASSESSMENT: Comprehensive analysis of environmental impacts of projects.
5. ECONOMICS: A brief introduction to environmental resource economics.
6. GAME THEORY: With applications to the management of common resources.
7. PREFERENCES: Elicitation of decision-maker values.
8. WILLINGNESS TO PAY: The value of environmental resources and services.

This part of the book deals with basic concepts of environmental decision-making and economics, and provides a framework and a structure for analysis of environmental management-problems. Chapter 3, especially, is a comprehensive introduction to how you may proceed if you want to perform your own analysis, serving as a reference for the rest of the book.

Chapter 1

INTRODUCTION

"It is our choices, Harry, that shows what we truly are, far more than our abilities" Dumbeldore to Harry in J.K. Rowling: *Harry Potter and the Chamber of Secrets.*

1. A MULTIDISCIPLINARY TEXT

Ecology and environmental sciences are quantitative disciplines. Also, making decisions involving the environment requires that the decision maker makes value judgements on behalf of others. This book brings together material from three major knowledge-spheres: Decision making, environmental technology and economics. The intention is to provide students of environmental management with necessary knowledge and skills in the area of environmental decision-making.

To make good environmental decisions requires an understanding of the complexity of the different impacts; one has to consider main effects as well as side effects. For example, the use of pesticides is obviously good for the crop, and for this reason laudable, but side effects involve human health hazards, decline in bird abundance, and even species extinction. For good decision-making, all these things have to be understood and evaluated.

The authors believe that a framework for bringing relevant information together for evaluation is quite helpful if one wants to make wise decisions, and this is what we want to provide. The book merges two approaches; it is an environmental decision-making manual as well as an ordinary textbook with a useful curriculum. It is intended for students taking a course in environmental decision-making.

2. CHAPTER STRUCTURE

The chapters are generally organized this way:

- An *Introduction* presents the typical decision problem, with an overview of the values and costs that are involved.
- A *Basic information* section summarizes basic concepts and terms.
- A *Quality standard* section outlines the relevant current standards, such as critical loads, admissible loads and tolerance limits.
- A section called *Natural and man-made impacts* deals with environmental conflicts and the rationale for changing the environment.
- A section on *Dose response functions* presents models for calculating effects of human activity on the environment.
- A section on *Mitigation measures* suggests some common ways of dealing with the problems.
- Finally, there is an *Application section* where decision analysis is applied.

Throughout the text, we give rules-of-thumb to make educated guesses. There are several calculation examples. They are made as realistic as possible, but at the same time kept simple. Some examples are called "demonstrations". These are examples which can be "played" in the classroom to see how certain theories function. For example, one demonstration explains how tradable pollution permits can be distributed among students, and students are asked to buy and sell permits.

3. CASE

The best way to learn environmental decision-making is to work with a case. Our advice is therefore that students at the beginning of the course select their own environmental decision-making problem that is complex enough to be really challenging. It is important that the best decision is not obvious at the outset – it should be a hard problem requiring careful analysis and judgment. Three or four students should work together to formulate an objective, finding relevant information, identifying alternatives and making judgments to arrive at a recommended decision. In Chapter 3, we will work through a model case that shows how this can be done.

4. APPLICATIONS

Several chapters have a section called "Application" at the end. These are real environmental decision problems, where we propose a structured approach based on the content of the book, and with cross-references to relevant chapters. Hopefully the abundant use of cross references also show that the material of the book belongs to an interconnected body of knowledge that is necessary for making well founded and wise decisions.

Table 1-1. List of applications.

Chapter	Theme
3	Oil pollution combat. Decision with uncertainty.
4	Siting of paper mills. The life-cycle of paper.
6	The tragedy of the commons. How to avoid unsustainable harvesting.
8	Prioritizing among environmental categories in the Netherlands.
9	Land reform, land use and deforestation in the Brazilian Amazon. Sharing out land gives surprising effects.
10	Management of cultural heritage monuments. Harvesting a finite resource.
11	Valuation of aesthetic qualities. Disagreement among reasonable persons.
12	Protection of plants. The death of the Altai Hawk's beard.
14	Contaminated ground. Legacy of an innocent time?
17	Lake rehabilitation. What is water for?
19	The case of the black-footed ferret recovery. The value of the charismatic and the few.
20	Pesticide spraying. Disadvantage of good will.
22	Wind mills. Noisy energy with a view.

Chapter 2

DECISION-MAKING CONCEPTS

Introduction of basic concepts in environmental decision-making

We introduce basic concepts so that you can start on your first decision analysis in Chapter 3. We will return to most of the concepts in later chapters where they will be treated in greater depth. You will learn about:

- Consequentialism as a guiding ethical principle for choice.
- Conflicting objectives and trade-off.
- Decision rules under certainty and uncertainty.
- Utility functions and risk aversion.
- Multi criteria utility functions.
- Rationality as a three-dimensional concept.

●

"Tis not contrary to reason to choose my total ruin, to prevent the least uneasiness of an Indian". David Hume, 1748.

1. GOALS AND TRADE-OFFS

We find the first comprehensive formulation of national environmental goals in the American National Environmental Policy Act in 1969. It has six main sections as shown in Table 2-1. It takes into consideration aspects of ecology, human health and well-being, and preservation of culture; it focuses on sustainability as well as high standards of living. Sustainability in the sense of environmental quality may well be competing with the goal of high living standard in a market context.

These goals focus on human use of the environment. We are the masters of nature so to say, and should employ it for our purpose as long as we do not overuse it. We call this an **anthropocentric** perspective, since it puts man, *anthropos*, in the center. Other ethical viewpoints exist, however, that ascribe intrinsic value to nature, which is not supposed to be violated no matter what other benefits that might create. In this book, we shall follow the example set by the US act and adopt an anthropocentric perspective.

Table 2-1. US National Environmental Goals.[1]

1. Sustainability	Fulfill the responsibility of each generation as a trustee of the environment for succeeding generations
2. Safety	Assure all Americans safe, healthful, productive, and aesthetically and culturally pleasing surroundings
3. Beneficial use	Attain the widest range of beneficial use of the environment without degradation, risk to health and safety, or other undesirable and unintended consequences
4. Culture preservation	Preserve important historical, cultural and natural aspects of our national heritage and maintain, where possible, an environment that supports diversity and variety of individual choice
5. Standard of living	Achieve a balance between population and resource use that will permit high standards of living and a wide sharing of life's amenities
6. Renewable and non-renewable resources	Enhance the quality of renewable resources and approach the maximum attainable recycling of depletable resources

This means that we will be primarily concerned with what are good or bad consequences of our decisions, than with what is right or wrong. This is consequential ethics – or **consequentialism** – in practice. Keep your eyes open, however! Consequentialism is not necessarily always appropriate, and may some times be too narrow.

When we make decisions that will have impact on nature, some consequences will usually be good and some will be bad. A hydroelectric power plant may require submerging of land, but will produce beneficial energy. Using land for cattle grazing will change the local ecology – which is often considered bad, but provides beef (and some happy bulls for a while, but notice that bull happiness may not be covered by the US goals above). Examples are plentiful, and they all have in common that they involve a **trade-off** between consequences. To be able to decide, we must assign value to the consequences so that we can compare. This is a central task for environmental decision-makers, and it should be done in appropriate ways. The next chapter will show one way to do that.

2. DECISION RULES

The science of decision-making is partly **prescriptive** – prescribing how people should go about making decisions, and partly **descriptive** – observing how people actually make decisions. Such observations reveal that people often behave in strange – even self-contradictory – ways. This book is mainly prescriptive, but we will now and then discuss lessons from descriptive science to illustrate the difference between theory and practice.

A central task in prescriptive decision theory is to recommend suitable **decision rules** – rules that prescribe how to choose between alternatives.

Appropriate rules are easy to formulate in simple situations, but this becomes more complicated in the face of uncertainty and when there are several decision criteria. We therefore start with simple decision problems and proceed towards more complex ones.

2.1 One Decision Criterion and Certainty

The simplest situation imaginable is when you can choose between different sums of money. Say, you can choose between receiving €100, €200 and €400. Naturally, you choose €400 – you *maximize* outcome. In this case, we have only one decision criterion, money, and we have **certainty** – we know for sure what the consequences will be.

The decision rule is: *Maximize value*.

2.2 Uncertainty

It becomes more complicated if the payoffs are uncertain. Most environmental impacts cannot be predicted with certainty. What actually happens depends on such things as future climate, migration of species, or the true content of an oil well. This is also true when it comes to economical impacts. Such basic things as future interest rates and market prices are inherently uncertain, but will have impact on the economic consequences of our decisions. We need a common term; **future states** denote conditions that will affect the consequences of our decision today.

We shall make the important assumption that nature is disinterested; it does not try to harm us; it does whatever it always intended to do no matter what our actions are. This way, we can talk about probabilities of future states of nature. Notice, however, that we apply the term to nature as well as the economy.

We shall distinguish between uncertainty that we characterize by probabilities, and strict uncertainty. **Strict uncertainty** means that we have no idea whatsoever about the likelihood of the future states of nature. Many decision makers like this notion, for the simple reason that they find it hard to come up with probabilities. This is not a good excuse, however. Even in cases where it is next to impossible to estimate probabilities, decision-makers will – consciously or unconsciously – have subjective beliefs about how likely things are, and this belief can be elicited in a scientific way as **subjective probability**.

Let us give an example of how the dialogue might run: We want to measure how probable *you* think it is that the sea will rise at least 5 cm over the next 10 years. Even if you have no preconceived idea of this probability, we can elicit it by giving you a choice between two alternatives:

A: We will give you €1000 in 10 years if this has actually happened.
B: We will toss a die in ten years and give you €1000 if six eyes come up.

Which option do you choose? If you prefer A over B, we know automatically that you believe the probability of the sea level rising at least 5 cm is higher than $^1/_6$. Instead of a die, we could use a roulette wheel to pin down the probability more exactly. Such wheels can actually be bought for that purpose.

There is also a third sort of uncertainty. It arises when there are other conscious actors that might try to harm us, a competitor for instance. In such situations, probabilities may not be appropriate – and it becomes necessary to use other kinds of reasoning to develop decision rules. This is the field of game theory, which we will return to in Chapter 6.

2.3 Strict Uncertainty

Assume that you consider developing a pasture in the near desert and that nature has three different future states: *wetter*, *dryer* or *unchanged* climate. Further, you have little idea of the probabilities of the three states and therefore want to analyze the problem under assumption of strict uncertainty. The development will cost €5000 and the estimated net cash flows are shown in Table 2-2.

Should you develop or should you not? That is not obvious if you have no idea about the likelihood of the future climate. The most popular decision rule for this situation is Savage's **minimax regret**, which means that you should choose that alternative which gives you the least possible regret – you *minimize maximum regret*. If you develop the pasture, and the climate turns wetter, you will have no regrets of course; but if it turns drier, your regret will be worth 5000; and if it stays unchanged, your regret will be 1000. However, if you do not develop the pasture and the climate turns wetter, your regret will be 10,000; otherwise you did the right thing and your regret will be zero. Thus, you risk the highest possible regret if you do not develop the pasture, and therefore you should do it if you believe in Savage. It turns out that Savage's minimax regret criterion is a good approximation of what decision-makers tend to do in practice.

Table 2-2. Pasture development under strict uncertainty, consequences in terms of net cash

Future climate:	Wetter	Drier	Unchanged
Develop pasture	10,000	–5000	–1000
Do not develop	0	0	0

But unfortunately, human beings do not have a good record of being rational decision-makers. The minimax regret decision rule suffers from several weaknesses and can therefore not be recommended. One problem is rank reversal. If you have used the rule to rank two options, and then add a third, this can in some situations reverse the ranking of the first two.

Actually, it can be shown that there does not exist any consistent decision rule under strict uncertainty, and therefore the whole concept is better laid aside. We use probabilities instead.

2.4 Probabilistic Uncertainty

Assume that you have obtained probabilities for the three different states in Table 2-2, so that the consequence table now looks like Table 2-3. In principle, the probabilities reflect the subjective beliefs of the decision maker, and it is an advantage if these beliefs are based on some sort of evidence, like scientific estimates.

Table 2-3. Pasture development under probabilistic uncertainty, consequences in Euro.

Probability:	0.5	0.3	0.2
Future climate:	Wetter	Drier	Unchanged
A_1: Develop	10,000	–5000	–1000
A_2: Do not develop	0	0	0

A common decision rule for this situation is to **maximize expected value** (EV). The expected values of the two alternatives are:

$$EV(A_1) = 10{,}000 \times 0.5 + -5000 \times 0.3 + -1000 \times 0.2 = 3300$$

$$EV(A_2) = 0 \times 0.5 + 0 \times 0.3 + 0 \times 0.2 = 0$$

Thus, this decision rule tells you to develop the pasture – with a clear margin. Do you feel comfortable with this recommendation? You need not. The max EV-rule omits an important part of reality, namely *risk aversion* – or the counterpart; *risk preference*.

2.4.1 Utility

If €5000 is all the money you have saved, it is a sore thing to lose it. Perhaps the subjective utility derived from the €10,000 you could earn may be less than twice as much as the disutility of losing €5000. If that is the way you feel, you will behave as if you are **risk averse** – you will be less inclined to take chances that might make you lose a lot of money even if the potential gains are high. If that is the case, it is inappropriate to use maximization of expected value as a decision rule. Instead, we substitute the money scale with a so-called **utility scale**, where utility is a non-linear function of money. Exactly what form the function should have depends on the preferences – or feelings if you will – of the decision-maker, and can be elicited in a scientific way. Once we know the utility function, we can apply the famous **expected utility theorem** of von Neumann and Morgenstern, which says that the only consistent decision rule is to maximize expected utility: *max EU*. We shall illustrate how this works below, after we have elicited the utility function.

2.4.2 Elicitation of Utility-Functions

There are several ways to elicit a utility function from a decision maker. Let us demonstrate one method by eliciting your utility function of money gains in the example above. For a start, we can without loss of generality assume that the utility of the worst value is zero and that of the best value 1:

$U(-5000) = 0$ and $U(10{,}000) = 1$

Next, suppose you were given the choice between A and B:

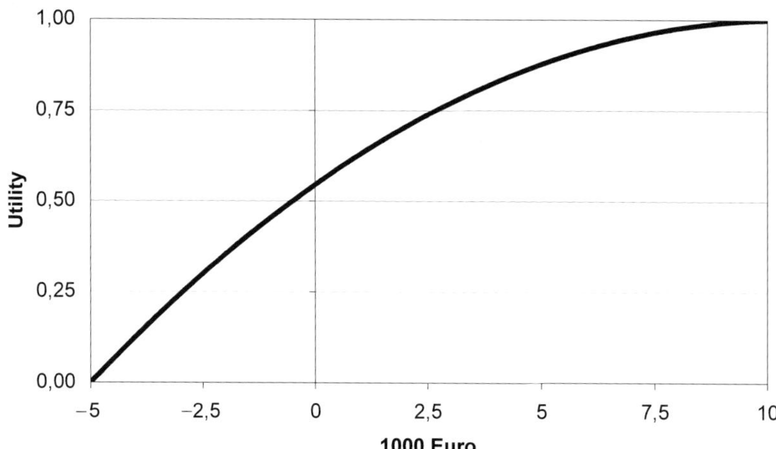

Figure 2-1. A risk averse utility function.

A: You get €2500 for sure;
B: You participate in a lottery where you will either gain €10,000 or lose €5000, both with probability 0.5.

What do you choose? Given that you are risk averse – as we discussed above, you would probably take the €2500 to avoid risking losing the €5000. If you agree to this and we assume that you maximize expected utility, we already know that *your* utility of A is larger than the expected utility of B. But we need to be more precise; we want to find the so-called **certainty equivalent**, which is a prize under A that makes you indifferent to the choice between A and B. Therefore, we continue to ask questions like the one above, but with lower values than €2500 for A. Let us assume that you feel that the certainty equivalent is approximately – €500. That means that you would just as well *pay* €500 as take the risk involved in the lottery where you may lose €5000. That is certainly risk aversion.

In utility terms, this means that your utility of – €500 is equal to the expected utility of the lottery in B. Equation 2-1 shows how to compute the expected utility.

$$U(-500) = EU(B) = 0.5 \times U(-5000) + 0.5 \times U(10{,}000) = \\ 0.5 \times 0 + 0.5 \times 1 = 0.5 \tag{2-1}$$

We have now three points on the utility function, and that is enough to draw a smooth curve as an approximation of your utility curve. In Figure 2-1 we have drawn a second order polynomial curve through the three points. Of course, we could have used other kinds of functions, and if we wanted to be more precise, we might have found more points on the curve by asking more questions the same way, but this will do for now.

2.4.3 Risk Aversive and Risk Seeking Behavior

The curve in Figure 2-1 is concave, meaning that there is a decreasing marginal utility – the utility increases at a decreasing rate. This is, according to utility theory, tantamount to risk aversion. If you are sorely hurt by losses, while high prices mean comparatively little to you, you will not be tempted to take chances – you are risk averse. Risk aversion is a common attitude, among investors as well as environmental protectionists, but there are exceptions. People who gamble do so for two reasons, the thrill of suspension and increasing marginal utility of money. The thrill of suspension is not considered rational by the theory and is not part of it, but an increasing marginal utility of money means that the utility function is *convex* as shown in Figure 2-2.

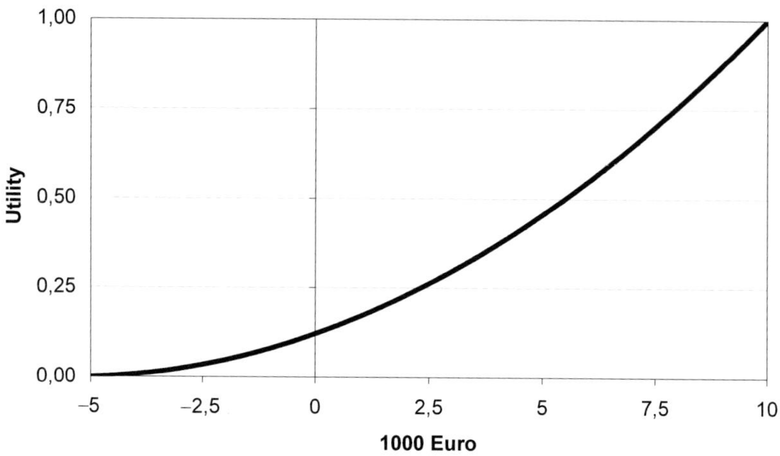

Figure 2-2. A risk prone utility function. The curve is convex, with an increasing marginal utility of money.

2.4.4 Maximization of Expected Utility

We are now ready to apply the maximum expected utility – *max EU* – decision rule to our little problem. The first thing we have to do is to substitute the monetary values in Table 2-3 with utilities, which we can read off from Figure 2-1.

Table 2-4. Utilities have replaced the monetary scores in Table 2-3.

Probability:	0.5	0.3	0.2
Utilities	Wetter	Drier	Unchanged
A_1: Develop	1	0	0.45
A_2: Do not develop	0.54	0.54	0.54

Calculating expected utility from Table 2-4, we now find:

$$\begin{aligned} EU(A_1) &= 0.5\times 1 + 0.3\times 0 + 0.2\times 0{,}45 = 0.59 \\ EU(A_2) &= 0.5\times 0.54 + 0.3\times 0.54 + 0.2\times 0.54 = 0.54 \end{aligned} \tag{2-2}$$

We see that development is still recommended, but now the alternatives come much closer. The reason is that we have accounted for risk aversion, and since development is the riskier alternative, its rating has gone down.

2.5 Several Decision Criteria

So far, we have considered decision rules where we have only one decision criterion – net cash flow, which we measured on a monetary scale. Most important decisions concerning environmental management have impacts in several dimensions, however, which means that we have to take into account several decision criteria. We saw for instance that the American National Environmental Policy Act specifies six main sections of concern.

To illustrate, let us continue with the example above and assume that development of a pasture in the near desert requires redirection of water, which in turn increases the probability of further spread of the desert. Now, we have two things to take into consideration, cash flow and risk of desert spread; and since these are in conflict, it becomes necessary to evaluate how important they are in relation to each other. But who has the authority to do that kind of evaluation? In other words, who is now the **decision-maker**? Above, *you* were the decision-maker – the developer who risked your own money in order to return some profit. But we can obviously not trust *you* with the task of solving the conflict at hand. What we need is a decision-maker who is entitled to compare the two societal goals: Value creation and Risk of desert spread.

This decision-maker can still apply the *max EU* decision rule. What is needed is to attach importance weights to the two criteria. Let us name the criteria x_1 = *value creation* (€1000) and x_2 = *probability of desert spread.* The total utility of the decision could then be computed as:

$$U(x_1, x_2) = w_1 u(x_1) + w_2 u(x_2) \qquad (2\text{-}3)$$

Here, w_1 and w_2 are importance weights ($w_1 + w_2 = 1$); u_1 and u_2 are the utility functions of the separate decision criteria. This is the simplest possible **total utility function**; more complex functions will be treated later. The decision maker must now find which one of the alternatives has the highest expected utility by taking the uncertainties into account. The analysis could proceed as shown in Table 2-5.

We now have to convert the consequences to utilities. Let us use the utility function of value creation (u_1) that is shown in Figure 2-1.

Table 2-5. Consequence table with two decision criteria. x_1 = value creation, x_2 = probability of desert spread.

Probability Future climate	0.5 Wetter		0.3 Drier		0.2 Unchanged	
Criterion	x_1	x_2	x_1	x_2	x_1	x_2
A_1: Develop	10,000	0	–5000	0.5	–1000	0.25
A_2: Do not develop	0	0	0	0.25	0	0

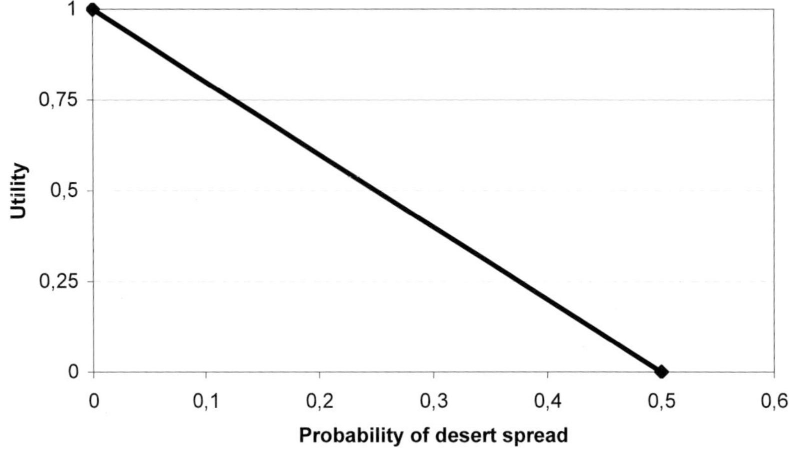

Figure 2-3. Risk neutral utility function for x_2 = probability of desert spread.

Let us further assume that the decision-maker is risk neutral when it comes to probability of desert spread. As shown in Figure 2-3, the utility function u_2 is then linear with $u_2(0.5) = 0$, $u_2(0.25) = 0.5$ and $u_2(0) = 1$. The result is Table 2-6.

Table 2-6. Utility table with two decision criteria. x_1 = *value creation*, x_2 = *probability of desert spread.*

Probability Future climate	0.5 Wetter		0.3 Drier		0.2 Unchanged	
Utility	$u_1(x_1)$	$u_2(x_2)$	$u_1(x_1)$	$u_2(x_2)$	$u_1(x_1)$	$u_2(x_2)$
A_1: Develop	1	1	0	0	0.45	0.5
A_2: Do not develop	0.54	1	0.54	0.5	0.54	1

Finally, let us assume that the decision-maker has determined that in this case it is twice as important to avoid the risk of desert spread as to obtain the value creation in question, meaning that $w_1 = 0.33$ and $w_2 = 0.67$. The expected utilities of the alternatives are:

$$EU(A_1) = w_1 \times 0{,}59 + w_2 \times (1 \times 0.5 + 0 \times 0.3 + 0.5 \times 0.2) = 0.54$$
$$EU(A_2) = w_1 \times 0{,}54 + w_2 \times (1 \times 0.5 + 0.5 \times 0.3 + 1 \times 0.2) = 0.75$$

Now, “no development” wins with a clear margin.

3. RATIONALITY

It is often assumed that people make rational decisions. The term rationality has many meanings, however, and we need a definition that is useful for our purposes. Some people take rationality to mean self-serving – that a rational person is only concerned with his own well-being in a narrow sense and do not think of others. Since we will be concerned with decision-makers who make decisions on behalf of others, this is evidently not very useful. It will serve our purposes better if we define rationality in a way that most people will regard as an exemplary standard for environmental management.

3.1 Definition

Rationality requires: [2]

1. Well founded beliefs
2. Well founded values
3. A consistent decision rule

Well founded **beliefs** means that the decision-maker has done a decent job at identifying and assessing all important **consequences** of the decision, economical and social, as well as environmental. Even if we never know exactly what will actually happen beforehand, we at least need to develop educated guesses as a basis for decision-making.

Well founded **values** means that we have assessed the relative importance of the impacts properly, either through information about people's **willingness to pay** to avoid damage, WTP, or – if no such information can be made available – through balanced subjective judgment by the decision-maker.

To maximize expected utility is a **consistent decision rule**, and by many regarded as the only one. This follows from the expected utility theorem of von Neumann and Morgenstern.

Notice that this definition of rationality says nothing about what kind of values you should have – that depends on personality and feelings; it says only that the values you have should be well established so that you are not led astray in the heat of the moment. This means that it might be perfectly rational to give away money to help other people or to save the environment, if that makes you happier.

It is important that you appreciate the difference between beliefs and values. This distinction lies at the heart of decision-making. Beliefs are about facts – what will actually happen in the real world. We form beliefs about consequences through observations, models and analysis. Beliefs can be proved wrong if they do not fit with reality, but not values. Values can be judged inappropriate, but not wrong in an objective sense.

3.2 Inconsistency

Here we first present an inconsistent decision rule and thereafter discuss theory and practice.

3.2.1 An Inconsistent Decision Rule

To give you an idea of what inconsistency entails, let us consider an example caused by so called intransitivity. The symbol $\succ$ means "preferred to". If a decision rule ranks the alternatives *A*, *B* and *C* so that $A \succ B \succ C \succ A$, then that decision rule gives intransitive ranking and is therefore inconsistent.

Example: A student has the choice of the job-alternatives *A*, *B*, *C* and *D*. She has identified four decision criteria and ranked the alternatives according to how well they score on each criterion. See Table 2-7. On salary, job *A* scores best, and job *D* worst, etc.

Table 2-7. Ranking of the alternatives on each criterion.

	Criteria			
Job	Salary	Prestige	Travel	Location
A	4	3	1	2
B	3	2	4	1
C	2	1	3	4
D	1	4	2	3

Let us investigate the following decision rule: "If a job ranks higher than another job on a majority of criteria, it is better." We see easily that *A* ranks higher than *B* on salary, prestige and location; while *B* ranks higher than *C* on salary, prestige and travel, etc. Therefore $A \succ B \succ C \succ D \succ A$, proving that the decision rule is inconsistent. This kind of inconsistency is called *violation of transitivity*, and transitivity is a fundamental assumption of rationality.

3.2.2 Theory and Practice

The problem with *prescriptive* decision-making is that it does not very well *describe* how people actually make decisions – prescription and description are sometimes worlds apart.

Utility theory fails to account for aspects that are important for many decision-makers: It does not incorporate joy of gambling; the reason utility functions are curved upwards is not that the decision-maker enjoys uncertain situations in themselves, but simply that the possibility of earning a lot of money may change life completely, and thus give a more than proportional return of value. The other is failure to accommodate regret – which is the feeling you experience when you know what you should have done. A decision rule that aims at avoiding future regret is very attractive for most people, but such rules can be shown to be inconsistent. Look forward, not backwards! The famous Allais paradox[3] describes a choice between pairs of lotteries where the possibility of regret is very much present, and where it is easy to demonstrate that most people make decisions that go against utility theory.

Thus, according to prescriptive utility theory, common human feelings like joy of gambling and regret lead to inconsistent behavior. So, why is utility theory still useful? If you make decisions on behalf of other people – which is precisely what environmental management is about – such things as joy of gambling and regret on part of the decision-maker should not enter the equation. In that case, prescriptive theory should prevail.

3.3 Bounded Rationality

The most famous theory of non-rational decision-making is Simon's concept of **bounded rationality**.[4] He observes "that the capacity of the human mind for formulating and solving complex problems is very small compared with the size of the problems whose solutions is required for objectively rational behavior in the real world – or even for a reasonable approximation of such objective rationality". Therefore, decision-makers seldom try to explore the whole range of consequences of all possible alternatives to find the optimum solution to a decision-problem, but instead try to identify an alternative that has good enough consequences for their aspiration level. This is called **satisficing behavior** and is quite common in practice. In our terminology, satisficing behavior means that you start with the most important criterion and identify all decision alternatives that have at least a satisfactory performance with regard to this criterion. Then, you shift your attention to the next most important criterion and eliminate those options that do not perform satisfactorily on this one, and you continue down the criteria list until only a single option remains. If you do not find an alternative that satisfies the most important criterion, you need to look for other alternatives in that direction.

The prescriptive theory presented here aims at improving rationality by considering all criteria simultaneously, something that requires paying attention to importance. To keep track of everything, we easily exceed what is possible by the unaided human mind and therefore need computer assistance. Simon's concept of bounded rationality has lead many to perceive satisficing behavior as a prescriptive model of decision-making – as a suggestion of how it should be done. However, Simon never intended the model to be prescriptive. Indeed, in the introduction to the third edition of his book Administrative behavior, Simon refers to "...human beings who satisfice because they have no wits to maximize".[5]

---

[1] US Congress, 1970
[2] Føllesdal (1982)
[3] Stigum et al. (1987)
[4] Simon (1982)
[5] Simon (1997)

Chapter 3

GETTING STARTED

A step by step outline of the rational decision analysis process

How do I in practise carry out the seven steps of decision analysis? That is, how do I:

1. Specify the problem.
2. Formulate objectives.
3. Find decision alternatives.
4. Display the result of consequence analysis.
5. Elicit utility functions.
6. Elicit importance weights.
7. Rank the alternatives.

"Values are what we care about. As such, values should be the driving force for our decision making. They should be the basis for the time and effort we spend thinking about decisions. But this is not the way it is. It is not even close to the way it is." Ralph Keeney in *Value Focused Thinking*, 1992.

1. A RATIONAL DECISION PROCESS

The first thing you have to bear in mind if you want to carry out a rational decision analysis is that you must distinguish sharply between **values** and **beliefs**. If you follow that advice, you will be able to penetrate a decision problem much faster and avoid the confusion that arises when people start to discuss what to do before they have determined what they should try to achieve.

Values are linked to goals, which are what we try to achieve; and how important values are is in a fundamental sense subjective. It is important to know how much we appreciate such values as health, environment or economic goods. As a decision-maker, it is your job to choose and apply such well-founded values, and there are several ways to approach that.

Beliefs are linked to facts – what you think will be the actual real world consequences of your decision. Again, it is your job as a decision-maker to acquire well-founded beliefs and determine how they relate to the problem at hand, but you may certainly delegate to other people the job of model building and data collection.

In this chapter, we outline the seven main steps of what we call the **rational decision process**. When you understand the steps, you should be able to carry out a decision analysis yourself. It will improve your understanding if you already have a decision problem in mind that you would like to analyze. We illustrate the process with a case and recommend that you choose a similar problem for yourself and try to apply the steps on the way. Table 3-1 summarizes the seven steps.

Table 3-1. The seven steps of the rational decision analysis process.

	Step	Activity	Deliverables
1.	Specification of the decision problem	**Define** and restrict the problem. Identify decision makers and stakeholders.	Specification, including geographic extent, time horizon, decision makers and stake holders
2.	Formulation of decision objectives	Identify the **goal** and explain it further in terms of subgoals. Specify decision criteria that measure to what degree the subgoals are attained.	A **goal hierarchy** and a list of decision criteria x_1, $x_2 ... x_k$ measured in natural units like kg and km^2.
3.	Identification of decision alternatives	Find all feasible **decision alternatives**, and select the most promising of these (Screening).	A list of decision alternatives
4.	Consequence analysis	Estimate the **consequence** of each decision alternative in terms of a score on each criterion. These are beliefs of what will actually happen.	A **consequence table** with a row for each criterion and a column for each alternative, containing the estimated scores
5.	Elicitation of utility functions	Elicit the utility function for each criterion as shown in Chapter 2, by interviewing the decision-maker. These are called **primary utility functions**.	A utility function for each criterion and a **utility table** that corresponds to the consequence table, but with utilities instead of scores
6.	Elicitation of importance weights	Elicit importance weight for each criterion through an appropriate value trade-off procedure with relevant decision makers.	A list of weights w for each criterion such that $\Sigma w = 1.0$
7.	Ranking of decision alternatives. Recommendation	Compute the (expected) utility of each decision alternative by using the total utility function and rank them accordingly. Discuss the results.	The expected utility of each decision alternative and a ranking. Discussion and recommendation.

Bear in mind that although we call it *the* rational process, this is just one out of several possible prescriptions of how to perform rational analysis. It is important that you are eclectic rather than dogmatic, meaning that you should always use the most suitable method available, depending on the problem and the circumstances.

A decision analysis requires the following:

- A nontrivial decision problem – a situation where the best course of action is not obvious.
- A number of decision alternatives – the decision problem is which one to choose.
- Sufficient knowledge about the possible consequences of the alternatives to make it meaningful to choose. Scenario development is a technique that can be of help when there is little factual knowledge about costs and environmental consequences.

A rational decision analysis will usually produce:

- A ranking of the decision alternatives.
- Supporting scientific facts and an assessment of which facts were most important for the ranking.
- Supporting information on the value of non-monetary goods and how it was obtained.
- Supporting assessment of the robustness of the ranking.

2. STEP 1: SPECIFICATION OF THE DECISION PROBLEM

2.1 What is a Decision Problem?

To have a decision problem is to be in a situation that requires action, and there are several options available. You can specify a decision problem by indicating why an action is required or what kind of actions you have in mind. Environmental management is concerned with water, air and soil as well as survival and proliferation of life on earth, including human beings and their welfare. Thus, environmental projects have consequences for many stakeholders, now and in the future. To restrict a decision problem is therefore important, but also challenging. On the one hand, all issues that the decision-maker, the stakeholders, and the scientist think are important should be included. On the other hand, the decision problem has to be manageable.

2.2 Restricting the Problem

Environmental management tasks usually cause chain reactions. Suppose that the decision problem is whether to log a forest, and one objective is to minimize the resulting environmental burden. Should the analysis be restricted to the burden caused by the loss of trees, or should it include the environmental effects of using heavy equipment in the forest? Should impacts of fuel consumption of trucks be included? Trucks emit green house gases, which contribute to global warming, which again causes the sea level to rise. Should all this be included? It is obvious that we have to stop somewhere. Consequences of actions follow many pathways that can lead to several possible decision criteria. The question is where to stop. There is no fixed answer. A guiding principle is that one should stop at impacts that have obvious value. Health and welfare are of obvious value, and so are species diversity, clean air, pristine nature, waterfalls and recreation.

It is also advisable to define the geographic and temporal extent of the problem. For example, in the case of the *Amoco Cadiz* oil spill in 1978 at the coast of Brittany in France, the costs were in principle calculated for the world, but thereafter subdivided into costs for the country (France) and the region (the coast of Brittany). The loss for the tourist industry was substantial in Brittany, because fewer persons spent money in the region. However, many of those who did not travel to Brittany in 1978, traveled to other parts of France, spending money at other recreational communities. Loss of revenue for the tourist industry after the *Amoco Cadiz* oil spill is shown in Table 3-2. The numbers assume that three fourths of the lost tourist revenues were incurred in France, but outside Brittany. The last quarter of the revenues was lost outside France.[1] The losses shown in Table 3-2 were incurred in 1978, but it is conceivable that this continued to a certain extent in the following year, therefore, a time horizon should have been defined, and present values of the costs computed accordingly.

Table 3-2. Tourist industry loss because of the Amoco Cadiz oil spill in 1978.

Cost category	Cost to Brittany	Cost to France including Brittany	Rest of the world	Total net social cost to the world
Column #	(1)	(2)	(3)	(4) = (2+3)
Million Franc	110–116	29	–29	0

2.3 Decision-Makers and Stakeholders

The decision problem is not specified until it is clear who the decision-maker is and who the stakeholders are. At the outset, the decision maker is whoever or whatever has the formal authority to make the decision, be it a government, parliament, council, public servant, or someone in a private firm. Advisors who make recommendations to the decision-maker often have considerable decision power. Such relations should be explicitly stated to make it clear who is making the necessary judgments in the process.

Stakeholders are all people and organizations that have an interest in the decision, such as authorities, shareholders, greens, residents, and consumers. Some stakeholders are users or potential users, and some just want to know that the environmental resource in question exists. In that case, we say that it has a use value as well as an existence value.

Stakeholders can also be categorized according to how local they are. We distinguish between local residents, residents of the country and members of the world community, which are often represented by UN organizations.

Value judgments will depend on the decision-maker, as well as who the stakeholders are.

2.4 Case: Oil Spill Combat, Specification of the Decision Problem

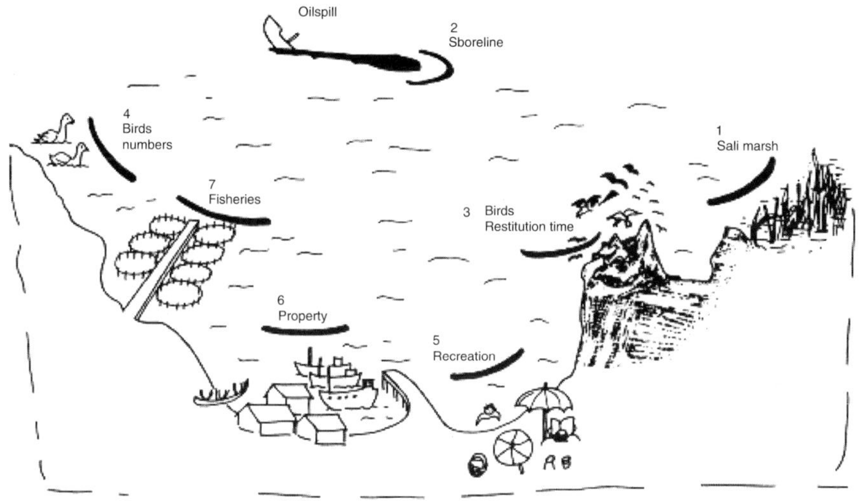

Figure 3-1. Oil Spill combat decisions.

The 1970s witnessed increasing oil production in the North Sea and Russia, which led to increasing traffic of ships in Norwegian costal waters with substantial risk of accidental oil spills as a consequence. The seriousness of the problem was estimated at around two spills per year in the 1980s. The Norwegian Parliament therefore decided to develop a strategic oil spill combat plan to be better prepared for situations like in Figure 3-1.

The main decision problem in 1980 was to identify an optimal level of preparedness against oil spills and to evaluate alternative combat strategies over the next 10 years. The problem was restricted to coastal waters, excluding blowouts on the shelf. The problem was also restricted to how to deal with spills when they occurred, not how they could be avoided in the first place. That would have included a much wider range of decision alternatives.

The most important stakeholders were the Norwegian Parliament, the oil firms operating in the North Sea, environmentalists, animal rights people, the coastal population, the tourist industry and the fishing industry.

The Ministry of Environment financed the decision analysis process. One problem was to identify the best strategy to contain the damage once a spill has occurred. There were three alternatives:

1. Do nothing.
2. Try to collect oil at the spill site as quickly as possible.
3. Try to direct the oil from the spill site to a convenient place on the coast.

The authors participated in the project, and we shall use it as a case throughout this chapter. This gives us the opportunity to share with you experiences that are useful in many decision contexts.

3. STEP 2: FORMULATION OF DECISION OBJECTIVES

A **goal hierarchy** organizes the objectives of the decision maker. It describes the main goal and explains it in terms of subgoals. Finally, at the bottom level, it shows how attainment of the goals will be measured. This is done by variables called **decision criteria,** which will be used to evaluate the decision alternatives.

Typically, the **main goal** is to maximize the combined effects of benefits and costs. Examples of benefits are industrial value creation, improvement in health and safety, and environmental protection. Examples of costs are economic expenditures, environmental degradation, and use of non-renewable resources. There are obviously conflicts between benefit achievement and cost reduction, and such conflicts should be highlighted through competing subgoals in the hierarchy. To make it clear whether something is conceived of as a benefit or a cost, one should always state explicitly whether goals are to be maximized or minimized. One should never use the word optimize in a goal hierarchy, since optimization requires value trade-off which is the subject of the subsequent analysis. The purpose of the goal hierarchy is just to declare what the values are, not to state how important they are.

3.1 Case: Oil Spill Combat, Goal Hierarchy

Figure 3-2 shows that the main objective related to oil spill combat strategies is to minimize social economic costs. These costs are both of non-economic and economic nature. Economic costs are public combat costs as well as loss to private industries. By separating combat costs and industrial loss (hotels and fishery), it was left to the decision-maker to weight the criteria differently according to presumed distributional effects.

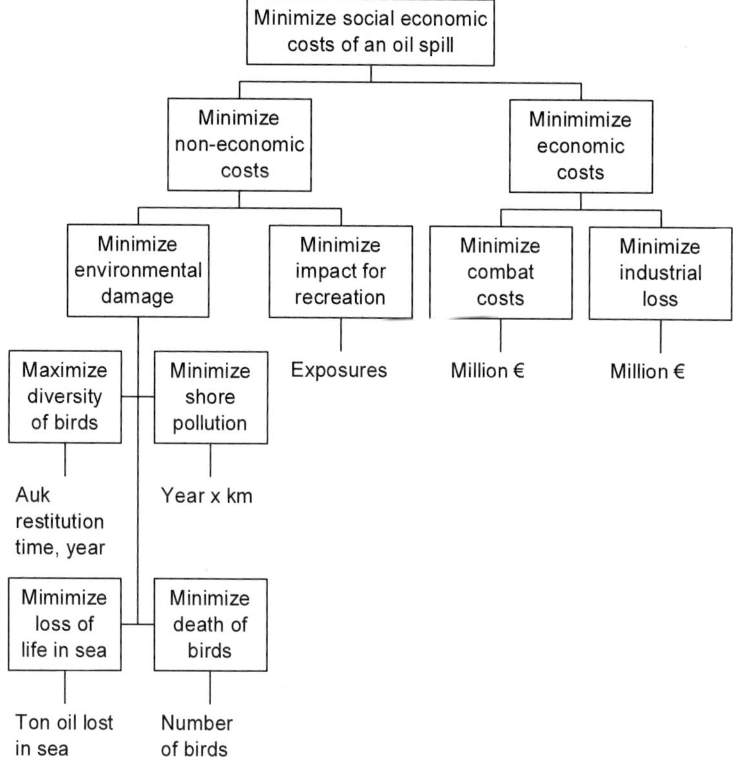

Figure 3-2. Goal hierarchy for the oil spill combat problem. Objectives are stated in boxes. Decision criteria are shown without frames.

Non-economic costs are environmental impacts as well as loss of use value in connection with recreation. The loss of use value was measured in terms of how many times persons would be exposed to the oil spill if they continued with their recreational habits. Environmental impacts include loss of diversity of bird life. Auks are especially vulnerable to oil spills. The impact was therefore measured in terms of auk restitution time, which is the time it takes before the auk population is 95% restituted.

Shore pollution is measured as the product of the length of the shore that is polluted (km) and the time it takes before it will be 95% cleaned by natural processes. Thus, one kilometer of shore polluted for five years is as bad as five kilometers polluted for one year, probably a fair assumption. The intention was to measure loss of life in sea in terms of kg benthic organisms per m^3, but it turned out impossible to predict. The **indicator** "ton oil lost in sea" was used instead. Finally, many people are concerned with animal suffering, and do not want birds to freeze to death, even if they are seagulls. The last criterion took care of that.

3.2 Developing a Goal Hierarchy

When you develop a goal hierarchy, you may use a *bottom up* or a *top down* approach. In a top-down process, you start from the main goal at the top and try to formulate subgoals. It is tempting to ask "How can this goal be achieved?" but this will lead you astray. A goal hierarchy is not a means-end diagram; it is a **value structure**. Instead, you should ask: "Which values are inherent in this goal?" or "Why is this goal important?" This is how the goal hierarchy of Figure 3-2 was originally developed. If you ask, "Why is it important to minimize damage to the environment?" the answer comes naturally that shore, birds and life in the sea are things we care about. If you ask, "How can we protect the environment?" you will be led astray and start thinking about actions instead of values.

It is usually a good idea to include stakeholders in the developing process. That makes it easier to achieve a consensus at the outset about the most important values. If this is done before you start to consider alternatives, people will probably have a more open mind than if they are defensive about the prospects of some specific alternative. This usually requires a **bottom-up process** where you make a list of subgoals, and ask people which ones should be included. Another method is to perform an open survey where decision-makers and stakeholders can formulate their own objectives. Questions in a questionnaire can be open or closed in the form of a fixed set of questions. Both methods have strengths and weaknesses. A question that scores highly in a questionnaire may be hardly mentioned when the respondents have to come up with it themselves.

3.2.1 Fundamental Values versus Instrumental Values

A **fundamental value** is something that is of obvious value in itself; another word for it is **end impact**. That a child gets enough to eat is an obvious fundamental value – no further questions need to be asked. Whether a new work place is of value in itself, is debatable. Some people hate to work – but work is an **instrument** or a **means** for family income, therefore workplace is an instrumental value. Is family income a fundamental value? Well, one does not eat money – it is an instrument for even more fundamental values.

This line of reasoning shows that when you develop a goal hierarchy, you have to stop somewhere. We advise you to try to carry through to end impacts, because their importance is usually easier to assess than if you stop at instrumental values. On the other hand, instrumental consequences are often easier to predict. If you look at Figure 3-2, to prevent oil from disappearing in the sea is obviously an instrumental value, while life in the sea is more like a fundamental value, at least for divers – but also for most other people. In the oil spill case, ton oil that disappears in the sea was used as an instrumental decision criterion to *indicate* the real concern, which is life in the sea. The variable is therefore also called an **indicator**.

3.2.2 Balance

People involuntarily put more emphasis on those parts of a problem that are described in great detail, than on parts that are described summarily.[2] In our context, this means that the goals that are divided into the highest number of subgoals tend to get more weight. On the other hand, when people are exposed to too much information, they may develop their own filter and discard information without assessing it.[3] It is hard to say which one of the processes is the most effective, but we need to beware that people's value judgment is susceptible to innocent looking things like the number of subgoals in the different branches of the goal hierarchy. You should therefore try to avoid unbalanced hierarchies.

The goal hierarchy in Figure 3-2 is technically unbalanced, with much more detail under environment. But then, if this is why we combat oil spills in the first place, maybe it is not unbalanced after all.

3.2.3 Monetary Criteria

At least one criterion will almost always be monetary; project cost is one example. If there are cash flows involved, you should use the present value, computed at a suitable discount rate. However, if there are several possible monetary criteria, you should make a careful analysis of which ones you can lump together. Money that flows in and out of the same account should be lumped together. You do not need one criterion for revenues and another for costs; you need only one for profit. But we do need one criterion for oil spill combat costs and another for industrial loss, since different stakeholders incur them.

Another problem with monetary criteria is that one does not eat money. Incurred combat costs are not end impacts; end impacts are loss of other good things that could have been bought with the money. Again, there is the familiar problem of where to stop, and of course practical considerations make combat costs a natural end point. However, when this case was actually carried out, combat costs where explained in terms of hospital facility equivalents for the decision makers so that they could better appreciate alternative use of the money.

Environmental goods are usually not bought and sold in a market place; they are called **external goods**, meaning that they are external to the economic system. Environmental goods therefore usually require non-monetary criteria. However, if reliable surveys are available that show how much people are willing to pay for them (WTP), they could be included in the goal hierarchy as monetary criteria. Such information was not available in the oil spill case, however.

3.2.4 Desirable Properties of a Goal Hierarchy

A good goal hierarchy is of great help, pedagogically as well as operationally. But there are many ways to draw a goal hierarchy, some are better and some are worse. Here is some advice:

- *Conflict*: To have a decision problem means that there is a conflict present. This has to be reflected in the goal hierarchy by criteria that are in conflict. In Figure 3-2 there is an obvious conflict between combat costs and environmental damage.
- *Completeness*: The goal hierarchy must reflect all values that are important to the stakeholders in the current decision problem. That includes “soft” values – those that are easily neglected by economists – which should be included on par with “hard” values.
- *Non-overlap*: The same value should not be represented by several criteria. That would produce double counting. In Figure 3-2, there is an apparent double counting since both recreation and shore pollution are included. This was done on purpose, however, since they represent two different values; recreation represents what we call a **use value**, while clean shore is what we call an **existence value** – clean shores are important even if we do not use them.

- *Consistency*: Subgoals must reflect the superior goal. Do not suddenly introduce something entirely new that could not be deduced from the objective above.
- *Direction*: It must be clear whether the score of a criterion is to be maximized or minimized. A criterion with an optimal score – like the amount of cake you eat – does not reflect a fundamental value, and should be avoided. It can usually be substituted by two other criteria – like pain and indulgence. You would probably like to minimize pain and maximize indulgence, and the optimum would be a proper balance.
- *Balance*: Avoid a very unbalanced goal hierarchy. Do not split some goals much more than other goals if they are of similar importance.
- *Parsimony*: Include criteria that are of real importance. Too many unimportant criteria clutter the picture without affecting the outcome.
- *Relevance*: Include only criteria that are relevant for the decision problem. Avoid criteria that are not affected by the decision alternatives, even if they may be important in other contexts.

3.2.5 Measurement Scales

Decision criteria should preferably be measurable on a numerical scale. Sometimes a scale is difficult to find – especially for soft values. Then it may be advantageous to use an **indicator**, which is a measurable criterion that only indirectly represents the value, such as ton oil that disappears in the sea in the oil spill case. If the criterion is hard to measure, you may construct a scale where you explain carefully what the different scores mean. Table 3-3 shows a **constructed scale** for the criterion "Biological impact in an area".[4]

Table 3-3. A constructed scale for the criterion "Biological impact in an area".

Score	Effect level
0	Loss of 1 km^2 agricultural or urban area. No loss of "natural areas"
1	Loss of 1 km^2 of primary agricultural land (75%) and loss of secondary vegetation (25%). No loss of wetland areas, or areas with endangered or threatened species
2-7	Increasing loss of "natural habitat areas"
8	Loss of 1 km^2 of mature forest and or wetland area and or habitat for endangered species.

4. STEP 3: IDENTIFICATION OF DECISION ALTERNATIVES

Only imagination of decision-makers, stakeholders and experts limits the number of decision alternatives. Alternatives that are clearly inferior may be eliminated from the beginning. An alternative that scores worse than another alternative on all criteria, is called **dominated**, and will never be chosen and need therefore not be included. In some cases, the number of alternatives is theoretically uncountable, as for instance if you were to decide on the optimal water level of a reservoir that could be any number in a range.

Brainstorming sessions are often useful to identify promising decision alternatives. The idea is to encourage creativity by uncritically listing all proposals that come up during the initial session. In subsequent sessions the list of possible alternatives are screened and reduced to a **Pareto set**, which is a set that consists only of decision alternatives that are not dominated by other alternatives in the set.

In the oil spill combat project, three clearly different strategies were visible from the start: to do nothing, to collect oil at the site, to try to direct the oil to a suitable place.

5. STEP 4: CONSEQUENCE ANALYSIS

Rational decision-making requires well-founded beliefs about consequences, and this can be achieved through systematic **consequence analysis**, which is performed to predict the impact of the decision alternatives on the criteria. The result of consequence analysis can be condensed to an $n \times m$ **consequence table**, when there are m alternatives and n criteria.

Consequence analysis often requires model building as well as information gathering. There is no unified theory for this, however; each decision problem presents its own challenges.

5.1 Case: Oil Spill Combat: Consequence Analysis

The consequence analysis in the oil spill combat case was performed through a mix of field observations, experiments and model building.

A particular place outside the west coast of Norway was chosen as the site for an experimental oil spill. The bird fauna was observed, and the number of individuals and the mix of species registered. Stretches of shore on the coast were classified as rock, sand or wetland. Based on these observations, computer models were developed that could estimate the impact of oil spills on different classes of shore and on birds in terms of restitution time of bird colonies. Tourism connected to hotels, camping grounds, and summer cabins were also registered, as well as revenues from fishing. One central objective was to learn how well the available equipment for containing and collecting oil would work.

Table 3-4. Consequence table for a given oil spill and three different combat strategies.

Criterion	Unit	No action	Collect oil	Direct oil
Recreation	1000 exposures	200	200	1000
Industrial loss	Million €	0.45	0.45	0.34
Shore pollution	Year km	600	600	300
Bird deaths	1000	1000	1000	700
Auk rest. time	Year	12	12	10
Oil lost in sea	Ton	10,000	8000	8000
Combat costs	Million €	0	1.25	1.50

Real experiments were performed where oil was poured in the sea. The results were not particularly promising, as is evident in Table 3-4. We see that the difference between trying to collect oil at the site, and doing nothing, is dismal. The only thing they achieved was to collect some oil, which therefore was prevented from disappearing in the sea, but this did not mitigate other impacts. It is obviously far more effective to direct the oil to a suitable point on the coast and deal with it there than to try to collect it at the site. Whether the strategy of directing oil is cost efficient, however, depends on the valuation of the impacts.

Table 3-4 is produced according to scientific standards. It is a summary of well- founded beliefs about what will actually happen if any one of the three strategies is carried out. The table provides a good basis for decision-making, but since none of the three alternatives dominates in the sense that it is better than the others on all accounts, something more is required to single out one option as better than the others. We see for instance that whether it is better to collect oil than to do nothing depends on whether it is worth 1.25 million euro to prevent 2000 ton oil from disappearing in the sea. That is a question of values, and the subject of steps 5 and 6.

6. STEP 5: ELICITATION OF UTILITY FUNCTIONS

Let us assume that we have obtained a consequence table such as Table 3-4. The scores are basically a technical or scientific matter. They are supposed to be objective – or at least subjective "best estimates" when factual knowledge was missing. Now, however, we turn from beliefs to values.

In this step, we include subjective preferences of persons with such authority. In principle, this is the decision-maker; but decision-makers may want to solicit the preferences of other stakeholders, such as the public.

Two types of subjective input are needed in decision analysis, utility functions and importance weights. We described the principles of utility function elicitation in section 2.4.2, and in this chapter we shall discuss how the process was conducted in the oil spill case. Elicitation of importance weights follows in step 6.

Proper handling of utility functions and importance weights requires careful consideration of scales. It is good practice first to define a *worst* and a *best* scenario with scores that lie somewhat outside the range of the estimated outcomes and then elicit utility functions and weights relative to those ranges. Let us look at the oil spill combat case.

6.1 Case: Oil Spill Combat: Elicitation of Utility Functions

The Norwegian Parliament was the formal decision maker, but it was obvious that preference information had to come from other sources. The most important stakeholders were the Government, the North Sea Oil Developers (NSOD), and NGOs. The most prominent NGO was the Norwegian Society for the Conservation of Nature (NSCN). It was decided to appoint a **decision panel** from each of these stakeholders. Each panel consisted of three persons, and the panels worked independently. The State Pollution Prevention Agency (SPPA) represented the Government.

Utility functions were elicited from the panels with the method described in section 2.4.2. That produced essentially risk-averse utility functions, where risk aversion was most pronounced for the environmental variables. The exception was *number of dead birds*; it was argued that each bird should count the same, and therefore a linear utility function was deemed appropriate. In this presentation, however, we shall simplify and use linear utility functions everywhere, although that does not do credit to the serious concern for major impacts that was displayed by some of the panels.

To elicit utility functions, we first have to specify the range of each criterion. In this case, it is natural to assume a best case where all consequences are zero. Given that an oil spill actually has occurred, zero damage is of course impossible, but the intention now is just to specify the minimum points on the scales. The maximum points can be specified as the worst scores of the alternatives, or somewhat outside that. This changes Table 3-4 to Table 3-5.

Table 3-5. Oil spill case: Consequence table with best and worst scenarios.

Criterion	Unit	Best case	No action	Collect oil	Direct oil	Worst case
Recreation	1000 exp	0	200	200	1000	1000
Ind. loss	Million €	0	0.45	0.45	0.34	0.50
Shore poll.	Year km	0	600	600	300	600
Dead birds	1000	0	1000	1000	700	1000
Auk rest.	Year	0	12	12	10	12
Oil in sea	Ton	0	10,000	8000	8000	10,000
Combat c.	Million €	0	0	1.25	1.50	1.50

With linear utility functions, it is easy to transform Table 3-5 to the utility Table 3-6 with the formulae:

$$u(x) = \frac{x^0 - x}{x^0 - x^*} \quad x^0 \text{ is the worst and } x^* \text{ the best value in the range} \quad (3\text{-}1)$$

Table 3-6. Oil spill case: Utility table based on table 3-5 and linear utility functions.

Criterion	Unit	Best case	No action	Collect oil	Direct oil	Worst case
Recreation	1000 exp	1	0.8	0.8	0	0
Ind. loss	Million €	1	0.1	0.1	0.32	0
Shore poll.	Year km	1	0	0	0.5	0
Dead birds	1000	1	0	0	0.3	0
Auk rest.	Year	1	0	0	0.17	0
Oil in sea	Ton	1	0	0.2	0.2	0
Combat c.	Million €	1	1	0.17	0	0

Above we have assumed that all primary utility functions are linear, and that the worst case has the largest criteria value. However, some of the criteria may best be expressed as nonlinear.

For example, if you were more concerned about the last dead bird than the first because repopulation by birth takes longer, you would use a risk-averse utility function for dead birds as shown to the left in Figure 3-3. On the other hand, the first kilometer of polluted shore may create more concern than subsequent kilometers, because you have to go somewhere else anyway. In that case, a risk-prone function is what it takes, as shown to the right in Figure 3-3.

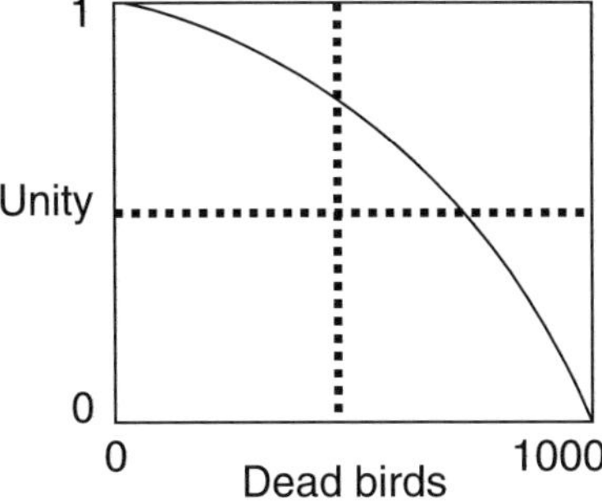

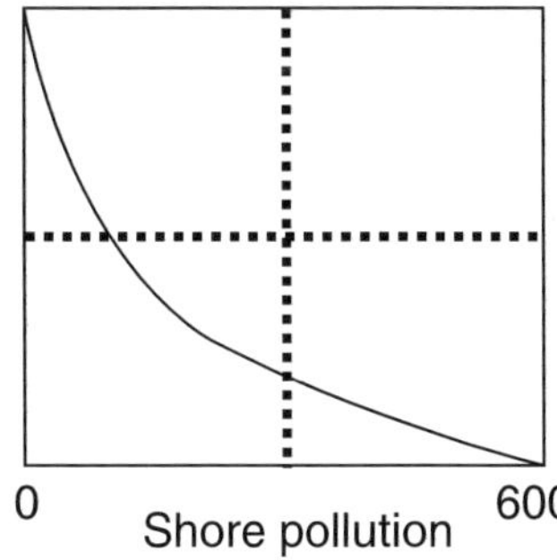

Figure 3-3. Non-linear primary utility functions for bird and shore damage. Left: risk-averse utility function. Right: risk prone-utility function.

It is usually a good idea to sketch a primary utility first function to consider possible non-linearity. You can do that by comparing the first and the last 10% losses in utility and see if the ranges are different.

In some cases you might want to use a utility function that first increases and then decreases. For example, if the amount of food additive is a criterion, small amounts may be advantageous, but large amounts may be toxic. In that case, you should split the criterion into two criteria, one representing the positive effect – preservation, and the other the negative effect – toxicity. That way you obtain one monotonically increasing and one decreasing utility function.

7. STEP 6: ELICITATION OF IMPORTANCE WEIGHTS

A crucial step in any decision analysis is to attribute importance weights to the decision criteria. There are several ways to do this. According to good practice, the simplest way is the method with swing weights. There are several other methods, however, and some of them require the assistance of a computer program. We will return to that in Chapter 7.

With **swing weighting**, the decision maker is first asked to consider the worst case. She is then told that she is allowed to improve the situation by giving one criterion its best score – this is called a swing, and the question is which criterion she would prefer to improve. The criterion she selects is assumed to be the most important criterion and assigned a temporary weight of 1.0. It is important to realize that the question is a pure value question. It does not matter if it in practice is impossible with a top score on this criterion and bottom on all the others. The question is hypothetical: "If you could move one criterion to its best value, which one would you choose?"

Notice that this way **importance** depends on the **intrinsic value** of the criterion as well as the range. Suppose for instance that combat costs were considered most important. This does not necessarily mean than it is more important to save one euro in combat costs than to avoid one euro in Industrial loss (intrinsic value). It means precisely that it is more important to save 1.5 million euro in combat costs than to avoid 500,000 euro in industrial costs – the whole range must be taken into consideration.

The next question runs as follows: Assuming again the worst case, which is the next criterion you would prefer to move to its best score, and how important is this as a percentage of the importance of the most important one? This way, we continue through the list of criteria, all the time selecting the criterion that the decision maker deems is next in importance.

7.1 Case: Oil Spill Combat. Elicitation of Importance Weights

In the oil spill case, a computer program was used to facilitate the elicitation process. Each of the three decision panels went through an interview process consisting of two sessions that each lasted about three hours, where the panels compared pairs of criteria in an effort to arrive at well founded values – that is, weights that the panels felt comfortable with. A crucial part of the process was discussions within the panels, which as a rule produced useful clarifications and insights, something that would not have been possible in ordinary willingness-to-pay surveys. The intention behind the use of three different panels with presumably different value systems was to come up with a range of weights that would contain the sentiments of the population.

To illustrate the method of swing weights, let us describe how the process might have proceeded. Table 3-7 is an excerpt from Table 3-5, showing only the worst and best scenarios.

Table 3-7. Case: Oil spill combat. Worst and best scenarios.

Criterion	Unit	Best case	Worst case
Recreation	1000 exp	0	1000
Industrial loss	Million €	0	0.50
Shore pollution	Year km	0	600
Dead birds	1000	0	1000
Auk restitution time	Year	0	12
Oil in sea	Ton	0	10,000
Combat costs	Million €	0	1.50

DP is the decision panel, and F is the facilitator:

- F: “Look at the worst case in Table 3-7. You may alleviate the situation by giving one criterion the best value. Which criterion would that be?”
- DP, after extensive internal discussion: “We would first of all spare the coast from pollution. 200 km of shoreline polluted for three years will lead to serious complaints” ($200 \times 3 = 600$).
- F: “OK, I assign a temporary weight of 1.0 to shore pollution. Assuming again the worst scenario, what is the next criterion you would give the best value?”
- DP: “A restitution time of 12 years for auks is serious for such an endangered species. This is therefore the one we would have chosen next if possible. But that would render the auks unhurt while at the same time killing one million birds, and this does not make sense.”
- F: “You have a point. The best situation would be if all criteria were independent. However, in real situations this is almost impossible, so let

us consider just how valuable criteria are, and not worry about whether individual scenarios might be technically possible! I have noted that auk restitution time is the next most important criterion, but how important do you feel it is compared to combat costs – can you give a percentage between 0 and 100?"
- DP: "We feel it is nearly as important as combat costs, let us say 90%."
- F: "OK, that means that the temporary importance weight of auk restitution time is 90%. What is the third most important criterion, and how important is that compared to shore pollution?"
- DP: "Actually, since it is a question of as many as one million dead birds, we feel that bird deaths are just about as important as the auk population."
- F: "Are you certain that you now only think of the suffering of birds, and not of the sustainability of the bird populations?"
- DP: "Yes, we have talked through that. We only think of the individual birds."
- F: "Then I assign a temporary weight of 0.9 to dead birds as well."

The interview session proceeds until all criteria have been assigned temporary weights. The temporary weights are then normalized to make the sum equal to 1. A possible result is shown in Table 3-8. Notice that to avoid industrial loss of one euro is considered more important than saving one euro in combat costs, since the former scale range is only 1/3 of the latter, while the weight is ½.

In reality, the panels from the State Pollution Protection Agency and the North Sea operators had comparable weights, the SPPA being somewhat more in favor of birds. The NGO, however, displayed a willingness to pay that was approximately 10 times higher than the two other panels, and they also had a stronger WTP to avoid oil disappearing in the sea. The possible free-rider problem was mitigated by comparing combat costs to operating costs of hospitals, and we believe that the panels expressed honest preferences at this stage in the process.

Table 3-8. Case: Oil spill combat. Ranks and weights according to importance.

Criterion	Unit	Rank	Temporary weights	Importance weights w
Recreation	1000 exp	4	0.7	0.14
Ind. loss	Million €	7	0.3	0.06
Shore poll.	Year km	1	1.0	0.20
Dead birds	1000	2	0.9	0.18
Auk rest.	Year	2	0.9	0.18
Oil in sea	Ton	5	0.6	0.12
Combat c.	Million €	5	0.6	0.12
	Sum		5.0	1.00

8. STEP 7: RANKING THE ALTERNATIVES

8.1 Total Utility

The decision alternatives can be ranked in order of preference simply by computing the total utility of each alternative. Depending on the preferences of the decision-maker, there are several ways to do this, and we shall return to a fuller discussion in Chapter 7. The simplest possible way, however, is to use an additive total utility function where we compute the weighted sum of utilities for each alternative. The formula for two criteria was presented in section 2.5, and is readily extended.

8.2 Case: Oil Spill Combat. Ranking of Alternatives

Table 3-9. Oil spill case: Total utility for the decision alternatives.

Criterion	Weight	Best case	No action	Collect oil	Direct oil	Worst case
Recreation	0.14	1	0.8	0.8	0	0
Ind. loss	0.06	1	0.1	0.1	0.32	0
Shore poll.	0.20	1	0	0	0.5	0
Dead birds	0.18	1	0	0	0.3	0
Auk rest.	0.18	1	0	0	0.17	0
Oil in sea	0.12	1	0	0.2	0.2	0
Combat c.	0.12	1	1	0.17	0	0
Total utility		1.00	0.24	0.16	0.23	0.00

Table 3-9 shows the computed total utilities for the decision alternatives in Table 3-5. For instance, with an additive utility function, the total utility of No action is computed this way:

$$U(\text{No action}) = 0.14 \times 0.8 + 0.06 \times 0.1 + \ldots\ldots + 0.12 \times 1 = 0.24$$

We see that No action comes out first, then Direct oil, and last Collect oil. However no action and directing oil utilities are almost equal. The natural decision would therefore be to stay put and not try to do anything with the oil spill. On the other hand, to collect oil at the site appears clearly inferior, so at least one lesson was learned from the analysis. To direct oil instead of collecting it at the site is often a superior strategy. This was illustrated by the incident with the Bahamian tanker Prestige, which in 2002 leaked oil close to the shore of Spain. To lessen the pollution impact, it was brought out further from land, but then broke in two and leaked 77,000 tons of oil that polluted an extensive shoreline. It was afterwards realised that the damage would have been much smaller if it had been brought close to shore instead of the opposite.[5]

8.3 Uncertainty Analysis

A ranking of alternatives based on total utilities is obviously not the last word. One has to consider uncertainty, of which there are essentially two major sources: uncertainty with regard to consequences and uncertainty with regard to values – or importance weights. Both of these have to be considered carefully. Uncertainty with regard to values can best be dealt with by retaining a protocol of the elicitation process, and using that to investigate whether changes in responses where the panel was doubtful would alter the ranking of the criteria.

In Chapter 2, we showed how we could take statistical uncertainty about consequences into account by assigning probabilities and computing expected utilities. The impact of uncertainty about consequences can be actually be analyzed very efficiently in a spreadsheet. There are add-inns to spreadsheet programs that make it easy to point at different cells and specify the degree of uncertainty in terms of a standard deviation and a probability distribution for the number in the cell. Then one points at the target cells, which in our case are the total utilities. The add-in will then perform a simulation and show how uncertain the targets are.

If suitable software is unavailable, you should at least construct reliability intervals for the most promising alternatives. For each alternative, identify the most uncertain criteria and change their scores within a reasonable range. In a spreadsheet you can immediately observe the effects on the utility of that alternative. Enter your results in a diagram as shown in Figure 3-4. Here, you see that alternatives *b* and *c* are rather similar, while alternatives *a* and *d* are less desirable than *b*. Recommend alternative *b*.

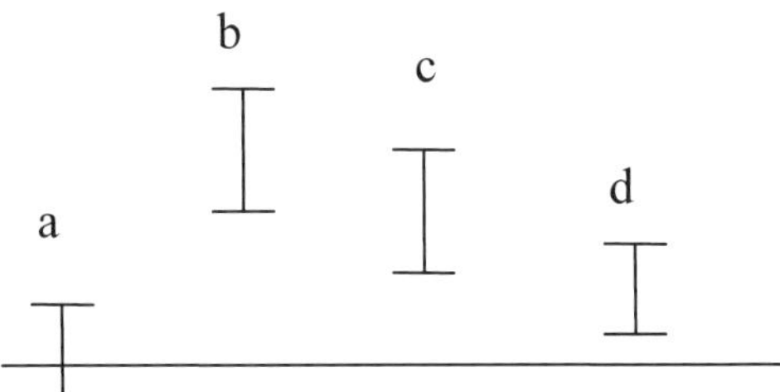

Figure 3-4. Uncertainty analysis: Four decision alternatives with reliability intervals.

If none of the alternatives are obviously better than others, then other criteria than those included in the analysis may be used. However, this should be stated explicitly. Sometimes new information emerges during the analysis; but if it is too late to include it, it should be treated as a separate piece of information together with the analysis itself. A formal decision analysis cannot be performed in isolation, but must be seen in relation to political realities.

---

[1] After Grigalunas (1983)
[2] Weber (1988)
[3] dos Santos (1988)
[4] Keeney (1980) p. 55
[5] www.cnn.com/2002/WORLD/europe/11/19/spain.oil/

Chapter 4

INTEGRATED ASSESSMENT

Comprehensive analysis of environmental impacts of projects

Integrated Assessment is an established term for comprehensive environmental analysis of projects. We discuss three central issues:

- How to make an Environmental Impact Statement?
- How to select good monitoring points and performance measures on the impact pathway?
- What are the key concepts in Life Cycle Analysis?

●

"But what happens when you come to the beginning again" Alice ventured to ask. "Suppose we change the subject", the March Hare interrupted, yawning. Lewis Carroll: *Alice's adventures in Wonderland*

1. INTRODUCTION

Smoke related cancers may develop tens of years after its causal agent has ceased to act; and pollution may drift across continents before they affect a target. Integrated assessment (IA) is a comprehensive approach that includes a bundle of tools for handling project effects that occur over long distances in time and space. A **project** is an activity that starts at some point in time, and is assigned time slots and resources; in our context a project is also assumed to have environmental impacts. An **integrated assessment,** IA, of a project thus includes a study of all impacts, on the environment as well as humans. It applies a life cycle perspective, considering production, consumption, as well as disposal and reuse. Impacts from different stages may be quite different for different kinds of goods or services.

IA is structured around four main themes:

- *Objectives*: Who is the decision maker and what are the end impacts or ultimate goals?

- *Industry*: Industrial, agricultural and socioeconomic activities that produce goods and services.
- *Ecology*: Dilution processes and links between industrial activities and impacts on humans and the environment.
- *Economics*: Monetary assessment that links material cycles to man's short and long-term strategies and goals. This includes ecological economics, as well as utility analysis and psychometric methods that deal with valuation of goods and services.

Environmental Impact Analysis (EIA) is more restricted than IA. It ends up with an *environmental impact statement* and consists of the following elements:

- *Environmental inventory*: A description of the existing environment at the site where the project is proposed according to a checklist of descriptors of physiological, chemical, biological, cultural, and socio-economic environments.
- *Environmental impact assessment*: A systematic identification and evaluation of potential impacts on the environmental inventory of the proposed project, program, or legislative action.

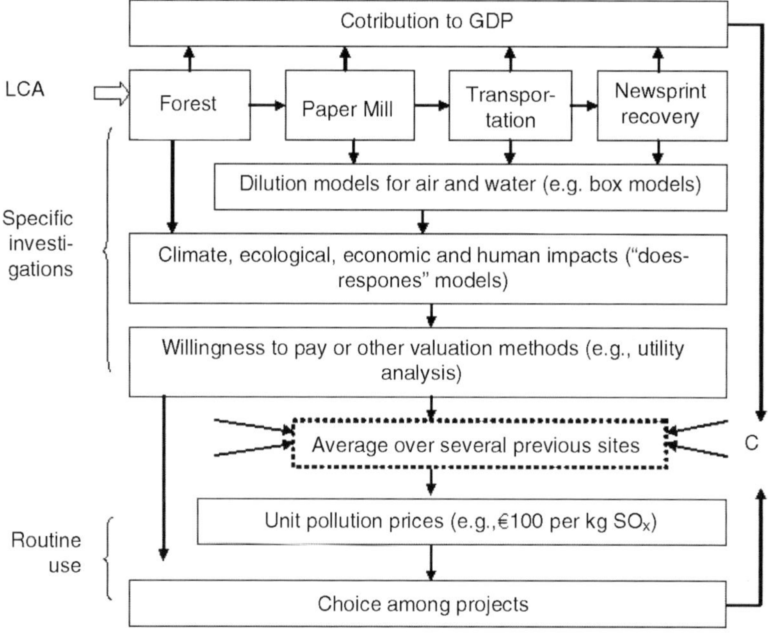

Figure 4-1. Integrated assessment of a paper production and paper use project. At C there are opportunities for comparisons.

Integrated assessment is illustrated in Figure 4-1, with the life cycle of paper as an example. The IA project starts with a life cycle analysis, LCA. The life cycle of paper consists of timber logging, production of paper at the mill, transport and recovery of waste products. Usually, an environmental impact statement will be required when a paper mill is built or upgraded, i.e., the second step in the LCA of Figure 4-1.

Common forest related damages are destruction of natural habitats during logging operations. The paper mill discharges pollutants, chronic or accidentally, to water and air. As part of the IA, one develops spread and dilution models to describe their impacts. To evaluate alternative mitigation measures that reduce the impacts, one needs to evaluate them, either through decision analysis or – more commonly – by using standard unit prices attached to the pollutants. In routine assessments, LCA and development of dilution models and dose- response models are expensive activities, and often replaced by unit prices, which are assumed to be average damage cost to the society of emitting one unit of pollution.

The cost of the environmental load of a project can be put into perspective by comparing its contribution to the gross domestic product, GDP.

2. EIA STEPS

Many countries require EIA by law. The details depend on the type and size of the project and the vulnerability of the environment. In general, EIA involves six steps some requiring considerably more attention than others.

2.1 Step 1: Screening the Environmental Problem

The first step is to perform a screening procedure to examine whether an impact assessment is at all required, or if it belongs to a category that is exempted. Renovation, maintenance, and minor extensions are usually exempted. It is often useful to describe environmental threats along two dimensions: *severity of impact* and *probability that the impact will occur.* Impact severity is categorized in four classes, and the combined effect of severity and probability is attributed a level of significance. The highest significance category – very significant – consists of those projects that probably will cause extremely large damage. Recurrent flood is an example. The significance of nuclear power fallout and collapse of very large dams depends upon the probability of failure. The highest significance category may also include projects where legal thresholds are surpassed or where damage is certain. In cases where a stakeholder firmly believes that both severity and probability are higher than the experts' risk estimates, the precautionary principle suggests that the higher significance category be applied.

An overview is shown in Table 4-1. If the probability is less than 10^{-6}, the combined effect is usually deemed insignificant, and the project is exempted. If the impact is "significant", an EIA should be carried out.

Table 4-1. Defining significance.[1] p is impact probability.

Impact severity	Impact likelihood category A: $p > 0.5$	B: $0.05 < p < 0.5$	C: $10^{-6} < p < 0.05$	D: $p < 10^{-6}$
very high	very significant	significant	significant	not significant
high	significant	moderately significant	moderately significant	no further study
moderate	moderately significant	not significant	no further study	no further study
low	not significant	not significant	no further study	no further study

If it is "moderately significant", an EIA may be required and a screening test should explore the problem. "Not significant" means that a screening study should be undertaken, but that a full EIA is not likely to be required. "No further study" suggests that no further consideration is required.

2.2 Step 2: The EIA document

The EIA document is to be included in the project proposal, which is submitted to the authorities. It normally contains the following:

1. Relevant government rules and regulations.
2. Benchmark information, that is, information about the environment without the project.
3. Estimates of the impacts of the project, as well as an outline of elements to be monitored before, during, and after the construction period. By comparing predicted and observed data, one can assess the accuracy of the predictions.
4. An evaluation of the predicted impacts, for example by comparing them to established standards. An example is transparency in waters designated for swimming, which should be more than 2m – the lowest required transparency in certified swimming ponds.
5. Identification of possible mitigation actions.
6. Writing the document.

Step 2 takes often 5%–15% of the budget.

2.3 Step 3: Multidisciplinary Team Management

EIA requires several types of expertise; engineers assess emissions and discharges; atmosphere and hydrology scientists predict how these spread in air and water; toxicologists assess the resulting levels of toxicity, and biologists and health workers assess the impacts on ecosystems and people. Output from one discipline is therefore input to another. Since researchers in all disciplines tend to stress the uncertainty of their own evaluations, the chained reasoning typically produces a large number of scenarios. It is the task of the team manager to reduce it to something manageable.

Allow 5%–15% of the budget for step 3.

2.4 Step 4: Public Participation

The public is an important stakeholder in environmental matters, and often wields considerable decision power. The purpose of public participation is to exchange information, elicit local values, establish credibility, and observe peoples' "right to know". There are several levels of public decision power:

Citizen control: Citizen control means that citizens without stakes in the project have the majority in decision-making bodies, or that a referendum decides the outcome. If a general referendum is not possible, it is necessary to select representative citizens.

Delegation of power: Decision making power may be delegated. The choice of delegates is important since many projects have failed, in particular in developing countries, because the project managers did not address the concerns of the people that were directly affected, or used a decision making process which was unfamiliar to the people concerned.[2] There are several ways to choose delegates:

- The proponent of the project may select members to the delegation that balance different interests and represent all stakeholder groups.
- The proponent may select members to a voting committee, which in turn selects the decision-making body. Alternatively, the voting committee may be elected by the people.
- The proponent may identify stakeholders and allow them to select their representatives. The stakeholders will often be invited to an initial meeting where the project is explained.[3]

Allocate 2–5% of the budget for step 4.

2.5 Step 5: General Study

The general study is the topic of the next section. It consists among other tasks of goal hierarchy construction and consequence analysis.

2.6 Step 6: EIA Review

This step is to ensure that the EIA has been adequately completed as agreed. An independent group should perform this important task.

Allocate 10% of the funds for step 6.

3. MONITORING POINTS ALONG THE IMPACT PATHWAY

In this section, we follow up the discussion of goal hierarchies from the previous chapter. There we defined an end-impact as a consequence that affects fundamental values – something we care about, or want to avoid. End-impacts are represented by decision criteria in the goal hierarchy. At the same time, an end-impact is an **end-point** in a causal chain of environmental processes. There is for instance a causal link from a production process to emission of pollutants to discharges through a chimney to concentrations in the air to end-point impacts like coughing in children.

The challenge is that we have freedom of choice when we design goal hierarchies. They are supposed to systematize values or end-point impacts that are at stake in connection with the project. If possible, the choice of end-impacts should reflect fundamental values rather than instrumental values. Suppose for example that you are concerned about water pollution from phosphorus. You could use *ton nutrient discharged* as a decision criterion, but this is only an indicator of your real concerns. You could therefore go further down the impact pathway and perhaps measure the impact as *water surface area with above guideline concentration*, and use that as a decision criterion. The guideline could be related to **water quality class**. Pollution control authorities define such classes according to whether the water is used for consumption, industry, recreation or other purposes.

Pollutants are emitted, spread, and do damage. The reason we recommend focusing on end-point impacts – the real damage – is that damage is easier to valuate. We feel it or see it, and are willing to pay to avoid it. On the other hand, prediction of damage effects is hard. While it is rather easy to predict emissions from a project, it is harder to predict ambient concentrations, and hardest to predict end-point impacts. Therefore we have to compromise.

Table 4-2. Pollution monitoring point and measurement.

Monitoring point and measure	Examples of decision criteria
End-of-pipe emissions or discharges	Gap between ambition and achievement such as ton emission above threshold per day Fraction of total emission and discharges Pollution fee in euro per ton emitted
Ambient conditions Area that is contaminated above a threshold	Damage area (km^2)
End-point impacts	Number of exposed persons Cases of increased risk of heart attack

We must therefore decide at which one of the three stages along the impact pathway effects can be most conveniently monitored and valuated. Each stage gives different decision criteria, as shown in Table 4-2. Each stage has its own pollution severity measures as shown in the right column of the table. Naturally, all methods require that the effects of the pollutant can be isolated from other sources. Each stage has advantages and disadvantages with respect to evaluation and implementation in an industrial context. To minimize the damage area, for instance, is an instrumental objective, and to evaluate it, one must assume that there is a causal relationship to the end-point objectives. If the goal is to minimize phosphorus load in a lake, it is because one assumes that the phosphorus causes end-point environmental effects by being converted to algal biomass. Sometimes we are able to model the relationship between emission and end-point damages by generally accepted causal chains, although uncertainties may be large.

The chain between emissions, polluted area and damage to valuable resources is illustrated in Figure 4-2.

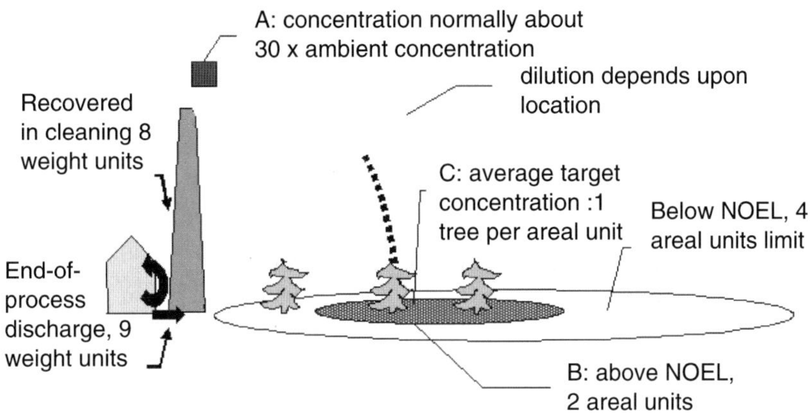

Figure 4-2. A typical impact pathway: emissions, polluted area and end-point impacts.

The left part of the figure illustrates the dilution processes; the right part illustrates the effect of pollutant concentrations on resources. The figure shows an example where the end-of-process pollution yields nine pollutant units. Through recycling and purification, one unit is left and discharged through the chimney. End-of-pipe discharges are thus one pollutant units. The consequence is two area units with concentrations greater than NOEL (NOEL = No observable effect level), and four area units below NOEL. The average resource concentration is one tree per area unit, thus the expected damage affects two trees. For a worst-case damage, one could assume that the emission becomes contained in a pocket of air and transported to a place where there is a maximum concentration of susceptible resources. For example, the package reaches two area units where the maximum intrinsic resource concentration is three trees per area unit, and thus the worst-case damage would be six units. The probability of a reasonable worst-case damage should be higher than 10^{-6}. If one of the hurricanes that hit the southeastern coast of USA hits Manhattan, a substantial disaster would occur.[4]

3.1 End-of Pipe Monitoring

Industrial emissions are monitored by the EPA and reported in the form of an inventory list of emissions and discharges. The so-called Toxic Release Inventory, TRI, assesses the environmental burden, EB, imposed by major industries through a list of 517 potentially toxic chemicals. (The 2002 TRI list, released 2004.)

In project evaluation, it is also desirable to give information about the severity of the pollutants. This is done in the form of an index that summarizes the total burden of the pollution. If the index is below a certain level, the project may be carried out; if it is above, the project should be cancelled. There are at least four indices: "Total release", "Toxicity weighted", "Distance-to-target" and "Fractional emission".

3.1.1 Total Release Index

This index is based on emissions of potential toxic chemicals from a list of chemical categories in terms of discharged tons. A Dow chemical facility in La Porte, Texas reported in 1987 that 1730 thousand pounds of toxic compounds were released and transferred, and this included release of monoclorobenzene to air and water. In 1993, the release was reduced to 960 thousand pounds.[5]

$$EB_1 = \sum_i^n X_i \tag{4-1}$$

X_i are emissions measured in ton per time unit. In some cases the burden is assessed by only one discharged chemical, $n = 1$, presumably the one with largest impact on the environment.

3.1.2 Toxicity Weighted Index

The concentrations of released products, although all rated as potentially toxic, may have different threshold values before they cause toxic effects. Threshold Limit Values, *TLV*, have been reported ranging from 0.37 to 1780 $\mu g\ m^{-3}$ for a limited selection of chemicals on the TRI list.[6] Total phosphorus – the sum of all chemicals that contain phosphorus, total nitrogen and a series of other substances affecting the environment are not included in the list. The index EB_2 is obtained by dividing the amount of toxic material (on the list of 517 potentially toxic materials) by the threshold concentrations of the respective products. Since *TLV* has a threshold value of 1 $\mu g\ m^{-3}$ for sulfuric acid and phosphoric acid, EB_2 expresses the burden as a transformation of the emitted products into sulfuric acid equivalent emission. With EB_2 as pollution index, the ranking of firms on the TRI list changes considerably.[7]

$$EB_2 = \sum_i^n X_i \times TLV_{ref} / TLV_i \qquad (4\text{-}2)$$

Exercise: With reference to Table 4-3, estimate the environmental burden caused by paper mills A and B. Which factory is the worst?

Solution: According to the Total Release Index it is factory A, but according to the Toxicity Weighted index, it is factory B, as shown in Table 4-3.

Table 4-3. Emissions in kg per 1000 kg product from paper mills A and B. *TLV* is Threshold Level Value.

Factory	SO_2	NO_x	Sum $SO_2 + NO_x$	Sum $SO_2 + NO_x$ (weighted)
A	0.3	0.72	1.02	0.0071
B	0.2	1.00	1.20	0.0077
TLV, 24 hr $\mu g\ m^{-3}$	100	175		

3.1.3 Distance to Target Indicator

This indicator relates pollution emissions to political targets for emission reductions. The idea is that it is easier for people to appreciate environmental performance regarding a certain emission if it can be reported as a fraction of the target, or distance-to-target.[8] The gap closure principle is applied when actual end-of-process discharges are compared to reduction goals:

$$I = \Delta E / (E_{present} - E_{goal}) \tag{4-3}$$

For example, if a national goal is to reduce nitrogen discharges from 28000 ton in 2005 to 14000 ton in 2010, a reduction of 1000 ton gives I = 1000/14000 = 0.07 over the time period.

3.1.4 Fractional Emission

The fourth method is to calculate emissions or discharges as a fraction of total discharges in a country or a region. This method corresponds to the distance-to-target method when national goals are zero emission.

3.1.5 Discussion

The Toxicity weighted index has been criticized because TLV has to refer to one type of toxicity, for instance with regard to human health. However, pollution damage occurs also at lower levels. For example, corrosion from SO_2 concentrations in the air occurs at much lower values than health effects. On the other hand, "safe" pollution levels for health effects may at closer examination be higher than reported, because excessive safety factors have been used.[9] Another uncertainty lies in subtle, long time effects that may not have been discovered. The distance-to-target index circumvents this critique, because it merely expresses current political concerns.

3.2 Ambient Pollutant Concentrations

Damage can be evaluated by reporting the environmental burden in terms of area with ambient concentration above TLV.

$$EB_3 = \sum_i^n A_i \tag{4-4}$$

A_i is the area (in hectare) with concentration above TLV with regard to chemical i.

A version of this method is the Habitat evaluation method. In this case, the concern is about the quality of a habitat. The quality may be degraded because of air pollution or toxic discharges, or because of noise and human activities that affect populations of the habitat. The areas are weighted with respect to their habitat quality, HQI, ranging from 0 to 1.0, where 1.0 is best. Thus, the environmental quality is measured as "prime quality equivalent area".

$$EB_4 = \sum_i^n A_i \times HQI_i \leq \sum_i^n A_i \qquad (4\text{-}5)$$

An underlying assumption for reporting the damaged areas, EB_3, or EB_4 is that most people have an idea of the extent of an area. That is, they can intuitively relate to 10 km^2 of degraded land. The Method described in 3.2 is used to calculate the maximum area damaged. Area methods reflect the toxicity of each pollutant, as does the TRI index.

3.3 End-Point Impacts

There are many different end point impacts, and we only give some examples below. We start with birds; their wings make them especially challenging since they move easily.

3.3.1 Birds that Move and Communities with Long Recovery Times

An oil spill may pollute large areas of the ocean, but there is not necessarily a direct connection to the number of birds that die. Thus, for natural resources that move, like birds, some kind of estimate of concentrations at the time of a discharge must be available. For some recipients, it is also relevant to include the time it takes for the environment to return to normal state. This was done in the oil spill case in Chapter 3. If the **restitution time** is long, the effects are worse than if it is short, other parameters being equal. For example, if an oil spill hits a colony of kittiwakes (*Rissa tridactyla*) and kills 1000 birds, this may be considered as a smaller environmental damage than if the oil spill kills 1000 Guillemots (*Uria alge*). The reason is that it takes much longer time to restitute the guillemot population than it takes to restitute the kittiwake population.[10]

3.3.2 Box Calculation of the Polluted Area

This method can be used to calculate the polluted area by assuming that the pollutant is confined for a certain time period to a box with a certain area and height. We have to know i) the amount, Q, of the pollutant emitted over a time period, ii) the mixing height, h, in air, water or soil, depending upon the receiving medium, and iii) the No Observable Effect Level, NOEL, of the concentration of the pollutant. NOEL is measured in per mille concentration or $\mu g\ m^{-3}$. For contamination of water by nutrients, an upper threshold level defining the best "Water class" can be used.

End-of-process discharges can sometimes be obtained from **Life-Cycle Analyses**, LCA. LCA typically computes discharges per "functional unit", for instance 0.01 kg SO_2 per 1000 kg newsprint paper produced. The total amount of pollutant, Q, can then be found by multiplying with the production volume over a time period corresponding to the residence time of the pollutant in the receiving medium. Typical residence times for pollutants in air are two hours to 24 hours, and in water one day to one week. The height of the box corresponding to the minimum mixing height, can be set to 5 m in air and 1 m in water although 100 m and 10 m would be more typical minimum mixing heights.

The polluted area is calculated this way:

$$A = \frac{10^9 Q}{hc} \tag{4-6}$$

Q is discharge (kg) during the residence time – 24 hours in air and a week for water pollutants. *A* is area in km^2, *c* is the NOEL (24 hour) concentration of potential pollutant in $\mu g\ m^{-3}$. The factor 10^9 converts from μg to kg. *h* is height of equivalent air volume or water volume, 5 meters for air and one meter for water.

3.3.3 Resource Damage Calculation

This method estimates the damage within an area, *A* (km^2) caused by a certain degree of pollution. For air pollutants, we use as input the concentration of natural resources or people susceptible to the pollutant in the area. For health effects, we would for example use the density of people living in, or passing through, the area considered. Calculate the resource damage as:

$$D = C_R \times F \times A \tag{4-7}$$

Here, C_R is the density of resources, such as the number of people per km^2. *F* is the fraction of the resources that is susceptible. The damage index unit is then number of people. (The average density of people in 29 cities in the Baltic drainage basin is 10,000 km^{-2}.)[11] If there are several resources or several areas with different levels of pollutant, they have to be divided into areas with approximately uniform distributions.

3.4 Valuation of End-Impacts

To assign monetary values to end-impacts is probably one of the greatest challenges in environmental decision-making. It is a necessary step if monetary gain from a project is to be compared directly to environmental loss. As we discussed in Chapter 2, valuation of environment and health is a question of subjective preference, and there are no clear answers regarding the best way to perform valuation. Experts may do it on behalf of the public, or it could be done by popular votes, or assessments based on answers to questionnaires.

Table 4-4. Estimated costs of NO_x emissions from a project.[12]

Impact	Damage cost, euro kg^{-1} 1994
Acidification of water	0.0025
Acidification of forest	0.0625
Health impacts	12.0250

We will discuss methodological issues in depth in Chapter 8 on Willingness-to-pay for environmental amenities (WTP).

In several other chapters in the book we also include some information on observed WTPs. It is very practical, especially in Integrated Assessment, if some information about estimated monetary costs of emission of one unit of a polluting agent can be included in the report. Such estimates must be based on predicted damages, such as in a Norwegian study where the cost of 1 kg NO_x emission was estimated. The result is shown in Table 4-4. NO_x emission was assumed to have three end-points, water, forest and human health, and damage costs judgmentally assessed by the Norwegian pollution control authority. Note that damage caused by acidification is valuated much lower per kg NO_x than damage to human health.

If it is not known which target the pollutant hits, one has to make assumptions about probabilities, and compute an expected damage cost based on the target specific unit costs.

4. LIFE CYCLE ANALYSIS

Life cycle analysis, LCA, deals with the environmental burden caused by a product or a service during its life span "from cradle to grave". Nowadays, one should perhaps say "from cradle to cradle" since materials are often reused. LCA analysis is most often conducted with the help of commercial LCA software, like SimaPro[13] and the Carnegie-Mellon system.[14] A full description of LCA that complies with the ISO 14040 standard can be found elsewhere.[15] Here, we shall only give a brief account of the method.

4.1 Three Central Concepts

Three terms are important in LCA. The first is **functional unit**, which is the unit of analysis in LCA. A functional unit is a product that has certain functions and causes certain environmental impacts. For example, assume that you want to use LCA to compare the environmental burden of glass and plastic used as packing material. Then you have to choose a unit of analysis that makes comparison possible. You could for instance define the product as material that will pack 1 liter of wine. This product is the functional unit.

The second important term is the **boundary** of the system supporting the good through its life. A good system boundary definition is important in LCA, and Figure 4-3 shows a scheme for that purpose.

The horizontal streams in Figure 4-3 illustrate the flow of the good as it moves through production and usage. The upper stream is the production stream, and the lower the reuse-stream. The vertical streams symbolize the accumulated burdens during the lifetime of the product. (1) is pollution associated with production; (2) is pollution associated with scrapping of production equipment; (3) is pollution associated with product use; (4) shows the reuses stream, (5) shows end-of-life waste depots, and (6) shows the pollution distantly related to the production of the product

The third important term is **inventory analysis**. This is the step in LCA where energy sources, raw material, and environmental loadings for the entire life cycle of the product or process are identified.

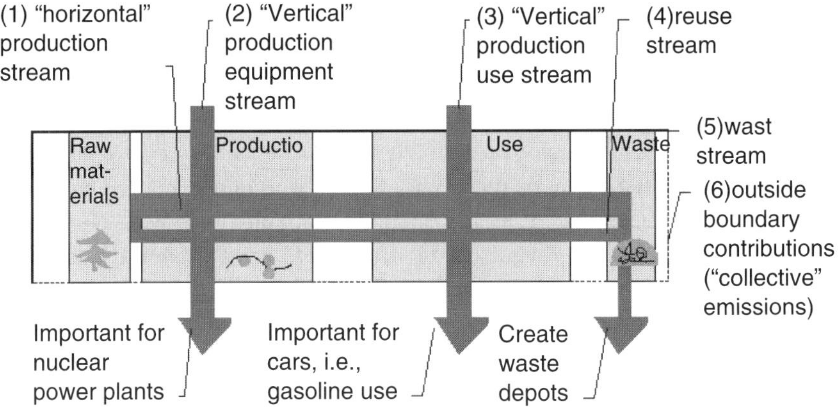

Figure 4-3. System boundaries and material streams.

Often, LCA stops here. The result is a list like the following for a washing machine costing 1000 USD (the Carnegie Mellon system): 700 kWh electricity used, 12 GJ energy used, 1 ton CO_2 equivalents released, 0.4 ton ores used, 4100 kg SO_2 and 600 kg PM_{10} emitted.[16,17]

4.2 Impact Analysis

A next step, **impact analysis**, or risk analysis, characterizes the effects on environment and human health. To relate end-of-pipe to end-points is difficult and laborious. There are, however, several shortcuts available, depending upon the purpose of the study. One approach is to represent all effects in monetary units by using unit prices. A second approach is to apply models that are generic for a geographical region. To illustrate: The result of an LCA could be that in a particular region, PM_{10} emitted from a factory that manufactures washing machines will become diluted with resulting air concentration around 0.1 $\mu g\ m^{-3}$. By using dose-exposure models, this finding could translate into a probability of 0.009 per year that a person in the region acquires respiratory symptoms. A single washing machine would then be responsible for part of this risk.

In some industries, such as nuclear energy, the costs of factory scrapping and site remediation are substantial. More than 30 years ago, such costs were seldom included in LCA; industrial sites were simply abandoned. It is estimated that there are about 1500 toxic waste sites in the US, which is about 0.6×10^{-5} sites per capita.

4.3 Outside System Boundary Contributions

To save time and efforts, the life cycle analyst may specify narrow LCA boundaries, and take into account contributions from outside the boundary in a more cursory way. She may for instance decide to consider only the 10 presumed highest cost items,[18] or 0.1% of the environmental load. It is also important to take into consideration that many products, like petrol, now have added a "green" tax to its sales price. The environmental burdens of a product from outside boundary activities, I_{ob}, can be estimated by assigning a proportion of countrywide total discharges, I_{tot}, as they are reported in official statistics.[19] However, there are arguments that suggest that this procedure introduces larger errors than by not using it.

$$I_{OB} = I_{Total} \times F_{OB} \times AV / TAV$$

Here, AV is added value from the industry in the study – it can be found in standard governmental statistics, F_{OB} is the fraction of the industry's added value from outside system boundaries, and *TAV* the total value added from the main segments of the industry division.

5. SELLING INTEGRATED ASSESSMENT

Having completed an IA, we perhaps include an LCA, and an accompanying decision analysis to evaluate alternatives; the task now is to convince public or private decision makers about the merits of the work. If the decision makers accept the results, the work is finished.

Very often they do have objections, however, and for various reasons. They may find the consequence model inadequate; they may find the study incomplete, or disagree about the choice of decision criteria. And that does not complete the list. There are also at least five less tangible reasons the decision makers may have for being reluctant.

1. *Legitimacy*: The legitimacy and the quality of the study may be questioned.[20] Did those that judged the criteria weights have sufficient authority to do so?
2. *Risk*: The decision-maker goes for a project with less expected return than the one you recommend. This may be because risk may not have been dealt with adequately. If it is possible that things could turn out much worse than anticipated in the models, the risk should be made explicit. The standard way is to use risk-averse utility functions, which translates into a risk premium, meaning that the decision-maker is willing to sacrifice some the expected returns from a risky project and go for a less risky project with smaller expected returns. The size of the risk premium is a subjective matter, however, and depends on the feelings of the decision-maker.
3. *Stakeholder conflict*: The receiver of the benefits may not be the same as the one carrying the cost. Examples are long-range transport of pollutants, or a government that only looks to its own monetary income.[21]
4. *Discount interest rate*: The benefits may have been discounted according to an official interest rate, but the decision maker, in rejecting the project may unconsciously have used a higher rate. For example, many people would use a low discount rate to calculate the present value of future lives saved.[22]
5. *The "dumb farmer" syndrome*: This syndrome refers to a situation in which the environmental assessment presupposes that everything is constant except the direct consequences of the project. Decision-makers understand, however, that people normally will do things to improve the situation. Dumb farmers do nothing, but real farmers shift to other crops.

6. APPLICATION: SITING PAPER MILLS AND THE LIFE CYCLE OF PAPER

Where should paper mills be sited to minimize adverse environmental and socio-economic impacts of production, use, and disposal of pulp and paper? Figure 4-4 shows a typical paper mill. The emissions in this case may be water vapor, but for many mills there are significant chronic emission and discharges, and for all mills accidental pollution incidents. A third concern is the possible detrimental effects of unidentified chemicals.

Below, we describe a study where two projects were compared, a paper mill at a polluted site, and a paper mill in a pristine environment. It was assumed that benefits are the same regardless of location; thus only the externalities of the paper mill will affect the decision. [23]

Figure 4-4. Paper mill.

6.1 Decision Analysis

The impacts of paper production and use are not restricted to the site of the paper mill, which may have a lifetime of 100 years. The forest where wood is cut for pulp production and the waste dump or incineration plant, may be somewhere else, are affected as well – and it also matters where the paper finally ends up.

Decision-maker and stakeholders: The decision maker is the paper company's board of directors, and that the Local Ministry of Environment, NGOs and forest owners are important stakeholders (Chapter 3, section 2.3).

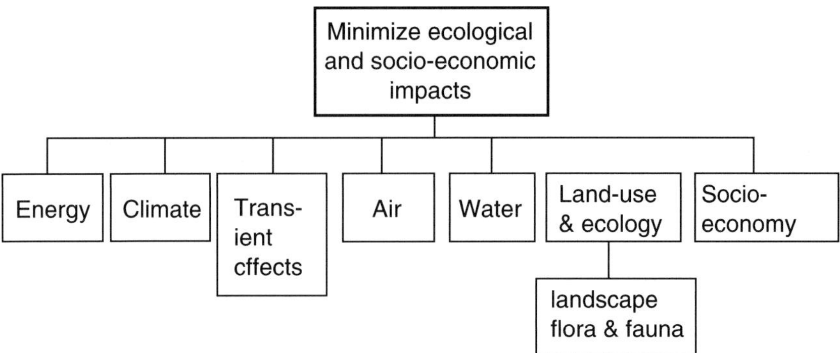

Figure 4-5. Generic goal hierarchy for an industrial product.

If the paper mill were built in a developing country, forms of ownership – tenure – may differ from that of developing countries, and discussion on how to manage the site or the forest where the timber is cut may be conducted in a discursive format where the weakest part is empowered (Chapter 9, section 6.3).

Decision objectives: Figure 4-5 shows a generic goal hierarchy, which had previously been used in studies that involve LCA of paper products.[24] End-point criteria were selected according to the main transportation medium or "vehicle" bringing the pollutant to the affected target. In addition, three categories not directly related to any vehicle, and one category that summarizes transient effects related to the construction period, was added. Thus, there were seven subgoals as shown in Figure 4-5.

Adding decision criteria that measure the achievement of the subgoals completes the goal hierarchy. Here is a list of the seven subgoals and 22 end impact categories that need to be further developed into decision-criteria; the details will depend on the particular setting:

1. *Energy*: Fraction of the world's non-renewable energy resources used. Other effects of non-renewable energy use, like emissions of toxic substances, can be categorized according to the vehicle that dilutes and transports them, e.g., water and air.
2. *Climate*: Global warming potential, Ozone depletion potential.
3. *Transient effects* during the construction period can be aggregated and described with one criterion.
4. *Air*: Human mortality, Human sub-lethal effects, Noise, Odor, Corrosion on buildings and monuments, Winter fog.
5. *Water*: Eutrophication, Water use.

6. *Land use and ecology*: Vegetation, Wildlife, Endangered species, Storage of waste.
7. Socio-economic damage: Recreation.

How to structure objectives and goals are discussed further in Chapter 3, secton 3.

Consequence analysis: To assess goal attainments for the two alternative projects, one has to consider expected effects, how probable they are, and the potential for natural or artificial restoration. The case study was based on a study of paper mills in Norway and central Europe, and scores were assessed for "typical" mills at the respective sites. LCA was used to describe processes such as logging of wood, transportation, production of pulp and paper, distribution of paper products and finally their disposal. Simple models such as the box model described in this chapter, section 3.3, were used to describe discharges into water and air. Gaussian dispersion models described in Chapter 21, section 4.3 could be used to estimate dilution caused by large chimneys. Emission of carcinogenic substances must be dealt with through toxicity slope factor calculations, whereas non-carcinogenic emissions can be dealt with through the hazard quotient method, both in Chapter 14, section 4. Noise impacts could be described by estimating the number of people exposed to noise above 55 dB(A), or one may use the noise impact index in Chapter 10, section 4.4. Logging in forests may cause 5% to 10% of a natural forest to be in non-attainable state, (Chapter 19, section 3). In Chapter 19, section 2.5 one will also find data to estimate forest area required to support the paper mill with raw materials. Water quality criteria are supplied in the chapters for the lake and river environments respectively (Chapters 17 and 18).

Preferences: The decision-maker determined the criteria weights, within constraints imposed by environmental authorities, and based on legislation. It may also be valuable to solicit the preferences of stakeholders with methods described in Chapters 7 and 8.

6.2 What Happened?

No single criterion appeared to dominate in the current case, in the sense that it was judged much more important than the others, although previous decisions often have emphasized closeness to timber resources and possibilities for discharging process water. The consequence analysis showed that end-point damage ranged from 3% to 23% of "worst case" damages for 9 out of 12 decision criteria if sited in a pristine region. However, to compromise the area of pristine environments of a region is a political decision and requires methods to solicit political judgments, such as the utility method or the pair-wise comparison method (Chapter 7, section 2.2.).

The external cost of paper production was estimated to range from 7% to 17% of the contribution to the gross domestic product of paper production, this chapter Figure 4-1. Methane production at the dumpsites for paper contributed the most.

7. CONCLUSION

"Integrated assessment" is a term that describes how a decision problem is embedded in a framework that includes all potential effects of the project. However, the term "potential" indicates that not all effects are included in the analysis, and good judgment has to be used when deciding what to include and what to exclude. LCA is a major contribution to the science of integrated assessment, with useful concepts like "functional" unit, and "system boundary". "Monetary unit price" for pollutants are only rough estimates of the damage the pollutants cause in a given region, but they are convenient when more detailed information is lacking.

---

[1] Based on Canter (1996) p. 24
[2] Davis and Wali (1994)
[3] Keeney (1992) p. 96
[4] Michaelis, Malmquist et al. (1997)
[5] Greer and Sels (1997)
[6] Horvath, Hendrickson et al. (1995)
[7] Horvath, Hendrickson et al. (1995)
[8] This method was used by Seip and coworkers (2000)
[9] Power and McCarty (1997)
[10] Seip, E. Sandersen et al. (1991)
[11] Folke, Aa. Janson et al. (1997)
[12] Data Renskaug (1998)
[13] Sima-Pro (2003)
[14] Carneigie-Mellon (2003) CM
[15] Guinée, Gorrée et al. (2002)
[16] Carnegie-Mellon (2003)
[17] Matthews, Lave et al. (2002)
[18] Matthews, Lave et al. (2002: 854)
[19] Scc Seip, Betele et al. (2000)
[20] Wenstøp and Seip (2001)
[21] Cifuentes (2000)
[22] Poulos and Whittington (2000)
[23] Seip et al. (2000)
[24] Seip, Betele et al. (2000)

Chapter 5

ECONOMICS

A brief introduction to environmental resource economics

We present a model of the free market with consumers and producers that maximise their own benefit, and show how the market fails to create an optimal solution when the industry pollutes. We then discuss how different policies can improve the situation. In particular, we discuss:

– The laws of supply and demand.
– Externalities to a free market.
– Taxes as a means to curtail production and thereby reduce pollution.
– Incentives to cleaner production.
– Voluntary green instruments.
– Land tenure forms.
– Discounting as a means to incorporate the future value of the environment.

●

"Every individual ... intends only his own gain, and he is in this … led by an invisible hand to promote an end which was no part of his intention. By pursuing his own interest he frequently promotes that of the society more effectually than when he really intends to promote it."
Adam Smith in *The Wealth of Nations,* 1776.

1. INTRODUCTION

In market economies, most goods and services are priced through the market by supply and demand. According to economic theory, free markets are efficient in the sense that they maximize the utility of all actors, as if the "invisible hand" of Adam Smith were operating behind the scene. For a product or a service to be valued in the market, however, there has to be a seller and a buyer, and a limited amount of the good. Natural resources are for the most part not traded in a market because there are no well-defined buyers or sellers. Take clean air, for instance. Nobody owns it; therefore, nobody can sell it. It does not have a market price, and that makes it possible to "use" it by polluting it without having to pay for it. A commodity like clean air is therefore called an **externality** to the market economy, and must be taken care of by other means. This chapter discusses different ways to do that. It describes ways to assess the cost of pollution and how that information can be used to **internalize** externalities, so that we can benefit from the efficiency of the market place. For example, if a factory produces a certain good and pollutes a nearby river by discharging effluents, people downstream might have to pay the price by having to live with the polluted water. However, if that price were included in the price of the factory goods, the external effect would have been internalized and the polluters would pay a share of the price. – Of which "price" are we talking? The price is the cost of environmental pollution, and that cost is by definition equal to the population's **willingness to pay** to avoid it.

The theory of environmental economics describes how to identify:

- An optimal solution, if it exists.
- The instruments that are required to reach the solution.
- Negotiations that are required to implement the instruments.
- Means to enforce agreements.

2. BASIC CONCEPTS

In this section, we give a brief outline of some relevant concepts of economic theory.

2.1 The Free Market

The so-called **free market** consists of a uniform good that is produced and traded, and thousands of producers and consumers who are so small that none of them can influence the market alone. Since nobody can change the price of the good, all participants take the market price as given, and adjust their own supply and demand accordingly.

The free market is an abstract concept that does not exist in its pure form in real economies, although some markets – like the international money market – come close. Its simplicity, however, gives us a useful model to understand how externalities can be internalized in a real market.

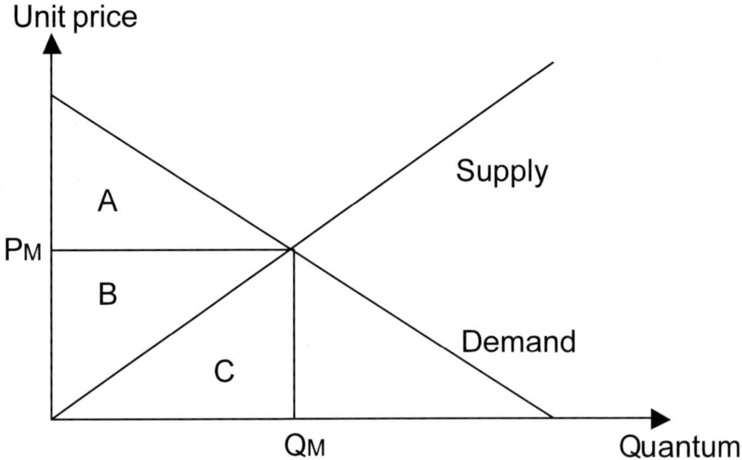

Figure 5-1. Demand and supply curves. The market price is determined by the intersection of the two curves. The x-axis is quantity traded, and the y-axis is price per unit. The intersection Q_M, P_M is the market equilibrium. The curves are *marginal* in the sense that they show how much the price would have to change if an extra unit were to be supplied or demanded. A is consumer surplus and B producer surplus. C is the cost of production. See text.

2.2 Supply and Demand

In the theory of supply and demand, the **unit price** of a good is the price consumers pay per unit to the suppliers for all units that are produced and sold. The price is determined by how much the consumers are willing to pay for the good – the demand function, and how much it costs to produce – the supply function. The price and quantity that is actually traded is found where the two curves intersect. This is the **market equilibrium.** See Figure 5-1.

2.2.1 Demand

The curve that slopes down in Figure 5-1 is called the **demand curve**. Imagine that the good in question is bread. At a certain price per bread, you might conceivably buy seven breads per week; but if the price were lower, perhaps you would buy more. If this is the case, you have a decreasing marginal utility of bread: The more bread you already have consumed, the less is your need for another bread, and therefore you are unwilling to pay more for it. The demand curve shows the aggregated quantum Q demanded by all consumers in the market. If all have a decreasing marginal utility, it is easy to see that at lower prices more bread would be demanded, thus making the demand curve slope downwards. This is not always the case, however. Some goods may become so cheap that the consumers lose interest and buy less at lower prices. Such goods are called inferior goods.

When we draw a demand curve like the one in Figure 5-1, we tacitly assume that everything except price and quantity is constant. If, however, people suddenly got more money, so that they could buy more of the good at the same price, the demand curve would shift upwards. Changes in taste could produce similar effects. The demand curve is in a sense an abstraction, since we cannot observe it directly. If everything else is constant, the price is also constant, and all we can observe is one single point on the curve.

The **consumer surplus** measures how well off the consumers are. Assume that you are prepared to pay €5 for your first bread, but only €4 for the next, and only €3 for the third. The market price is €3, and therefore you buy three breads. Your surplus as a consumer is then €12–€9 = €3. Likewise, in Figure 5-1 at equilibrium the consumers' total willingness to pay for the quantity Q_M is the area $A+B+C$, but they are only paying $B+C$. The consumer surplus is therefore the area A.

2.2.2 Supply

The **supply curve** describes the behavior of the producers of the good. It is a result of how much they want to produce at different prices when all other things are constant. The producers take the market price as given, and decide on the optimal quantum to produce. If the cost of producing one extra unit is less than the sales price, they will produce more. Usually, the unit production cost increases with the production quantum, making the supply curve slope upwards, meaning that more is supplied at higher sales prices. If the cost of producing one more unit is higher than the price, the production will not be increased. The reason unit costs normally increase with the quantity produced, is that factories typically have physical limitations to how much they can produce, and to go beyond this limit requires procurement of new production capital. If production cost for some reason increases, the supply curve will shift vertically upwards with the same amount. The result will be that the market equilibrium will move along the demand curve to the left, and the consequence will be less production – and therefore less pollution, and a higher market price.

The **producer surplus** is a simpler concept than the consumer surplus. The area $B+C$ in Figure 5-1 is the sales price for the quantum Q_M, while C is the cost of producing the same quantum. Thus, B is the total profit for the producers, which is also called producer surplus.

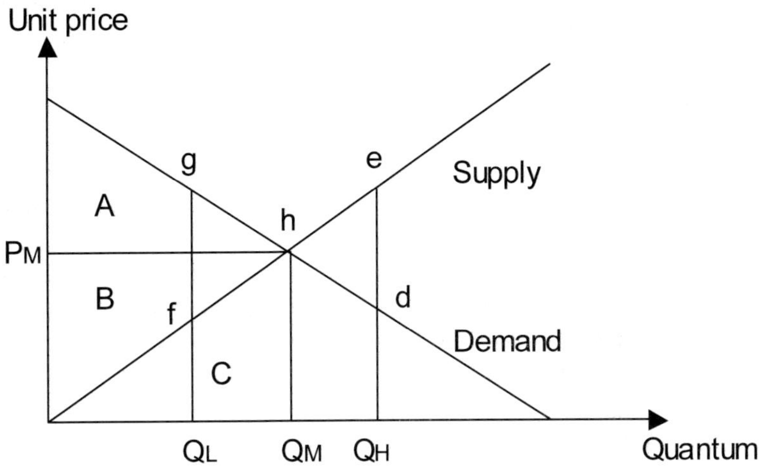

Figure 5-2. Efficient and non-efficient allocations: Society's net benefit is the area (A+B+C) under the demand curve less the area (C) under the supply curve to the left of quantum traded, Q_M. Q_L and Q_H are two alternative production volumes, which both are inefficient since the differences are smaller than in Q_M.

2.3 Efficient and Inefficient Allocations

An **allocation** is a distribution of outcomes between the actors in a market. What characterizes a good distribution? This is of course partly a political question where issues of fairness and equality as well as economic surplus are important. Let us focus on surplus and define some concepts that characterize different types of distributions.

In Figure 5-1, we identified the consumer and producer surpluses with the triangles *A* and *B*. The sum of these two areas is the **net benefit** to the society – the same as the value of the products less the cost of production. The value of the products is what people are willing to pay for them – or the area under the demand curve. The cost of production is the area under the supply curve. Figure 5-2 shows that Q_M maximizes net benefit; since if Q_L is produced, the difference decreases by the area *hfg*. On the other hand, if Q_H is produced, the difference decreases with the area *edh*, since the difference on this side is negative. The quantum in excess of Q_M has to be sold at a lower price than the cost of production. The equilibrium point therefore maximizes net benefit. It is called an **efficient allocation**, since no reallocation can increase net benefit. Net benefit is maximized when the marginal benefit equals marginal cost.

In an inefficient situation where the net benefit *can* be increased, somebody may benefit from a reallocation without anybody else being worse off. From an ethical point of view, it is presumably better if some become better off and nobody worse off, than if everybody stays the same. This leads to a criterion for **social optimum**, which was proposed by the Swiss economist Vilfredo Pareto at the turn of the twentieth century. An allocation is said to be **Pareto optimal** if no reallocation can benefit some without others losing. An efficient allocation is in other words Pareto optimal.

Not all accept the criterion, however. From an ethical point of view, one could argue that it is better if everybody are reasonably well off, than if most are reasonably well off and some extremely well off. In other words, not only wealth, but also distribution of wealth is important. This perspective can be included by adding distribution of wealth as a societal welfare criterion, but the economics of this is outside the ambitions of this book.

So far, we have discussed the theory of the free market without taking into account the fact that production produces pollution. The market equilibrium we have found maximizes the sum of benefits to consumers and producers and is therefore a social optimum – but only in a limited sense, since we have not yet studied the external costs of pollution.

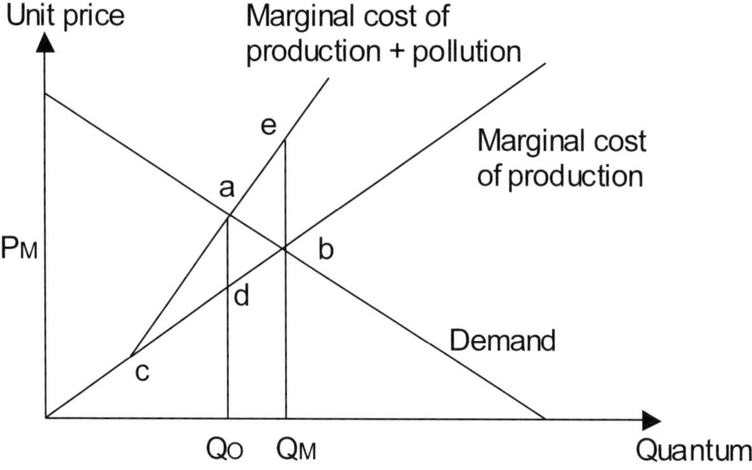

Figure 5-3. Demand and supply curves with the cost of pollution included.[1]

2.4 The Free Market and the Cost of Pollution

Figure 5-3 shows a situation where the cost of pollution is included in the demand and supply diagram. The marginal cost of production is what it costs to produce one extra unit, so this curve is identical to the supply curve of Figure 5-1. The free market equilibrium is as before Q_M. In addition, there is a curve starting in *c*, and sloping upward more steeply than the supply curve. The distance between this curve and the supply curve is the **cost of pollution**; or more precisely, what society is willing to pay to avoid the pollution caused by the production of one additional unit. How to measure that is an important question to be discussed in Chapter 8 on Willingness-to-pay.

2.4.1 Social Optimum with Pollution

We see that if production is less than at the point c, there is no pollution problem, but then it becomes increasingly more severe with each additional unit produced. The total cost to the society is now the *private* cost of the producers plus the society's cost of pollution. The same reasoning as before shows that Q_M is no longer the social optimum, but rather Q_O, which gives higher price and less production. However, a free market would continue to produce Q_M and thereby cause a total pollution cost equal to the area *cbe*, while it would only have been *cde* under the new social optimum at Q_O.

Our reasoning allows us to make the following observations:

- In the new social optimum Q_O, we still have a certain level of pollution. Including the cost of pollution introduces a trade-off between how much we appreciate the good that is produced, and how much we are willing to pay to avoid the associated pollution. A pollution free social optimum would imply an infinite marginal cost of pollution, corresponding to a vertical line through point c in Figure 5-1, but this is not realistic.
- In a free unrestrained market, the production will be too high. The **loss of welfare** caused by the unrestrained free market, is the area *abe*. This is the higher cost of pollution *adbe* less the area *adb*, where the consumers are willing to pay more than the production costs.
- Who is to blame? Ethical reasoning has made the so-called Polluter Pay Principle (PPP) become popular. The culprit should clean up! Who is the polluter here – is it the producer? The producer is only producing what the costumers are willing to buy. Is it the consumers then? Well, the consumers would not have bought what they did if the producers had not produced it. We have to conclude that the actors are just doing what they are meant to do in a market economy. The free market itself is to blame! It is therefore necessary that authorities adopt policies to amend the free market so that a social optimum can be created.

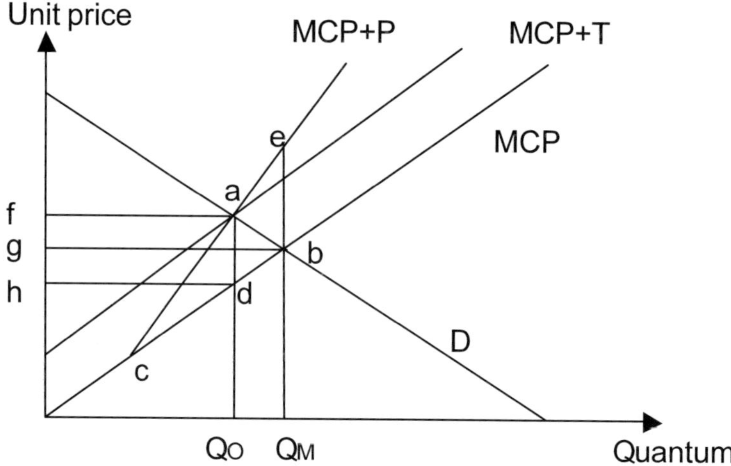

Figure 5-4. Internalizing pollution cost with a tax $T = a - d$ per unit produced. *MCP* is marginal cost of production. *MCP* + *P* is marginal cost of production and pollution. *D* is demand. Q_O is the social optimum. *MCP* + *T* is marginal cost of production with a tax added.

2.5 Internalizing the External Cost of Pollution

The problem with the free market is that neither producers nor consumers take the cost of pollution into account. Fortunately, however, the market can do this for us, if the authorities impose a tax on production.

2.5.1 Tax on Production to Reduce Pollution

Figure 5-4 shows a situation where a tax equal to the distance between *a* and *d* is imposed on each unit that is produced. That shifts the supply curve upward by the same amount, and if the producer and the consumers optimize their own behavior as before, the new social optimum in point *a* will automatically be reached.

We see that it is in principle easy to induce the market to produce optimally in the face of pollution. The only information the authorities need is the optimal production quantum Q_O, the marginal cost of production and the consumers' marginal willingness to pay for the good at that quantum. It is another question whether this is also a practical method. We shall return to that question.

2.5.2 Winners and Losers

Who are the winners and losers in the new social optimum compared to the original market equilibrium? Let us look at Figure 5-4 again.

- The consumers have reduced their surplus by the area *fabg*.
- The producers have reduced their surplus by the area *gbdh*.
- The taxpayers have reduced their tax by the area *fadh*, assuming that the taxes paid by the producers are credited to the taxpayers.
- The victims of the pollution have their cost reduced by the area *adbe*.
- Finally, the net welfare gain is the area *abe*!

3. POLICIES FOR CLEANER PRODUCTION

Above, we saw that the authorities may induce producers to reduce pollution by imposing a tax on each unit produced. The producers will respond by reducing production – and thereby pollution, the consumers will pay a higher price; and if the tax is of correct size, a socially optimal solution will establish itself through the operation of the market forces.

The shortcoming of tax on production, however, is that it gives no incentive to adopt cleaner production technology. If a single producer tries this, his production cost will increase above that of his competitors, and if this cannot be compensated for through higher willingness-to-pay for "green" products, he will be forced out of business. Incentives for cleaner production must therefore focus directly on emissions of polluting agents. In this chapter, we will discuss policies to make producers reduce emissions. There are three important kinds of policies:

- *Emission standards*: Each factory is permitted to discharge a certain amount of pollutant (end-of pipe control).
- *Emission fees*: Each producer has to pay a certain amount of money per unit of emission.
- *Transferable emission permits*: A permit gives the right to emit a unit of the pollutant, and they can be traded between the producers.

3.1 Performance Criteria

We will evaluate the policies with two different performance criteria:

- *Allocation efficiency*: We defined this concept in section 2.3 where we discussed the optimal allocation of benefits and costs to producers and consumers. In this chapter, we will discuss allocation of benefits from production and pollution cleanup and pollution damage costs.

- *Cost effectiveness*: In this case, we do not know the pollution damage costs. Instead, we use a politically determined target that represents an acceptable pollution level. The objective is to minimize production costs for the given target.

The two terms, **efficiency** and **effectiveness** have different meaning. Efficiency means that we get as much benefit as possible for the least cost – both benefit and cost are variables, and the benefit/cost ratio is as large as possible. A cost effective policy, on the other hand, minimizes the cost of production for a given pollution target – here, the only variable is cost.

3.2 Allocation Efficiency

3.2.1 Efficient Quantity of Emissions

Consider a firm that produces certain goods and consequently emits Q units per week of a certain pollutant. We shall assume that the firm keeps the production volume constant, and that cleaning technology that reduces emission is available.

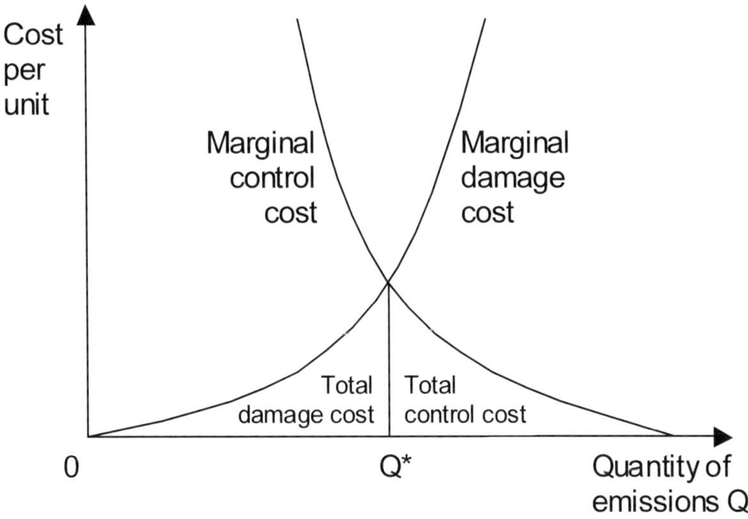

Figure 5-5. Efficient control of pollutant emission.

The **marginal pollution control cost** is what it costs the firm to reduce pollution by one unit. Normally, the costs are small when emissions are large because "Off-the-shelf" technologies are available. Below that level, expensive technologies must be bought, or new technologies invented. The marginal control cost curve therefore tends to decrease with increasing emissions of pollution as shown in Figure 5-5.

Emissions cause damage, and such damage tends to increase more than proportionally, as suggested by the marginal damage curve in Figure 5-5. **Marginal damage cost** is what the society is willing to pay to reduce the emissions by one unit. From a societal point of view, how much should the firm clean? If the emissions Q are higher than Q^* in Figure 5-5, it costs little to reduce emissions by one unit, while the benefit to the society is larger. If Q is less than Q^*, it costs more than it is worth to reduce Q by one unit. The efficient emission of pollution is therefore Q^* where marginal costs are equal to marginal benefits. At this point, the sum of total control and damage costs is at minimum. You may be surprised that the optimal emission of the pollutant is larger than zero. Why is it optimal to pollute? Why can we not get rid of pollution altogether? The answer lies in the shape of the marginal control cost curve. It costs very much to remove the last remnants of pollution. We therefore should accept a certain level of pollution, and use environmental management to find out what that level should be and the right instruments to implement it. Think of pollution emitted by cars. It is possible to reduce it, and new technologies for that purpose are constantly being developed. The most *effective* policy, however, would be to simply ban all cars, but that would not be an *efficient* policy, since it would create huge costs for transportation.

If the authorities do nothing, the producer in Figure 5-5 will emit high quantities of the pollutant, with a correspondingly high environmental cost. The authorities, however, can use two conceptually simple alternative strategies to achieve the efficient cleaning Q^*.

- *Command and control with emission standards*: The producer is simply not allowed to emit more than Q^*.
- *Internalizing the damage with emission fees*: A fee on each unit of emission is imposed on the producer. The fees could be identical to the marginal damage curve, or constant and equal to that curve in the point Q^*. In both cases, the producer would maximize his own profit by reducing emissions to the point Q^*.

3.2.2 Emission Standards

The traditional pollutant control strategy is to impose emission standards for end-of-pipe emissions. Discharge into water may be subject to discharge permits or end-of-pipe permits that allow a certain amount of pollutant per time unit such as one day or 30 days. In addition, one could require the producer to use Best Available Technology (BAT) or Best Economically Feasible Technology (BEFT). A third possibility is BATNEC, which means Best Technology Available Not Entailing Excessive Costs.

Standards are often set after an iterative procedure where the government usually knows the best available technology (BAT) for a particular industry, and the effect of that technology on the emissions. Thereafter, the factory is permitted to discharge a certain volume of a pollutant the following year according to three criteria; the known BAT efficiency; the production volume, and the capacity of the receiving environment.

3.2.3 Emission Fees

This fee is paid to the government for each unit of actual end-of-pipe pollution discharge. The total fee is the fee per unit of discharge times the volume of the discharges. Since the fee implies an extra cost of production, a factory will always try to reduce its emission so long as the control costs remain less than the fee.

3.2.4 Assessment

Both strategies require knowledge of the producer's control cost curve as well as the damage curve, and the emissions would have to be monitored for control purposes. However, control authorities usually do not have this kind of knowledge, not even for one producer. In practice, there are many producers and the policies would become quite infeasible unless one makes the pragmatic assumption that all producers are similar and therefore impose the same standards or identical fees. Such gross simplifications mean that the policies are not efficient in reality. Think of a paper mill that discharges into a small brook and another into an open fjord. For the one discharging into the fjord, strong pollution control may be a waste of money. The methods are therefore not necessarily "fair" in the sense that they allow different factories to pollute to different degrees according to how much damage they actually cause.

Emission standards are probably more frequently applied than fees. Politicians tend to prefer command and control, since it gives them a feeling of being effective, and in many countries politicians are skeptical of market forces on ideological grounds.

3.3 Cost-Effectiveness

Many countries, including the US, have adopted a pragmatic strategy that requires less information than needed to establish emissions standards or fees. For a given pollutant, they establish a maximum level of emissions based on consideration of threshold values of what humans and the environment can sustain, and endow this level with a legal standing. The next question is how that level can be attained *at least possible cost.* This criterion is called **cost-effectiveness**. It implies that one disregards marginal damage cost, which requires information that is not easily obtainable, and instead specifies an appropriate pollution target based on considerations of health, environment and sustainability. This usually requires political processes, and the results will not necessarily depend primarily on scientific facts, but political feasibility. The Kyoto protocol is an eminent example. Once a target has been reached, however, the scientific question becomes how that target can be realized at minimum cost. Figure 5-6 shows two producers that each emits six units of a pollution. Both of them can reduce their emission, but it is more expensive for producer 2 to do so. The curves show the marginal cost of reducing emissions. Imagine that the pollution control authority has determined that they are allowed to emit only six units together.

Any point on the *x*-axis of Figure 5-6 is a solution, but how can the target be achieved in the most cost effective way? If producer 1 reduces emissions by four units, the cost will be equal to the area *A*; and if producer 2 reduces emissions by two units, that cost will be area *B*. The total cost is *A*+*B*. Is there a more effective allocation?

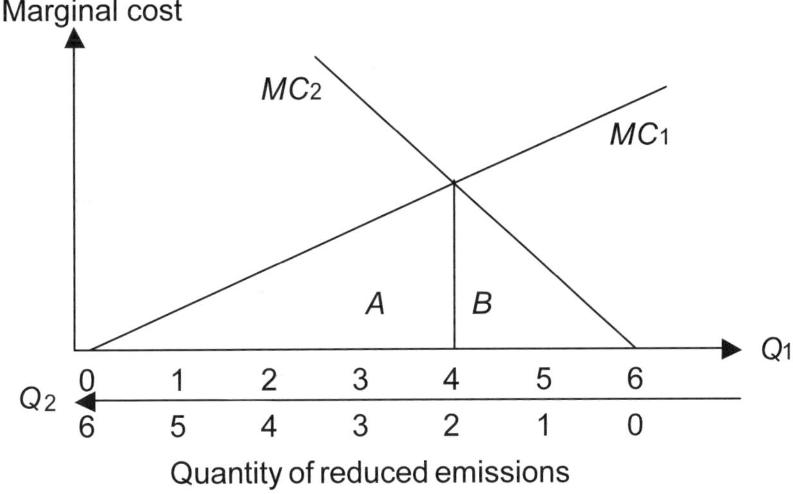

Figure 5-6. Cost-effective allocation of emissions between two producers with different marginal costs of emission control.

Consider what will happen if they both reduce emissions by three units. Then it is easy to see that *B* will increase much more than *A* decreases, making total costs larger. Likewise, if producer 1 reduces by five units and producer 2 by one unit, *A* will increase much more than *B* decreases. The cost effective solution is therefore where the two producers' marginal costs of pollution control are equal.

The simple example in Figure 5-6 has only two producers, but illustrates an important theoretical principle that is valid for any number of producers, as long as the marginal control costs increase with the amount of reduction: *The total cost of achieving a given reduction in pollution emission is minimized when the marginal costs of control are equalized for all emitters.*

This points to a very simple policy instrument of emission control: The authorities can simply calculate how much each producer would have to reduce emissions to reach a given target in a cost-effective way, and then give each a corresponding quota. The problem with this, of course, is that it requires knowledge of each emitter's control cost function, and that is seldom the case. A much simpler solution is to let the producers solve the question themselves by buying and selling emission quotas. We shall look at how that mechanism works in the next section.

In some cases, however, we do know something about cost control functions. Figure 5-7 shows marginal control or pollution abatement costs for carbon emission reductions as a function of the quantity of reduction required to meet the Kyoto commitments in Japan, EU, (EEC in Figure 5-7), USA, and other OECD countries, OOE. The expression "shadow price of carbon" corresponds in the present case to the marginal costs of abatement.

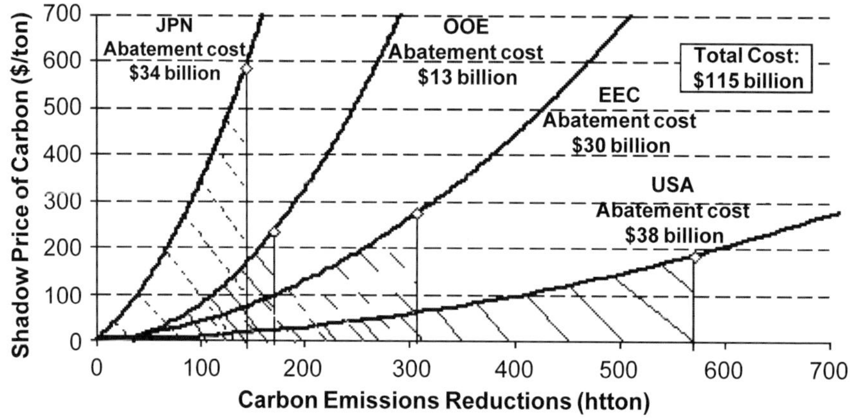

Figure 5-7. Control costs of CO_2 abatement in selected countries.[2]

The diamonds show the 1999 level of reductions required by the Kyoto commitment on the *x*-axis, and the no-trading marginal costs of abatement on the *y*-axis, e.g., US$584 ton^{-1} in Japan. The areas under the curve represent the total costs, e.g., US$34 billion for Japan.

3.3.1 Tradable Permits

A tradable emission permit is a permission to discharge a certain amount of pollutant into the environment, which may be traded among producers. The permission is initially issued or sold to potential producers who can use it for its own discharge, or sell it to other potential polluters.

Let us see what would happen if our two producers in Figure 5-8 were given an initial emission quota of three units each, and were allowed to trade them as well. Since they originally emitted six units each, both of them will now have to reduce by three units and the situation would look as in Figure 5-8 before any trade took place. We see that the control costs of producers 1 and 2 are A = €2 × 3/2 = €3 and B = €4 × 3/2 = €6, respectively.

Now, if producer 2 could emit one more unit, he would only have to reduce emissions by two units from the original six, and the control costs would drop to B = €2 × 2.7/2 = €2.7.

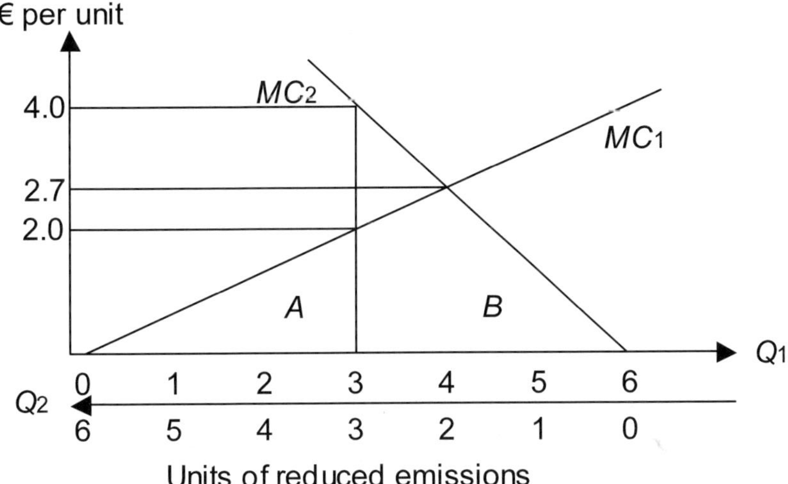

Figure 5-8. Tradable emission permits. From the situation shown, producer 2 will buy the permit to emit one more unit from producer 1, thereby reducing cleanup costs more than the cleanup costs of producer 1 will increase.

He would therefore be willing to pay up to €3.3 for this extra permit. Would producer 1 sell? That would mean that he had to reduce by one more unit, bringing his total control costs up to €2.7×4/2 = €5.4 – in other words, an increase of €2.4. As a rational actor, he would certainly sell, as long as they can agree on a price somewhere between €2.4 and €3.3.

If you study this example a bit, you will probably realize that all initial splits, except the cost-effective solution where the two marginal cost curves intersect, allows for profitable trade. This simple exercise with two producers therefore illustrates another important theoretical principle: *Cost effectiveness will be achieved automatically through the market if the producers may buy and sell emissions permits freely, assuming no transaction costs*. There is therefore no need for the control authorities to know the individual control cost functions.

It is an interesting experience to try out this exercise in a classroom situation. Two students acting as managers of the two factories are endowed with initial emission permits, and encouraged to trade them. Sometimes, transactions will halt before cost-effectiveness is obtained because one of the participants wants a better deal than the other is willing to give, although both would benefit from continued transactions. This is a strange, but common property of human nature, which will be discussed further in Chapter 6 on game theory.

3.3.2 Initial Distribution of Tradable Permits

Tradable permits require that the control authorities perform an initial distribution of the permits, and how this is done is not without consequence. There are three main methods:

- *Quotas are free of charge*: The producers get permits according to their discharge at a certain date. This is called "grand-fathering" and has been used in many cases. However, it punishes those who have been "nice" and already reduced their discharges, while it gives advantage to those who adopted a "laisser faire" attitude. Another version of the same principle is to issue permits according to historical production volumes before a certain date, or a historical production value. This was done for the former Soviet Union states, but they only got "hot air" to sell, because the emissions were well below the 1990 emission target at the time the agreement was put into effect. A third version is to give everyone equal quotas.
- *Quotas are sold*: Ideally, the price should be set a little lower than the cost of abating the discharges, because that is the price the market would be willing to pay. However this would require knowledge of the producers' control costs, or at least a lower bound on them.

- *Quotas are auctioned*: This means that the factories can buy quotas if they find the price acceptable. The price will reflect the cost of pollution control.
- *Quotas are allocated*: Quotas are given to industries in areas that need development.

3.3.3 Tradable Permits in Practice

Tradable permits are used in the US to regulate discharge of CO_2. Tradable water permits issued to farmers would allow them to allocate water efficiently since permit holders often sell some of their permits However, price-schemes for water are considered regressive and unfair to low income farmers in many countries.[3] Other types of tradable permits that have been issued are tradable fishing permits and tradable housing developing permits. By setting up a "clearing house" which monitors the price of the permits, it is possible to infer the actual price of the commodity being traded, like the price of pollutant abatement measures. In some cases, this has shown that it costs less to reduce CO_2 emissions than earlier believed. Indirectly, one also obtains a minimum estimate of goods damaged by emissions.[4]

Tradable permits cannot be used for all kinds of emissions of polluting agents. Here are some criteria for using tradable permits:

- A common recipient, like the atmosphere or a common lake.
- Knowledge of the recipients' capacities to find acceptable discharges.
- Knowledge of the actual discharges at end-of-pipe.
- A market or "clearing house" for tradable permits.
- A mechanism for initial distribution of permits.

3.3.4 Standards, Fees and Tradable Permits

In addition to tradable permits, it is also possible to use emission fees and emission standards as policies to approximate cost-effectiveness. In this section, we shall use a worked-out example to demonstrate the consequences of the three policies.

Imagine that we have two factories that presently emit altogether 30 kg of pollutants. Factory A's marginal cost of cleaning is €1000 per kg when 10 kg is cleaned. The marginal cost function increases linearly with the amount that is already cleaned. Factory B's marginal cost of cleaning is twice as high. The factories know these cost functions, but the pollution control authority does not. It has determined that the capacity of the environment is 15 kg. The objective is therefore to induce the factories to reduce their combined emissions by 15 kg.

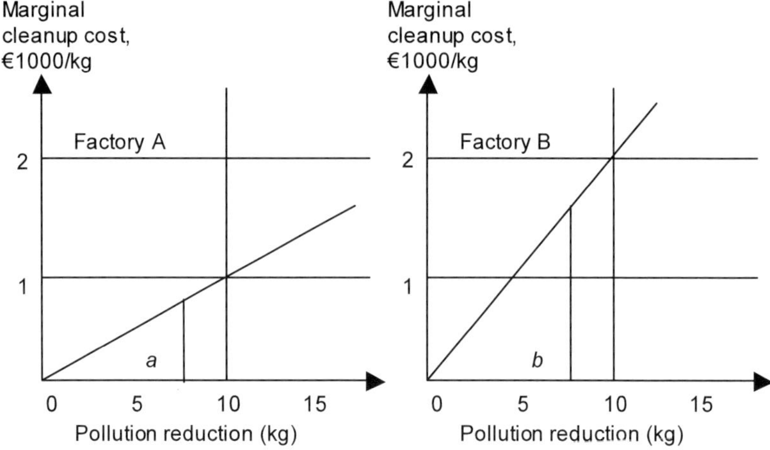

Figure 5-9. Emission standards: Each factory is allowed to emit 7.5 kg and must reduce by the same amount. Cleanup costs are the areas *a* and *b*; the sum is approximately €8400.

- *Emission standards*: Since the control authority does not know the marginal cleanup costs, the most sensible policy is simply to tell both factories to clean up 7.5 kg each. The resulting cleanup costs for the factories are shown in Figure 5-9. They are €750×7.5/2 = €2800 for A and twice as much, €5600 for B (approximately).
- *Emission fees*: If the control authority does not know the marginal cleanup costs, it is uncertain what the effect of different fees will be. Let us say that a fee of €800 per kg of emission is imposed. Then factory A will respond by cleaning up until cleanup costs become higher than the fees, which happens when the reduction is 8kg. The rest, which is 7kg, will be emitted and a fee of €5600 paid. Factory B has twice as high cleanup costs and will only reduce emissions by 4 kg, and must pay fees of €8800 for the 11 kg still emitted. We see that the imposed fee was too low to achieve a total reduction of 15 kg of emissions. The total emission will be 18 kg, and a corresponding fee of €14,400 paid. From Figure 5-10, we also see that A's cleanup cost is *a* = €3200 and B's cleanup cost is *b* = €1600. Total cleanup costs are therefore €4800.

If we know the marginal cost curves, it is actually easy to see from Figure 5-11 that a fee of €1000 would have been correct, since it would have induced A to clean 10 kg and B 5 kg, but a general problem with emission fees is that they are usually not known.

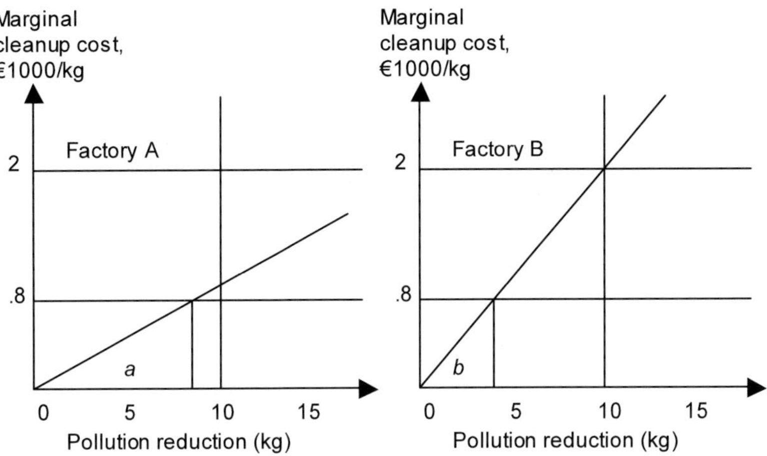

Figure 5-10. Emission fees: The factories must pay €800 per kg emitted. They clean up until marginal cleanup fees exceed this amount, which happens at 8 kg for A and 4 kg for B. Cleanup costs are $a =$ €800×8/2 = €3200 and $b =$ €800×4/2 = €1600 for B. Fees of €15,000 are paid for the remaining 18 kg emissions.

– *Tradable permits*: Let us assume that permits are given for a total discharge of 15 kg, and that each factory is initially allotted 7.5 kg. With that distribution, both must clean up 7.5 kg, and that will cost A €750 × 7.5/2 = €2800 and B twice as much, or €5600, just as we found under emission standards. B will therefore want to buy emission permits from A, who will gladly sell. The trade will continue until an equilibrium point is reached, and this is, as we have found before, when the two marginal costs are equal. This happens at a marginal cost of €1000, as we see in Figure 5-11.

This example shows that emission standards and fees generally do not produce cost-effective results. When emission standards are imposed equally on all producers, the cleanup costs will usually be unnecessarily high since producers with high cleanup costs must clean as much as the others. Emission fees can easily be set too low – and result in too much emission; or too high, with needlessly high cleanup costs as a result. Tradable permits is the only policy that, at least in theory, guarantees a cost-efficient result. Tradable permits are not all that popular, however. Politicians may feel they lose power when so much is left to market forces, and NGOs often do not like the concept of "buying the right to pollute"

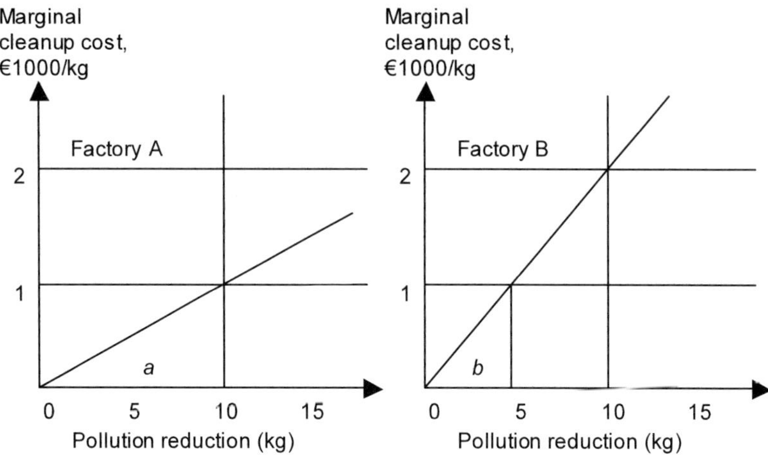

Figure 5-11. Tradable permits: From initial emission permits of 7.5 kg each, B will buy permits from A until the marginal costs are equal, which happens at €1000. A cleans up 10 kg and B 5kg, with cleanup costs a = €5000 and b = €2500. 15 kg is emitted.

4. VOLUNTARY GREEN INSTRUMENTS

The previous section described different policies that the authorities can use to induce the producers to clean up polluting emissions. However, we also witness a proliferation of voluntary arrangements that are becoming increasingly popular among producers who want to appease important stakeholders. In this section, we discuss three promising green instruments, which are called "Environmental credits", "Mitigation banking" and "Eco-labeling".

4.1 Environmental Credits

A production process does not always have to harm the environment; some may actually be beneficial. For example, an industry may discharge process water that is cleaner than the water it uses as input to the process. The environment may also be enhanced in ways that are more direct. An industry – or the government – may for instance purchase land for habitat preservation which otherwise might have been degraded. An industry that does this may obtain a "preservation credit" in the form of an increased

Habitat Quality Score (HQS) – a measure of land quality in terms of natural habitat suitability.

For example, planting of forest by the paper industry may add to a "preservation credit" score for the paper industry and its product. Power stations and other stationary sources emitting CO_2 gas may achieve recognition by planting new forest that sequesters CO_2. For example, a group of electric utilities including Wisconsin Electric is spending US$3.5 million to protect about 2500 hectares of forest in Belitze.[5] A third example is mitigation measures implemented when hydropower dams are constructed. Natural landscapes are "remodeled"; rivers diverted to new courses, sometimes with higher HQS; cultural artifacts are excavated, preserved, and historically interpreted, thus adding a piece of information to the cultural heritage.

Preservation credit is an example of environmental credits, which can be used in mitigation banking, the subject of the next section. So far, environmental credits are used on local levels and often in an immediate exchange process. The HQS are then calculated in the context of a specific situation.

4.2 Mitigation Banking

Mitigation means that some action is taken to reduce unwanted side effects. It can include economic compensation and is a major issue in many development projects. Mitigation banking consists of transactions with environmental credits. It allows impacts of a project on one habitat in return for an improvement in another habitat. Mitigation banking may be a good concept for several reasons. It is important that wilderness areas are large, because extinction of some species is inversely related to the area of protected land. Thus, many small areas do not give the same protection as a large one of the combined size. It may also be advantageous to have several small development projects contributing to the preservation of one large non-developed area.

The major steps in setting up a mitigation bank is:

- Establishing program goals.
- Selecting an area which is to be protected or enhanced as a wilderness area or nature park area.
- Creation of a bank and doing banking transactions.

As with Environmental credits, mitigation banking is still in its infancy, but has the potential for offering cost efficient solutions.

4.3 Certification and Eco-Labels

Firms frequently employ eco-labeling of products to highlight environmental friendliness. There are different eco-certificates to choose from, with different profiles and characteristics. For example, eco-labels on tuna fish cans under the 1997 International Dolphin Conservation Act guarantees that the fishing method used does not harm dolphins.[6] Eco-labeling may increase sales (about 1% in the tuna case) or make it possible to charge higher prices than for similar non-green products. To make eco-labels credible, producers pay a fee to a third party for certification of product quality claims. Reputation and unverified information are not sufficient to convince costumers,[7] and it may even take several years before information and verification of eco-label claims are accepted.[8]

5. LAND TENURE

What kind of objectives a land manager has depends in significant ways on who owns the rights to use the land, and the natural resources in question. Below, we will discuss land management in connection with several different forms of land ownership – or **land tenure** – from open access through group tenure to private property.

5.1 Land Management Objectives

Efficiency, fairness and sustainability are central objectives of land management.[9] The natural resources of the land should be reaped with as little cost as possible, and in a sustainable way and the wealth accruing from it should be distributed fairly. It is hard to define fairness of course, but equitability is certainly a part of it.

Tenure forms are important determinants of how natural resources are used. One example is distribution of water. In regions where water is in short supply, wastefulness will curtail agriculture production. A charge on water use imposed by private or government water holders could therefore make agriculture more efficient, but water allocation policy is a sensitive issue and difficult to implement. Free use of water is traditionally by many considered a natural right, and any limitation of that right may be considered unfair. Fairness also has to do with people's right to communicate, to lobby lawmakers, and sanction one another if resources are abused.

If everybody has the right to use a resource it easily becomes over used. The results are seen as overgrazing in commons and overfishing on the high seas. The resources become damaged, with decreasing renewal as a result and consequent loss of sustainability.

5.2 Resource Characteristics

Resource characteristics influence how suitable different tenure forms are. Water, wildlife, fish, wood, and medical plants are typical resources that are reaped without well-defined property rights.

Resources may be characterized according to size and carrying capacity, as well as temporal and spatial distribution of the resource flow.

A second set of characteristics is storage property, whether the resource moves as do animals or is stationary like trees and how various harvesting technologies affect resource size and regeneration.

A third set of characteristics is the ease with which one can measure the size of the resource and predict its future development. For example, the amount of wildlife and fish stocks are not easily measured; they vary in space and time, and there are complex interconnections between natural factors and harvesting systems. Trees, on the other hand are easily counted, often uniformly distributed, and give predictable yield.

All these characteristics have implications for how tenure forms will function. For example, if several persons share common land, it may be easier to manage if resources are countable so that everyone knows what the others have taken out.

5.3 Tenure Forms

We distinguish between at least eight different land tenure forms, as shown in Table 5-1. Tenure forms determine how potential harvesters are included or excluded, as well as the efficiency of harvest control.

Table 5-1. Land ownership forms with an indication of prevalence in England and Wales in 1973.

	Type of ownership	Characteristics[10]	% Ownership[11]
1	Government	Owner regulated or subsidized use	9 (1)
2	Open access	Absence of enforced property rights	
3	Indigenous communities	Several sub-forms	–
4	Group tenure	Group of users can exclude others	
5	Private property- owner occupied	Individuals can exclude others	35
6	Private property – to let	Individual can exclude others	31
7	Mixed tenure		34
8	Tenure, dependent on use	Owner has to put the land to use	

(1) Including institutions

Let us look closer at three important tenure forms: open access, private ownership and group tenure, with regard to efficiency, fairness and sustainability.

We must bear in mind, however, that group size, family structure, cultural traits, religious beliefs, sanction systems and economic conditions also have important effects on the three objectives.

- *Open access*: In this tenure form, exclusion of harvesters through physical or institutional means is difficult and costly, and exploitation by one user reduces resources available to others. Open access land is also called **commons**, although many commons have informal access restrictions, harvesting rules and sanction systems. In tropical areas, the distribution of wealth may change abruptly if disease or drought hit one flock of animals and not another. Families therefore have institutionalized non-monetary insurance systems. Herdsmen hoard cattle in order to build systems of human bonds aimed at increasing security. For example, they lend cattle to poorer relatives or neighbors to ensure their help if needed later. Stocking out large herds also lessens the grazing pressure on the ground close to their village.[12] Thus, although informal, such arrangements may lessen the pressure on the common.
- *Private ownership*: There are different opinions about private land tenure. Economists will argue that setting up private or public ownership on previously un-owned resources creates an opportunity for efficient use. If private or public bodies owned all resources, and property rights were fully enforceable, then everybody would think twice before engaging in "use as you please".[13] However, it can be argued that if forests in Bolivia were privatized, investment in heavy machinery to take out as much as possible immediately would be the most probable outcome. The reason is that real interest rates in Bolivia are about 17%, whereas trees give typically less than 4% profit.[14]
- *Group tenure*: Groups of people who can identify with each other so that each member can draw on trust, reciprocity and reputation can develop norms that make resource use sustainable. This has in the past limited the size of groups who rely primarily upon evolved and shared norms. It probably does the same in our times because they have to exclude people who behave in a narrow, self-interested way and seldom cooperate. Resources with large stocks, and with easily predictable yields are best suited for this tenure form.

5.4 Empirical Evidence

Empirical studies of ownership forms show that no single tenure form is best suited for all resources and all land management objectives. So far, no theory can predict what tenure form is best for certain management goals. An important question is under what circumstances landowners themselves are able to change a badly working land tenure form. Evidence shows that resource users can change tenure form from open access to group tenure if resource degradation is evident.[15]

6. DISCOUNTING

Cost and benefits of a project usually accrue over a time period. A new hydroelectric power plant, for instance, needs investment capital to be repaid in fixed installments. Then there are income streams from the sales of electricity and operating costs.

Money streams in and out are called **cash flows**. To choose between two alternative projects, we need a method to compare future cash flows with cash flows today. In other words, we need a notion of the present value of future cash flows. We usually prefer to get something earlier rather than later and we need to take that into account.

Suppose you can choose between getting a gift of €1000 today or in a year. Most people would rather have the €1000 today. It would be fun to come home in a new suit or a new dress. You are impatient. Besides, the future is uncertain. You never know what will happen; it is therefore safer to get the money now than perhaps never. Growth, uncertainty and impatience are the main reasons our society has institutionalized banks that will pay you an interest if you deposit money, and charge an interest if you borrow.

If you put an amount N in the bank now, and leave it for n years at a constant annual interest rate r, you can then take out the following amount:

$$N_n = N(1+r)^n \tag{5-1}$$

We can turn this formula around: The **present value** of getting an amount N_n n years in the future, is:

$$PV = N_n/(1+r)^n \tag{5-2}$$

A project usually results in cash flows that come in or go out every year over a number of years. Typically, construction costs occur in the beginning of a project, while cash flows in from sales later. If N_i is the net cash flow in year i after the project has started, the **net present value** of the project is computed as:

$$NPV = \sum_{i=1}^{n} \frac{N_i}{(1+r)^i} \tag{5-3}$$

We use a positive interest rate r – or **discount rate**, since we prefer to have the capital now rather than in the future. To use a negative rate would imply that we preferred to have the capital in the future rather than now. This is not common, except perhaps among people with certain religious beliefs. The higher the interest rate is, the lower is the present value of future capital. If the future is far ahead, say 50 years, it turns out that even small discount rates make the present value very small. 5% and 10% discounts give present values of approximately 1:10 and 1:100, respectively. As a rule-of-thumb, 0.7 divided by the discount rate gives the number of years required to reduce the capital value by 50%. The present value of a constant sum N over infinite time is N/r.

There are two main types of discount rates:

- *Private time preference rate, PTPR*: This is the discount rate for private persons. The interest rate paid on debt is an indication of the discount rate of an individual.
- *Social time preference rate, STPR*: This is the discount rate levied on society as a whole. It has two components: the annual relative increase in consumption by society, c, and the pure time preference rate p, reflecting society's impatience. Long time bond rates are indications of c.[16]

$$STPR = c + p \tag{5-4}$$

6.1 Discount Rates for Natural Resources

Suppose you are evaluating a project that will damage a natural resource in the future. Think of a polluted lake. In the future, people would be willing to pay something, *WTP*, for a clean lake. This *WTP* can be regarded as a cost, similar to a cash flow out, which you can take into account today by discounting it. If you use a positive discount rate, the present cost will be smaller than the future *WTP*, and a high discount rate makes it even smaller. Therefore, the use of positive discount rates for environmental resources is often regarded as unfair or unjust: One discriminates against the future, paying too little attention to coming generations. It is argued that such practice may lead to extinction of species and unsustainable use of natural resources.

The important question is therefore which discount rate to use when projects with environmental impacts are evaluated. The pure time preference rate, p, is supposed to reflect the impatience of society. Another factor, however, also contributes to making p larger: Expected technological development holds a promise that we will be better equipped to deal with environmental problems in the future than now. Therefore, it is reasonable to leave some of the problems we create to future generations, thus allowing for a higher discount rate.

The World Bank has been using discount rates ranging from 3.5% to 12% for project appraisals in developing countries.[17] It appears, however, that theory is not always followed up by practice; discount rates that are different from that of society are often seen.[18]

The question is: which discount rates should be used for projects when natural resources are involved? Consider a valley where a hydroelectric power plant is planned. If the plant is constructed, the landscape will be largely lost as a natural resource. This loss must be compared to the benefit of the electric power from the plant. The net utility, *NU*, can be written as:

$$NU = U(D) - C(D) - U(R) \qquad (5\text{-}5)$$

U(*D*) is the utility of the hydropower. *C*(*D*) is the cost of constructing the hydropower station, and *U*(*R*) is the utility of the natural resource.

6.2 Opportunity or Compensation Cost

Assume that one effect of a project is future damage that can be amended or compensated for. Estimate the future cost, C_A, of reversal or compensation and calculate the present value of C_A with formulae 5-2 using the social discount rate, *r*. Add this cost to the present value of the project's costs, calculated at normal discount rates. Since you have the opportunity of amending the damage in the future, you can calculate the present day cost of compensations. Projects that have irreversible impacts on nature or would need extremely costly abatement measures with foreseeable technology cannot be treated this way. Examples are erosion damage, desertification, or damages to cultural monuments.

6.3 Resource Specific Discount Rates

The value of many natural resources is often seen to increase with time, and there are probably several mechanisms behind this phenomenon. Since the abundance of natural resources in the world decreases, they become scarcer and are therefore appreciated more. People also become more affluent and have more leisure time, thus increasing the demand for recreation activities. Our awareness of the importance of species diversity and gene-pool diversity also increases, and probably so does the weight on species' right to existence. These trends and others work in the same direction and make the value of natural resources grow. This growth rate can be represented by a growth rate, *g*, which according to Krutilla-Fisher[19] should be taken into account when calculating the present value of a project. They recommend using the standard financial discount rate, *r*, which represents the growth of the economy, but then subtracting from it the growth rate *g*, thus in effect using the discount rate *r-g*. Instead of just arbitrarily manipulating the standard discount rate – an approach economists do not appreciate – Krutilla and Fisher provide an argument for a way to assess a correct net discount rate, depending on the environmental resource in question.

The problem with market-based discount rates, *r*, however, is that they exist in variants, and are different in different times and different locations, due to monetary policy, saving habits, etc. They may also be corrected for all sorts of imperfections including expected inflation.

An alternative is to use a discount rate based directly on value considerations, which can be related to sustainability, where restricted exchange with other economic aspects is allowed and where different discount rates would come up.

The conclusion is that there are different sets of discount rates to apply, from private market considerations to the ones holding the point of view of certain stakeholders, and those reflecting collective value considerations.

---

[1] Adapted from Grønn (1990)

[2] After Ellerman et al. (1998)

[3] Syers et al. (1996) p. 486

[4] Romm et al. (1998)

[5] Moffat (1997)

[6] Teisl et al. (2002) p. 341

[7] Carson and Gangadharan (2002)

[8] Teisl et al. Teisl et al. (2002) p. 343, 352

[9] Ostrom et al. (1999) p. 278

[10] Ostrom et al. (1999) p. 279

[11] Ownership in England and Wales in 1973 in Blacksell et al. (1981) p. 26, Alston et al. (2000)

[12] Ruthenberg (1980) p. 335-336

[13] Kula (1994) p. 42

[14] McRae (1997) p. 1868 and Rice et al. (1997) p. 36

[15] Summary by Ostrom et al. (1999) p. 278 and Dietz et al. (2003)

[16] Lumley (1997) p. 73

[17] The world banks discount rent for project appraisals in developing countries, Ninan et al. (2001)

[18] Lumley (1997)

[19] Fisher and Krutilla (1975)

Chapter 6

GAME THEORY

With applications to the management of common resources

The Tragedy of the Commons is a well-known enigmatic problem in environmental resource management. Game theory provides a framework that makes the phenomenon easier to understand, perhaps suggesting ways to avoid the tragedy. We discuss among other things:

- How to identify situations where the actors make decisions only once?
- Extension of the theory to several actors and several decision rounds.
- More realistic situations that can only be studied with simulation.

●

"When two friends have a common purse, one sings and the other weeps." *English proverb*

1. INTRODUCTION

Grazing land for sheep or reindeer, fishing banks and the atmosphere are examples of resources with limited capacity that are used by numerous individuals or groups that we refer to as actors or players. If there is open access, everybody is tempted to exploit the resource to serve their own ends; and if all do so, it will eventually lead to depletion. To avoid that, players sometimes negotiate an agreement to share the resource in a sustainable manner. The problem, however, is that it is tempting to cheat; so the result may still be overuse and subsequent depletion, or what is known as *the tragedy of the commons*.

Game theory is a tool that is useful for studying such phenomena. One of game theory's many achievements in environmental studies is to explain why many biological resources are overused to depletion like anchovy off the Peru coast, or threatened by extinction like cod in the North Atlantic. Game theory has been used to determine the distribution of quotas among bluefin tuna fishermen in Australia.[1] It has also been applied to the gas trade and to the study of international pollution abatement.

Game theory makes a basic assumption that self-interested actors will try to maximize their own gain, and this assumption is often able to explain actual behavior well. But it often happens that predicted behavior contrasts with what we see in real life, demonstrating that the assumption of narrow-minded "rationality" is not always correct.[2] Game theory allows for simultaneous decisions by several decision makers, where the outcome for each depends on the decisions of all. A decision maker therefore has to consider what competitors will do, before making up her own mind. However, that depends on what she thinks they think she will do, etc. This can go on ad infinitum, although in reality we have found that most people think only two steps forward in a decision process.[3]

Game theory can be used to:

- Give advice in decision situations.
- Illustrate ethical dilemmas when two or more actors make decisions affecting everybody.
- Find optimum social solutions to complex decision dilemmas, such as how to control pollution by issuing pollution permits.
- Predict how people will react in a decision situation.
- Find sustainable solutions to repeated games.

1.1 Basic Concepts in Game Theory

The study-objects of game theory are strategies available to individual decision makers, and the resulting outcomes when all actors have made their decisions. The simplest kind of game has only two players, each one having the choice of one out of two strategies, and the choice is made only once.

1.1.1 Conventions

For convenience, we shall call the players *Row* and *Column*; Row has the choice between rows and Column the choice between columns. The standard assumption is that they have to choose independently and simultaneously, but sometimes we shall allow for communication between them. Both players know the payoffs, which are shown in the game matrix as two numbers, the first is Row's payoff and the second Column's payoff. Table 6-1 shows an example.

Table 6-1. Game with no conflict. The cells show first Row's payoff, then Column's payoff.

		Column	
		C_1	C_2
Row	R_1	12, 8	7, 5
	R_2	10, 2	4, 0

1.1.2 Choice of Strategy

If you were Row in Table 6-1, which strategy would you choose? This is not a hard question; you would probably choose R_1. But why? Think about that for a moment before you read on.

In conventional decision-making under uncertainty, we assign probabilities to the future states of Nature and select the alternative that gives the highest expected outcome. This makes sense when Nature is blind and does not pay attention to what we do – Nature does not try to hurt us (or help us). In games, the situation is quite different. Here, we have opponents who try to maximize their own gain, usually at our expense. Therefore, probabilities do not work; we need other principles to guide our choice.

There are three main principles. We recommend that you try number one first; if that does not work, try number two, and as a last desperate resort number three.

1. *Look for dominant strategies.* A **dominant strategy** is better than the other strategies no matter what the opponent does. Look again at Table 6-1. If Row chooses R_1, Row's outcome will be either 12 or 7 depending on whether Column chooses C_1 or C_2. This is better than 10 or 4, which are the possible results of choosing R_2. Thus, R_1 dominates R_2. In the same way, Column will find that the outcomes 8 or 2 from C_1 dominate 5 or 0 from C_2. Therefore, if both players choose their dominating strategy, they end up in state R_1,C_1 with payoffs 12 to Row and 8 to Column. Sometimes it is possible to analyze games through **iterative analysis** of dominant or dominated strategies: If your opponent has an option that performs worse for him than another of his options no matter what you do, you may assume that he will not choose that one and delete it from the matrix. Then, perhaps, one of *your* options may be deleted, etc.
2. *Make the worst that can happen as good as possible.* This is called the **maximin strategy**, and is a reasonable choice if there is no dominant strategy and your opponent is trying to hurt you. Look again at Table 6-1. If Row chooses R_1, the worst that can happen is that Column chooses C_2, which pays Row 7. If Row chooses R_2, on the other hand, the worst that can happen to Row is 4. Row's maximin strategy is therefore R_1. In the same way, we find that Column's maximin strategy is C_2.
3. *Choose your strategy at random with carefully computed probabilities, which we shall not delve into here.* This is called a **mixed strategy**, and may be used if maximin strategies do not lead to a stable solution. We will say more about that below.

These principles make sense, and most people are willing to accept them as guides for choice. They even describe pretty well what people actually do in such situations. However, they are just principles and can sometimes betray you. They should not be applied blindly, but only after careful analysis of possible consequences.

1.1.3 Analysis of Solutions

When both players make their choice, they end up somewhere in the matrix. All cells are therefore possible **solutions** to the game, and solutions can have different characteristics. The most important are:

- *Dominance*: A solution is **dominating** if no other cell is preferred by any player. In Table 6-1, the solution R_1,C_1 is preferred by both players. Thus, there is no conflict.
- *Pareto optimality*: A solution is **Pareto optimal** if no other cell is preferred by all players. Thus there is no room for negotiations once a Pareto optimal solution is reached, since somebody must lose by further moves.
- *Nash equilibrium*: A **Nash equilibrium** is a solution that no player would want to move from unilaterally. This means that if all other players stick to their strategies, they will be worse off if they alone do something else. A Nash equilibrium is therefore stable. There could very well be better solutions for all players, but only if they agree to move together.
  Let us classify the cells in Table 6-2.
- R_1,C_1: If both players follow their dominating strategies, the solution will be R_1,C_1. This solution is *not dominating*, since Row would rather be in R_1,C_2 and Column would rather be in R_2,C_1, but neither of them can persuade their opponent to make the required move. R_1,C_1 is *Pareto optimal*, since no other cell is better for both players. It is also a *Nash equilibrium*, since Row would lose 2 by moving to R_2, and Column would lose 3 by moving to C_2.
- R_2,C_1: This cell is *Pareto optimal* since at least one player must lose if they move together to another cell. It is not Nash *equilibrium*, since Row would gain by moving to R_1. The same classification applies to R_1,C_2.
- R_2,C_2: This cell is obviously neither dominating, nor *Pareto optimal* nor a *Nash equilibrium*.

Table 6-2. Game with weak conflict.

		Column	
		C_1	C_2
Row	R_1	12, 8	13, 5
	R_2	10, 9	4, 0

2. SYMMETRICAL GAMES BETWEEN TWO PLAYERS

In situations where there is open access to common resources, it is convenient to consider all actors as equal. Real harvesters of common resources are of course of somewhat different sizes – some fishermen have large boats, some have small boats. We shall, however, for game theoretic purposes assume that they have available exactly the same strategies, with equal payoffs, thus making the game matrix symmetrical. In spite of being simple, such games turn out to be surprisingly characteristic of a series of common situations in daily life, on the political arena, and in ecology.

With two players, there are essentially two options open to each, they may either cooperate with the other (C), or defect (D). The general game matrix is shown in Table 6-3.

Table 6-3. General representation of 2 × 2 games with symmetric gains.

		Column	
		C	D
Row	C	R, R	S, T
	D	T, S	P, P

It is perhaps surprising that much interesting can come out of something as simple as Table 6-3, but we shall show that the behavior of the players will depend crucially on the internal order of the payoffs, and that the situations sometimes become very complex. The letters are chosen for easy remembrance; R can often be interpreted as reward for cooperation; T stands for temptation to defect, S for sucker – the losers score; P is punishment for non-cooperation. Row's payoff is the first number in each cell; Column's payoff is the second number.

The four payoffs R, T, S and P can be ranked in 4! = 24 ways, giving rise to 24 different types of games. However, all of them except four are rather trivial with weak conflicts or none. The four remaining, on the other hand, have interesting applications. They are known as the *archetypical* games. When we represent the games, we shall assign the ordinal values 1, 2, 3 and 4 to the four payoffs. Thus, R = 4, T = 3, S = 2 and P = 1 means R > T > S > P, but the real payoffs would be quite different numbers.

2.1 Four Archetypical Games

2.1.1 The Tragedy of the Commons

Consider two fishermen who harvest from the same fish stock. They may cooperate so that each fishes a sustainable annual share with a consequent decent payoff for both. If one of them cheats and fishes more than his share, the cheater will make a gain, but at a considerable expense to the honest fisherman since the fish stock will eventually be depleted. If both cheat, the fish stock will be depleted even faster. The consequent payoffs are shown in Table 6-4.

Table 6-4. The tragedy of the commons, payoffs.

		Fisherman 2	
		Cooperate, C	Deplete, D
Fisherman 1	Cooperate, C	10, 10	1, 12
	Deplete, D	12, 1	2, 2

Seen from the perspective of fisherman 1, if he maximizes his own interest, it is better to cheat than to cooperate no matter what the other fisherman does. If fisherman 2 cooperates, fisherman 1 will earn 2 units more by cheating; if fisherman 2 depletes, fisherman 1 will still earn 1 unit more than fisherman 2 by cheating. In other words, to deplete is a dominating strategy.

The game is symmetrical, so the same goes for the other fisherman. Thus, if each maximizes his own interest, they end up both depleting the stock (D,D). Is this a Nash equilibrium? Yes, none of them has any incentive to change, not even if the other starts to cooperate. In other words, the solution D,D is stable. Is it Pareto optimal? Certainly, not! They both would be much better off by cooperating. But the tragedy is that while C,C is Pareto optimal, it is not a Nash equilibrium, and therefore unstable which means that both of them would gain by being the only defector, at least in the short run.

This classical dilemma is sometimes called **prisoners dilemma**, but it has implications for all types of provision of public goods.[4] What is the advice to a self-interested decision-maker? Communicate and make a strong commitment to equal sustainable shares, and then cheat! The more certain you are that the other will cooperate, the more certainly you will gain in the short run by cheating. We see here that some kind of environmental management is called for; we need a third party that can enforce an agreement.

Notice that it is a gross simplification to assume that the decision is only made once. In reality, the game will go on for years and one may question for how long one fisherman can harvest more than his share and the other behave responsibly?[5]

Table 6-5. The tragedy of the commons, ordinal payoffs.

		Fisherman 2	
		Cooperate	Deplete
Fisherman 1	Cooperate	3, 3	1, 4
	Deplete	4, 1	2, 2

Cast in the form of ordinal payoffs, the game looks like Table 6-5. It shows that the game is characterized by a temptation to cheat (T = 4) that is larger than the reward for cooperation (R = 3), which again is larger than the punishment for non-cooperation (P = 2). Lowest is the sucker's lot (S = 1).

2.1.2 Leader

If two drivers arrive at an intersection at exactly the same time, they may both "cooperate" by being polite and wait for the other to enter first – both incurring waiting costs. If, on the other hand, they both enter at the same time (D, D), they will collide and incur much higher costs. If one enters first (D) and the other waits a little (C), the result for both will be better. The game matrix is shown in Table 6-6.

Table 6-6. Leader, ordinal payoffs.

		Column	
		C = wait	D = enter
Row	C = wait	2, 2	3, 4
	D = enter	4, 3	1, 1

Each driver is obviously best off if he enters if the other waits, and waits if the other enters. In other words, none of them has a dominating strategy. In that case, the advice is to look for a maximin strategy. Obviously, the worst thing that can happen is if both enter and collide. Therefore, each driver's maximin strategy is to wait, thus C,C seems a natural solution. Is this a Nash equilibrium? No, both would benefit from defecting alone. Is C,C Pareto optimal? No, both would benefit from moving together to D,C or C,D. What is the advice to a self-interested decision-maker? Communicate by either looking aggressive or smile and wave your hand. This is very different from the tragedy of commons where communication is treacherous; here it is very useful.

The game borrows its name from the one who enters first, (D) being called leader. The one who waits is called (C) follower. In this game, both players are rewarded if one player defects, but the one who defects rewards herself more than the other.

Empirical studies have shown that approximately 78% choose C. The percentage depends on how bad C,C is. So far we have not seen applications of this archetype in environmental decision-making.[6]

2.1.3 Battle of the Sexes

This game is somewhat similar to the previous "Leader" game, but the one who defects (here: resigns) rewards the other more than herself. This example was constructed during the 1950s. In business it is called the Buyer-Seller game.[7] Ann and Kanute plan independently how to spend the evening. Ann would like to go to the opera, but perhaps Kanute will go to the Soccer match, and she would prefer being there with him, rather than at the opera without him. The worst that can happen to Ann is of course to be at the soccer match alone. Kanute has similar problems; also he prefers to be together. The game matrix is shown in Table 6-7.

Table 6-7. Battle of the sexes, ordinal payoffs.

		Ann	
		C = Ballet	D = Soccer
Kanute	C = Soccer	2, 2	4, 3
	D = Ballet	3, 4	1, 1

Since for Kanute it is best to be wherever Ann is, he has no dominating strategy; and vice versa. Thus, he applies his maximin strategy, which is to go to the soccer match. Ann does the same, and goes to the opera. The solution C,C is not a Nash equilibrium, however, and is therefore unstable. How can Kanute (Ann) get the solution he (she) wants? Send an SMS and say where he (she) is going, but being sure to be the first one to announce it. Alternatively, since C,C is not Pareto optimal either, they might negotiate one of the two Pareto optimal solutions D,C and C,D. Then the battle begins.

Empirical observations show that 82% use the maximin strategy, C. So far, we have not seen any application to environmental decision making in the literature.

2.1.4 Chicken

The name of the game comes from California where young car drivers approach each other with their left wheels on the dividing line, but it is more representative of situations where the process of solving a conflict has hardened; the negotiators stay with their claims, and no agreement is in sight. The archetype is shown in Table 6-8.

Table 6-8. Game of chicken, ordinal payoffs.

		Column	
		C = cooperate	D = stay
Row	C = cooperate	3, 3	2, 4
	D = stay	4, 2	1, 1

No player has a dominating strategy, and the maximin strategy for both players is to cooperate in order to avoid the risk of no agreement. The solution C,C is a Pareto point, meaning that there can be no mutual agreement to another solution.

How would you advise a decision-maker? Of course, reasonable advice is to cooperate. But the problem is that C,C is not a Nash equilibrium. Each player would gain by staying on track, and the negotiator with the worst reputation for ruthlessness would have a good chance of winning. Empirical observations have shown that the probability of choosing C strongly depends on how serious the worst outcome is. In one particular situation[8] 65–70% chose alternative C.

The game has been used to describe important international negotiations that have become deadlocked, such as reduction of CO_2 emissions. If there is no agreement to control CO_2 emission, it will continue and global warming may result in disaster. However, an agreement is only one prerequisite. As long as the temptation to break the deal (T) and expected reward (R) is greater than S (suckers outcome) and P (punishment), some players may be tempted to break the agreement.

The game has also theoretical applications in the analysis of stable evolutionary strategies.[9]

2.1.5 Summary of Archetypes

The tragedy of the commons is different from the other three games; a fisherman who cooperates, punishes himself, but rewards the other. If both cooperate, they will both be rewarded. The Leader game formalizes the situation when a leader uses his power to lead. It favors herself most, but also the people she leads. The battle of the sexes formalizes a situation where a person takes an initiative, but rewards the other more than herself. Both players, however, are better off than if no one takes an initiative. The chicken game describes the power of threats, the person that is the most daring risk-taker wins, the other loses. The game was used as a metaphor for the arms race during the cold war around the 1960s.

Tables 6-9 and 6-10 compare the four games.

Table 6-9. Order of payoffs and of actor labels in the four archetypical games.

Name of game	Order	Cooperator	Defector
Tragedy of the commons (1)	T > R > P > S	martyr	martyr
Leader	T > S > R > P	follower	leader
Battle of the sexes	S > T > R > P	admirer	hero
Chicken	T > R > S > P	victim	exploiter

(1) This game is also known as prisoners dilemma. In the prisoners dilemma one adds the condition R > (S+T)/2

Table 6-10. Pareto optimal solutions and Nash points of the four games. Pareto solutions are bold faced, and Nash points italic.

	Tragedy of the commons		Leader		Battle of the sexes		Chicken	
	C	D	C	D	C	D	C	D
C	**3,3**	**1,4**	2,2	***3,4***	2,2	***4,3***	**3,3**	***2,4***
D	**4,1**	*2,2*	***4,3***	1,1	***3,4***	1,1	***4,2***	1,1

3. MORE GENERAL GAMES

The set of symmetrical games between two players where each have two options, is of course very restricted . In this section, we shall look at some other games.

Games are often construed to look like real situations. The challenge is to include as much realism as required for the solution to be trustworthy.

A Contest Game is an encounter between decision makers in which the participants do not rank the possible outcomes in the same order; thus, there is a conflict of interest.[10] The four archetype games are all contest games. A **zero-sum game** is a game where one player's gain is exactly matched by the other player's loss; thus there is no occasion for cooperation. In a **flatland** game the players can only interact with the neighbors, such as in a community that consists of many small villages. Such games tend to allow more diversity in strategies over a longer time because it is more difficult to eliminate non-competitive strategies.[11] In a **non-bargaining game** the players make the decisions once and simultaneously; no binding agreements can be formed under the game and it is not allowed to threaten other players if they don't play as one wishes. It may seem strange that non-bargaining games, are important when bargaining is so commonplace. However, people do not bargain if the number of affected parties is high or when bargaining power is unevenly divided among the actors. Also bargaining is not anticipated when the expected gains and losses from cooperation differ widely among the actors, or when property rights are non-existent or not well defined. A third group of non-bargaining situations is when bargaining is about public goods, or, when for some reason the costs of bargaining are large relative to the gains expected from cooperation.[12] An intriguing argument is that most complex games demand more cognitive capabilities than most individuals can muster.[13]

Methods to find solutions become more complex as the problems become richer. Complexity increases when the number of alternatives, players and decision rounds increases. Whether information is symmetrically shared among the players, also affects complexity. A very common situation with asymmetrical information is covered in *Principal-Agent* theory. It arises when a principal hires an agent to do a job, and the agent knows how much work is needed, but the principal does not.

3.1 Several Players

The classical 2 × 2 game of Tragedy of the Commons can be extended to more than two players, by letting all players play pairwise 2 × 2 games. The tragedy is probably best demonstrated when many people run to the exit to save themselves, then congestion occurs in front of the door and it can not be opened. Below, you find examples of other real world multi-person N-person Tragedy of the Commons situations, and you are invited to try to make the interpretation.[14]

- Give wage increases that are higher than the inflation rate.
- Follow government advice to save the environment.
- Carry a hidden gun.
- Sell nuclear weapons.
- Stand on your toes to see a parade.
- Increase one's family in India and China.

Observations show that the more players who participate in the game, the more players choose the option Defect, i.e., not cooperate. The reason is probably that everyone thinks there is an enhanced probability that there will be a defector among the players – **a bad apple** – and if there is one who earns more than her share, why should not I try?

3.2 Several Rounds

Suppose that the Tragedy of the Commons game in Table 6-4 is played 10 times between the same two opponents, so that the opponents each time can choose between Cooperating and Depletion (i.e., Defect). You must assume that nature is restored between every game. What kind of strategy should a player use to maximize his or her own gain, when the choice each time may depend on what happened in the previous rounds? It can be shown that a strategy called **Tit For Tat** (TFT) is particularly strong. With the TFT strategy you cooperate in the first round. In the next rounds you always do as your opponent did in the previous round.

Suppose that the two players must choose between two 10-period strategies, either TFT or Always Deplete. The payoff matrix after 10 rounds becomes as in Table 6-11. If both chose TFT, they will start with cooperation, and then they continue to cooperate for ten rounds, ending up with a payoff of $10 \times 10 = 100$. We have made an assumption here, that, all rounds have the same potential payoffs. In reality, of course, a choice of depletion by one player will reduce the fish stock in the next round, thus reducing payoffs and making depletion strategies worse for both contestants than shown here.

Table 6-11. Payoff matrix for Row after 10 moves[15] - Column's payoffs are symmetrical.

		Column	
		Always Deplete	Tit for Tat
Row	Always Deplete	$10 \times 2 = 20$	$12 + 9 \times 2 = 20$
	Tit For Tat	$1 + 9 \times 2 = 19$	$10 \times 10 = 100$

We see that Row does not have a dominating strategy, and that the maximin strategy is to always deplete. But this is obviously not a good choice, when it is possible to earn 100 by playing TFT. The natural solution is therefore that both play TFT; this is both a Nash equilibrium and a Pareto optimal solution. In fact, TFT often wins in competition with other strategies. It has three characteristics that account for its impressive performance: it is inviting – it cooperates at the first move; it retaliates immediately by punishing defection in the prior round; and forgiving – it immediately returns to cooperation when the contestant starts to cooperate.[16]

3.3 Playing Games

It is instructive to simulate what may happen in real or artificial situations by playing games in a laboratory, simulating the games on a computer or observing actions among people in a small community where everyone can interact with each other. Even if an artificial setting takes away important aspects of reality there is probably enough left for useful learning sessions.

3.3.1 The Archetypes

All archetypes can be played. Each player may be assigned a personal trade-off table, or there may be a common table known to both. It is also instructive to assign roles to players, like "always defect". One interesting observation is that some players – usually male – apparently put more emphasis on beating the opponent, rather than maximizing his own payoff.

3.3.2 Testing for Altruism

The so-called *Ultimatum game* is a test for altruism. Player 1 is promised a certain amount of money, say €1000, on the condition that Player 1 gives a share of it to Player 2, and that Player 2 finds that share acceptable. Thus, before any money is given, Payer 1 and Player 2 have to negotiate which share Player 2 is going to get. If they cannot agree, nobody gets anything.

It is interesting to observe what share Player 1 offers to Player 2, as well as how much Player 2 requires for accepting the money. Results of observations in 15 small, indigenous societies indicate that when the players belong to a small group, Player 1 will offer a larger share to Player 2, who in turn will require a larger share, than in a larger group.[17] It has also been observed that some people would not accept large gifts because they feared unspecific strings attached.[18]

3.4 Strategy Composition

The combined gain of all players in a game – the societal gain – seems to be greater if players use different strategies than if all players adopt the same strategy.[19] Is there an optimum mix of strategies that provide high, long-term yield to the society, and what would that mix look like? This can be identified by simulation, but since that requires simplifying assumptions, it is hard to judge how realistic the results are: a particular mix of "forgiving" and "policing" strategies seems to give the highest societal yield.

Game theorists describe strategies with parameters. A (p,q)-strategy means that C (cooperate) is played with probability p after a C from the contestant, and with probability q if the contestant played D (defect). Thus (0,0) is an "Always defect" strategy, and (1,0) means Tit-for-Tat. Figure 6-1 shows a simulation result after 1000 "generations" starting with a uniform distribution of 99 strategies ranging from Always defect (the corner in the picture that comes out of the figure) to Always cooperate ($p = 1$, $q = 1$) in the corner furthest away.

"Generous Tit-for-Tat" (GTFT) is a strategy that cooperates on the first move and forgives defection more than twice. When a stochastic Tit-for-Tat strategy was added, GTFT dominated almost completely after 1000 generations, although there are tiny amounts left of other strategies, like ALLD, which is not seen in the figure. The figure is from Nowak and Sigmund (1992) and the development is described in a dramatic language.[20]

> "But the TFT-like strategy that caused this reversal of fortune is not going to profit from it; having eliminated the exploiters, it is robbed of its mission and superseded by the strategy closest to GTFT. Evolution then stops."

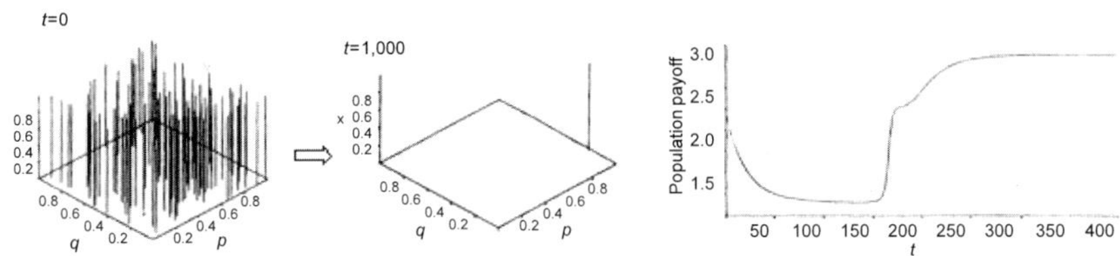

Figure 6-1. Game theory simulation of the fate of 99 strategies plus one strategy that includes a stochastic element after 1000 generations. The letters *p* and *q* measures the probability of cooperation and defection. The right figure shows the population payoff, or the societal gain.[21] Details in Nowak and Sigmund (1992).

Games can be used as metaphors for real life situations. Simulation shows that there are strategies that do well when chosen by only a few players, but bad when played by many. The important question is whether the game-theoretical caricatures of reality describe reality sufficiently well to draw realistic conclusions.

3.5 Rational and Irrational Behavior

Players in laboratory settings are supposed to act self-interestedly, aiming to maximize their own payoff. One objective of game theory is to give advice to reach this goal – that is, to act rationally; another is to try to predict what people will actually do. Failure to act rationally wants explanation, and in many cases it can be attributed to lack of understanding of the game situation or to the real goals of the players differing from mere self-interested maximization. In such cases, a reformulation of the game will often show that the observed behavior was indeed rational.

In the battle of the sexes, players often tacitly assume that decisions will not be forgotten, but have a bearing on expectations next time. Players also often include as an objective a desire for equitable outcomes among themselves.

Similarly, the Tragedy of the Commons could be turned into Victory of the Commons if mutual trust were established. Some players do actually assume that the opponent is trustworthy and consequently choose to cooperate. This brings virtuous ethics into the game, making it quite different from just maximizing payoff.

The example of the game of Leader assumes that the cars approach the intersection at the same time, but this is an almost impossible ideal. One person usually sees the other first, and has the right of way.

The game of Chicken depends on the balance of personalities of the players. If one is more anxious or daring than the other, the game will flip from cooperation to one of the C,D-solutions. The player with a reputation for being most daring – or insane – has the greatest chance of winning. The relative sizes of the four payoffs determine which solution is most likely. A central concept is the **critical risk**, which is the highest risk for negotiation breakdown a player is willing to accept. The critical risks of the two players determine the relative strengths of the bargaining positions. The more important it is to win compared to losing, the stronger is your bargaining position.

As noted, social moral rules play a role in the archetypical games. The Golden Rule of Confucius is: "Whatever ye would wish that men should do to you, do ye even to them." Immanuel Kant said, "Act only on such a maxim through which you can at the same time wish that it would become a universal law".[22]

4. APPLICATION: AVOIDING THE TRAGEDY OF THE COMMONS

The game-theoretical representation of the Tragedy of the Commons explains why open access may lead to over-harvesting. Dairy farming with cattle grazing in the commons was typical farming practice in Europe even 300 years ago, and is still widely used all over the world (Chapter 20, section 2.4). It is tempting for a farmer to bring in extra cows since this increases milk production for that farmer, although it eventually reduces total milk production for all farmers. We have seen that if the decisions are left to the individuals, the result is very likely to be over grazing to the loss of all.

We therefore need management that changes the rules of the game. The question is how this can be done.

4.1 Decision Analysis

Decision problem: How can over-harvesting be avoided? The time frame for the problem is usually measured in decades, depending on the time it takes to ruin the pasture for grazing purpose.

Decision makers: Normally, the decision makers are the members of the local community, or their representatives.

Stakeholders: These are the users of the meadow and all others with an interest in it, including banks that lend money to the farmers. See Chapter 3, section 2.3 on decision-makers, Chapter 9, section 6.3 on environment and fairness, and Chapter 10, section 4.2 on chain impacts.

Decision objectives: The overall *decision objective* can be formulated as "Efficient and sustainable management of the common". There are at least five *subgoals*:

- Maintain sustainability.
- Preserve flora and fauna.
- Maximize the yield from the meadow.
- Minimize control costs.
- Maximize equity among the farmers.

To see more on how goals should be chosen see Chapter 3, section 3.2
Decision alternatives: Possible decision *alternatives* are:

- No control measures.
- Winter rule: no citizen may send more cows to the commons than he can feed during the winter.
- The field lays fallow every seventh year.
- Part of the field is allocated as a no grazing area, providing a refuge for fauna.

How to structure objectives and goals is discussed further in Chapter 3, section 2.

Consequence analysis: To predict consequences, one should first investigate whether one actually is facing a Tragedy of the Commons problem. This requires construction of a game matrix to see whether the inequalities $T > R > P > S$ are satisfied, this chapter, section 2. To calculate the sustainable level of harvesting, models based on precipitation data, Chapter 20, section 4 can be used to calculate primary production, and models from Chapter 12, section 5 to estimate grazer density. Some generic rules for sustainable harvesting can be found in Chapter 13, section 4 on harvesting and generic rules for maintaining terrestrial natural ecosystems in Chapter 19, section 3. Calculation of control costs depends upon local conditions. The balance between control costs and the utility of control measures is a topic in the economy chapter, (Chapter 5, section 3). The social cost criteria may be complex, including constraints set by subsistence, preservation of rural settlements, the intrinsic value of the "commons", etc.

Preferences: The decision makers use their preferences regarding commons in developed nations. In developing countries, where subsistence is an issue, a more complex decision process may be required (Chapter 9, section 6.3).

4.2 What Happens in the Case of Swiss Commons?

Alpine villages in Switzerland have faced the decision problem described above. They decided on the so-called "winter rule" by which no citizen is allowed to send more cows to the public meadows than can be fed during the winter. The rule was enforced through inspection and substantial fines. However, policing is expensive and it is sometimes hard to finance it. This may lead to a double tragedy: no funds for policing, no inspection, and then subsequent over grazing.[23]

When rules are enforced, the problem will switch from the Tragedy of the Commons to something else, since defection is no longer an option. An interesting question is what conditions favor sustainable use of commons. Apart from the obvious requirement that the harvest has to be smaller than the yield from the stock, it may be tempting to adopt a biological principle for policing: intermediate levels of relatedness among individuals facilitate policing, and as a consequence reduce the risk of over-harvesting. Other solutions for policing are described in the section on management of fisheries in Chapter 13.

---

[1] Klieve et al. (1993), and Simons et al. (1995)
[2] Sigmund et al. (2000) p. 281: Nowak et al. (2004)
[3] Coleman (2003)
[4] Gibbons (1992) p. 5
[5] A player who always obeys values and principles is called a UTOPIA player
[6] Coleman (1990)
[7] Hofbauer and Sigmund. (2003) p. 510
[8] Described by Colman (1990)
[9] Maynard Smith (1993) p. 193
[10] Maynard Smith (1993) p. 206
[11] Nowak et al. (1995; 2004)
[12] Perman et al. (1996) p. 318
[13] Hofbauer and Sigmund (2003) p. 509
[14] Colman (1990) p. 160
[15] After Maynard Smith (1993) p. 197
[16] Brembs (1996)
[17] Mace (2000), Hofbauer and Sigmund (2003)
[18] Henrich et al. (2001) showed that individuals in New Guinea demanded an equal share.
[19] Brembs (1996)
[20] Nowak and Sigmund (1993)
[21] Details in Nowak and Sigmund (1992)
[22] Kant (1983)
[23] This problem became famous with an article by Hardin (1968). It is treated in detail in Short and Winter (1999). The decision alternative used below is due to Hammerstein (1995), Ostrom et al. (1999) and Ostrom (2000).

Chapter 7

PREFERENCES

Elicitation of decision-maker values

We assume a decision-maker oriented paradigm – meaning that the decision-maker is the source of the values underlying the choice to be made – and discuss how her preferences can be elicited in a reliable and valid way. In particular, we look at:

- How far can we get at ranking decision alternatives without preference information?
- How to elicit importance weights with the swing method?
- How to elicit importance weights with the pair wise comparison method?
- How to use weights to calculate the performance of the alternatives?
- How emotion can enhance validity and reliability?
- How a simple spreadsheet program can facilitate weight elicitation and performance evaluation?

"C´est le temps que tu a perdu pour ta rose qui fait ta rose si importante." *Le Petit Prince, de Saint-Exupéry, 1997*

1. INTRODUCTION: THE SOURCE OF VALUE

The *raison d'être* of environmental management is concern about things that are valuable. Preferences are about the relative importance of values. A fundamental question is: whose preferences should be used in environmental decision-making? The answer depends on which **paradigm** you subscribe to. An economist would answer that valuation is the prerogative of the population; a decision maker would be inclined to rely on her own preferences. There are pros and cons for both viewpoints,[1] which we shall discuss in the chapter on ethics. However, we shall adopt the following pragmatic viewpoint: *If information about the population's Willingness-To-Pay (WTP) is available, it should be used; otherwise the preferences of the decision-maker should prevail.* In many practical situations this means that we use a combination; some decision criteria will have known WTPs while others must be attributed weights by the decision-maker.

In Chapter 3, we applied a **management paradigm**. We discussed how the preferences of a decision-maker can be elicited and made operational in a decision context. We demonstrated especially how we could use the *swing weight method* to establish the importance of the decision criteria in terms of weights.

The paradigm of Chapter 5, on the other hand, is that of an **economist**. We discussed the difference between cost-efficient and cost-effective pollution abatement policies. *Cost-efficiency* means that we clean up until the marginal cost of cleaning becomes equal to the marginal cost of pollution. According to the paradigm, the latter cost is how much the **population** is willing to pay (WTP) to avoid pollution damage. In practice that kind of knowledge is hard to obtain, but in the next chapter we shall discuss methods for estimating WTP. *Cost-effective* methods are more pragmatic. They assume that pollution reduction targets are set by the authorities, and investigate how they can be met with as little cost as possible.

In this chapter, we shall apply the management paradigm and have a closer look at preference models for decision-makers, which we briefly encountered in Chapters 2 and 3.

2. CONSEQUENCE TABLES

The **consequence table** is the basic object of study in this chapter. It is a matrix with expected consequences of the decision alternatives; rows are decision criteria and columns are decision alternatives.

Table 7-1, which is identical to Table 3-4, shows an example. The question in this chapter is: Given a consequence table, what should be done? Our task is to help our decision-maker decide between three strategies: *no action*, *collect oil* at the site, or *direct oil* to a place on the coast.

Table 7-1. Consequence table showing the consequences of different oil pollution abatement strategies.

Criterion	Unit	No action	Collect oil	Direct oil
Recreation	1000 exposures	200	200	1000
Industrial loss	Million €	0.45	0.45	0.34
Shore pollution	Year km	600	600	300
Bird deaths	1000	1000	1000	700
Auk rest. time	Year	12	12	10
Oil lost in sea	Ton	10,000	8000	8000
Combat costs	Million €	0	1.25	1.50

The problem facing the decision-maker is of course that it is cheapest to do nothing, but then nothing is achieved. To direct oil to the coast is worse for recreation than collecting it at the site because more people will be affected by it. But directing is better than collecting for most of the other environmental criteria. Thus, there is no clear-cut solution; it all depends on preferences. – Or does it?

2.1 Analysis Without Preferences

Decision tables can to some extent be analyzed without preference information. We can illustrate this with a two-dimensional graph, and have therefore created Table 7-2, which is a reduced version of Table 7-1. All criteria except combat costs are lumped into a single environmental damage index. (We have, in fact, used the preference information in Chapter 3 to get from Table 7-1 to Table 7-2).

Table 7-2. Compressed consequence table with only two criteria.

Criterion	Unit	No action	Collect oil	Direct oil
Environmental damage index	[0, 1]	0.87	0.84	0.73
Combat costs	Million €	0	1.25	1.50

2.1.1 Screening for Dominated Alternatives

Look at Table 7-2. Which alternative would you choose? The answer is not obvious. *No action* is cheapest, but worst for the environment. To *collect oil* is cheaper than *directing* it, but not so good for the environment. Thus, none of the actions strictly **dominate** another. If that were the case, we could just have deleted those that were dominated.

However, to *collect oil* is just a little cheaper than *directing* it, but near to *doing nothing* with regard to the environment. Let us have a closer look at that. The scores are shown in the graph of Figure 7-1. The closer to the origin an action is the better. The line between *no action* and *direct oil* is called the **efficiency frontier** since there are no projects on its inside. To *collect oil*, however, lies on the non-efficient side. Let us see what that implies.

The underlying task is to find the alternative that maximizes (7-1):

$$u(d,c) = w_d d + w_c c \tag{7-1}$$

Here, d means environmental damage and c combat costs, while w_d and w_c are the importance weights of the two criteria.

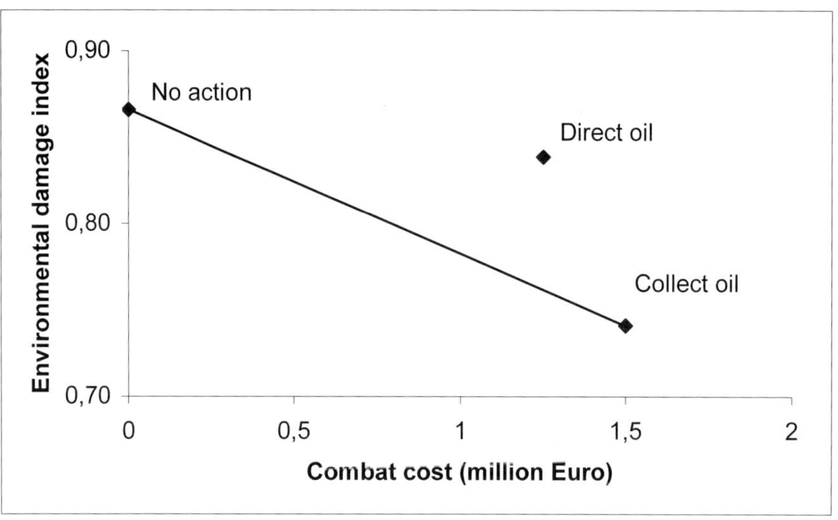

Figure 7-1. The performance of the three decision alternatives. The line between No action and Direct oil is called the efficiency frontier.

If w_d is large compared to w_c, *directing oil* will perform better than *no action*, and if it is the other way around *no action* will win over *directing oil*. However, since *collecting oil* lies on the wrong side of the efficiency frontier, it can never outperform both of the two others! You can easily verify this for yourself, by just trying out some weights; and it can be proven mathematically. Thus, one way to eliminate alternatives is to check in a diagram whether they are on the wrong side of the efficiency frontier.

This does of course not solve the problem of finding the best option; it only simplifies it, since we now have two instead of three candidates. To choose between the two remaining ones, we still need importance weights.

2.1.2 Data Envelopment Analysis

The procedure above works fine if we have only two decision criteria, but becomes intractable when there are many. A mathematical technique has been worked out, which is called Data Envelopment Analysis[2] (DEA). DEA requires software, however. A DEA program basically identifies all those alternatives as inefficient that have inferior performance regardless of weights. Alternatives that are better than the others for some weights are called efficient. This method is useful when there are many decision alternatives, but only as far as it goes; it has a tendency to produce many efficient alternatives – some of them being efficient only for very unlikely weights – and therefore one has to apply judgmental weights at the end anyway, to single out only one alternative.

2.2 Analysis with Preference Elicitation

We shall now look at weight elicitation. Let us return to the situation with the seven decision criteria in Table 7-1 and all three decision alternatives. Let us call the criteria $x_1, x_2, \ldots, x_7$. We want to identify the alternative that maximizes the **total utility function** (7-2):

$$u(x_1, x_2, \ldots, x_7) = w_1 u_1(x_1) + w_2 u_2(x_2) + \ldots + w_7 u_7(x_7) \tag{7-2}$$

The x-scores in (7-2) are known from the consequence table (Table 7-1), but the **primary utility functions** $u_1, u_2, \ldots, u_7$ and the weights $w_1, w_2, \ldots, w_7$ are unknown. They depend on the preferences of the decision-maker.

We showed in Chapter 2 how to elicit primary utility functions. When we do so, it is convenient to define a range for each criterion, with a worst value x^0 and a best value x^*, so that $u_i(x^0) = 0$ and $u_i(x^*) = 1$ for all i. With these definitions (7-2) implies:

$$u(x_1^0, x_2^0, \ldots, x_7^0) = w_1 u_1(x_1^0) + w_2 u_2(x_2^0) + \ldots + w_7 u_7(x_7^0) = 0 \tag{7-3}$$

$$u(x_1^*, x_2^0, \ldots, x_7^0) = w_1 u_1(x_1^*) + w_2 u_2(x_2^0) + \ldots + w_7 u_7(x_7^0) = w_1 \tag{7-4}$$

$$u(x_1^0, x_2^*, \ldots, x_7^0) = w_1 u_1(x_1^0) + w_2 u_2(x_2^*) + \ldots + w_7 u_7(x_7^0) = w_2 \tag{7-5}$$

Et cetera.

Thus, the total utility of a scenario where everything is as bad as it can get is zero (7-3). If one criterion is as good as it can be, the total utility is exactly equal to the weight of that criterion, (7-4) and (7-5).

2.2.1 A Closer Look at Swing Weights

In Chapter 3, we elicited weights with the swing method. We started with the worst-case scenario, which corresponds to (7-3) above, and asked which criterion the decision maker would most prefer to move to its best value. Let us say the answer was x_1. We then assigned x_1 the weight $w_1 = 1.0$. We again assumed the worst-case scenario and asked about the next most preferred criterion to be moved to its best value. Let us say the answer was x_2. Then the question was "how important is this compared to the first move?" We see from the equations above, that this is a question about the size of w_2/w_1, which is equal to w_2 when $w_1 = 1.0$. We proceeded in the same fashion and found all the remaining weights. The final operation was to normalize all weights to make the sum equal to 1.0.

This shows that the swing weight method is a *valid method* to estimate the weights in the total utility function (7-2). Validity means that we actually estimate what we want to estimate, and not something else. Whether it is a reliable method, is another question, however. For a method to be reliable, we should get approximately the same results every time it is applied, provided the real preferences are the same. But the problem with the swing weight method is that it considers extreme and very unlikely scenarios, and in addition it asks about degrees of importance. This is taxing for the cognitive skills of any decision-maker

2.2.2 Pair Wise Ordinal Comparison of Criteria

Pair wise ordinal comparison of criteria is a quick-and-dirty alternative that just asks whether one criterion is more important than another, without asking about degrees. It often comes close to the target, however, especially when there are many criteria. Let us demonstrate it by applying it to the oil pollution case. Table 7-3 shows the consequences with the best and worst scenario added, corresponding to Table 3-5.

The weight estimation procedure runs in the following fashion: Which criterion is most important, *Recreation* or *Industrial Loss*? To answer, the decision maker must consider the worst cases in Table 7-3 (in general, one has to consider the ranges from worst to best). The worst impact on *Recreation* is one million exposures, whereas *Industrial Loss* is at most €500,000. If she judges it more important to avoid one million lost recreation days than to avoid half a million euro in industrial losses, a score of 1 is entered in *Recreation's* column in Table 7-4 at the intersection with the *Industrial Loss* row, and a zero in the symmetrical intersection. If there is a tie, a score of 0.5 is entered in both intersections.

The procedure is repeated for all pairs of criteria; that is $n(n\text{-}1)/2 = 7 \times 6/2 = 21$ comparisons.

Table 7-3. Oil spill case: Consequence table with best and worst scenario.

Criterion	Unit	Best case	No action	Collect oil	Direct oil	Worst case
Recreation	1000 exp	0	200	200	1000	1000
Ind. loss	Million €	0	0.45	0.45	0.34	0.50
Shore poll.	Year · km	0	600	600	300	600
Dead birds	1000	0	1000	1000	700	1000
Auk rest.	Year	0	12	12	10	12
Oil in sea	Ton	0	10,000	8000	8000	10,000
Combat c.	Million €	0	0	1.25	1.50	1.50

It may very well happen that the decision-maker makes judgments that are inconsistent, and a nice feature of the method is that such inconsistencies may be discovered and corrected, thus improving the reliability of the method.

Table 7-4 is filled in with judgments that are consistent with the more demanding swing judgments that were elicited in Chapter 3. The judgment scores are then added together columnwise. The estimated weights are called weights_1. They are computed by dividing the column sums plus one by the sum of these sums. Do you expect these weights to come close to your more carefully elicited swing weights? Well, the swing weights are shown at the bottom of Table 7-4. They are pretty close, but there are also deviations of some importance.

Which method is best? That depends probably on the cognitive abilities of the decision-maker, but ordinal comparisons probably have a higher *reliability* than swing weighting. The *validity* of ordinal comparisons is, on the other hand, lower. If the weights_2 in Table 7-4 were the true preferences of the decision-maker, consistent ordinal comparisons could never come closer than weights_1.

Since Table 7-4 has symmetry properties, it is possible to make a simplified version by using half as many numbers. Then all criteria are listed in the first column, and the next columns list criterion 1 versus 2, 1 versus 3, 2 versus 3, and so on. Table 7-5 shows the simplified version for the three first criteria in Table 7-4. The preferences given to each criterion are thereafter calculated as the sum of entries in that criterion's row, plus 1 (for mathematical reasons). Thereafter the weights are normalized to make the sum equal to 1.0.

Table 7-4. Ordinal pair wise comparison of the criteria shown in Table 7-3. A "1" in a column means that the column variable is preferred to the row variable. "0.5" means indifference. Weight_1 are the weights from the ordinal comparisons. Weight_2 are the original swing weights from chapter 3.

	Recreat	Industry	Shore	Birds	Auks	Sea oil	Cost
Recreat.	0.5	0	1	1	1	0	0
Industry	1	0.5	1	1	1	1	1
Shore	0	0	0.5	0	0	0	0
Birds	0	0	1	0.5	0.5	0	0
Auks	0	0	1	0.5	0.5	0	0
Sea oil	1	0	1	1	1	0.5	0.5
Cost	1	0	1	1	1	0.5	0.5
Sum+1	4.5	1.5	7.5	6	6	6	3
Weights_1	0.14	0.05	0.24	0.19	0.19	0.10	0.10
Weights_2	0.14	0.06	0.20	0.18	0.18	0.12	0.12

Table 7-5. A simplified version of the pair wise comparison technique. The numbers are the same as in Table 7-4.

Number	Criteria	1 vs. 2	1 vs. 3	2 vs. 3	Sum +1	weights
1	Recreation	1	0	–	2	0.33
2	Industry	0	–	0	1	0.17
3	Shore	–	1	1	3	0.50
					6	1.00

Reducing the number of comparisons: Even though each comparison is cognitively rather simple, the sheer number of comparisons makes the process wearisome for the judge when the number of criteria is large. However, there is nothing sacrosanct about pairing each stimulus with every other in the series. It is often appropriate to select a limited number of criteria as standards. These criteria should be chosen at approximately equal weight differences – do a screening, and they should be among the least ambiguous of the lot.[3]

2.2.3 Rank Agreement Among Several Judges

It is convenient at this point to present Kendall's coefficient of rater reliability, κ (kappa), which in addition has statistical properties that allow for testing of whether the differences between judges are significant or not.[4]

Assume that there are K judges and that each one has independently compared n criteria and produced a table like 7-4, which we will assume contains only zeros and ones – indifference is not allowed. Let a_{ijk} be the score (0 or 1) in row i and column j of judge k's table. Kendall's κ is computed according to formulae (7-6):

$$\kappa = 8\frac{\sum_{ij}\left(\sum_{k} a_{ijk})\right)^2 - K\sum_{ijk} a_{ijk}}{K\times(K-1)\times n\times(n-1)} + 1 \qquad (7\text{-}6)$$

The summation index k runs over the judges, from 1 to K, but the summation over ij is made only below *or* above the main diagonal in Table 7-4. Kappa ranges from –1 for complete disagreement to +1 for complete agreement among the judges. Values around 0 indicate random differences.

Example: if we replace the "0.5"s by "1"s in Table 7-4, then for two judges that agree all the "1"s will be replaced by "2"s, and there are 13 of the "2"s below the diagonal, we get: $\kappa = 8 \times (13 \times 2^2 - 2 \times 13 \times 2) / (2 \times 1 \times 7 \times 6) + 1 = +1$. For two judges that completely disagree we have "1" in all entries below the diagonal because when one judge has "0" the other has a "1", and we get $\kappa = 8 \times (21 \times 1^2 - 2 \times 21 \times 1) / (2 \times 1 \times 7 \times 6) +1 = -8/4 + 1 = -1$.

3. PREFERENCE MODELS

Equation (7-2) is an example of the kind of preference models that are common in multi-criteria decision-making. It simply states that the total utility of a consequence is a weighted sum of the utilities that are derived from the individual decision criteria. It is important to notice, however, that this is just a **model** of how the decision-maker feels about the consequences. Models are simplified versions of reality, and decision-makers do not necessarily have such simple feelings. Why not multiply the utilities instead of adding them, for example? To answer such questions, it is useful to look at the underlying assumptions.

In this section we shall look at the assumptions behind two important models – additive and multiplicative total utility functions. There are other more complex models as well, but these two simple models have proved themselves especially useful.

3.1 Additive Utility Functions

We used an additive utility function (7-2) in the oil pollution example. The general form with n decision criteria is shown in equation (7-7). The primary utility functions in the formulae represent the decision-makers risk attitudes.

$$u(x_1, x_2, ..., x_n) = w_1u_1(x_1) + w_2u_2(x_2) + ... + w_nu_n(x_n) \qquad (7\text{-}7)$$

Just by looking at (7-7), three important assumptions are apparent:

1. *Utility independence*: Since the primary utility function of one criterion is independence of the scores (x-values) of the other criteria, the decision-maker's attitude towards risk does not depend on the scenario. It means for instance that the degree of risk-aversion with regard to deaths of birds is independent of the length of the shore that is polluted. This may sound like a rather strong assumption, but do not misunderstand: The number of dead birds in an oil spill is probably statistically *correlated* with shore pollution, but why should your willingness to pay to save another bird depend on the length of the shore that is polluted?
2. *Preference independence*: The weights in (7-7) are independent of the scores. In other words, the relative importance of the criteria does not change with the scenario.
3. *Decomposability*: The total utility is a sum of the contributions from each criterion; there is no preference interaction among the criteria. Suppose that your decision problem is to choose a job, and your two decision criteria are salary and length of vacation. Would the utility you derive

from a long vacation be independent of the salary? Probably not. A long vacation would be rather boring if you had to stay at home all the time, while it might be quite exiting if you had some money to spend. In such cases, the simple additive model would not be adequate, and you need something more complex – like the multiplicative model.

3.2 Multiplicative Utility Functions

The general form of the multiplicative utility function is quite complex, but it looks like (7-8) if we have only two decision criteria, such as x_1 = *salary* and x_2 = *vacation*:

$$u(x_1, x_2) = w_1 u_1(x_1) + w_2 u_2(x_2) + k w_1 u_1(x_1) w_2 u_2(x_2) \tag{7-8}$$

Notice that one new parameter is included, namely the **synergy coefficient** k. If k is zero, (7-8) boils down to the additive model, which therefore is a special case of the multiplicative model.

But what happens if k is positive? If the salary is low or the vacation short, the associated primary utilities will be close to zero, and so will their product. Thus, the multiplicative part of (7-8) has in those cases little impact on the total utility. But if the salary is high and the vacation long, the product will be substantial and give a boost to the total utility. Salary and vacation are called **complementary goods**. They have a positive synergy that is represented by a positive synergy coefficient in the multiplicative utility function.

There are other types of goods – like butter and margarine – that require a negative synergy coefficient, which subtracts from the total utility if you have a lot of both. Such goods are called **substitutes**.

Although theoretically important, multiplicative utility functions are seldom used in practice. One reason is that it is difficult to elicit reliable estimates of the synergy coefficients; another is that it makes the analysis less intuitive since it is hard to keep track of the synergy effects, making it more difficult to discover accidental errors. Actually, decision criteria that are complements or substitutes probably do not qualify as end impacts with intrinsic value. It is not really the length of the vacation that matters; it is its content. But then it is hard to measure content.

4. VALIDITY IN PREFERENCE ELICITATION

We have earlier in this chapter discussed validity and reliability in relation to weight estimation techniques, but we overlooked one important aspect that we address now: Validity requires emotion.

Validity means that we actually measure what we want to measure. And this is a key point: *What we want to measure is how much the decision-maker would like to live with the consequences of the decisions.* We measure this in an indirect way, by first eliciting preferences in terms of decision criteria weights. The reasoning is that this is easier than to evaluate complex consequences consisting of many criteria. To know the weights is also convenient, since you do not have to bother with limiting the number of alternatives; one can easily compute the utility of a whole range of consequences. This also makes sensitivity analysis simple.

4.1 Emotion

No matter how value elicitation is performed, it entails decisions. The decision-maker is repeatedly asked to state whether she prefers something to something else, or whether she is indifferent. But preference depends ultimately on feelings, and feelings depend on emotions.

In this context, **emotions** are defined as somatic events in the body, like rapid heartbeat, skin pallor or muscle contractions. Other people can usually observe your emotions, while the associated **feelings** are your own. Research has showed that decision-making without emotions is hazardous.[5] If you don't feel anything when you decide, you are not to be trusted as a decision-maker, since you are liable to make gross errors.

Notice that this is diametrically opposite to conventional ideas of rationality where a rational person is conceived of as a cold, calculating person. On the contrary, rationality requires emotion! How could you *calculate* how many birds you must save to compensate for one extra kilometer of oil polluted shore? That is of course impossible, the appropriate trade-off must be felt, it can in no way be calculated. To make certain that no misunderstandings creep in: The value of a bird for you as a decision-maker is how much you would be willing to pay to save the life of the bird by preventing it from freezing to death from oil on the feathers; is has nothing to do with how much you would pay to have it for dinner. Further, your willingness to pay for a kilometer of unpolluted shoreline has little or nothing to do with what it would cost to clean it.

Valid value elicitation requires that the alternatives under consideration are made *vivid* in a way that makes it possible for the decision-maker to *imagine* what it would be like to experience them in real life, and to feel something. This, of course, is a real challenge – a challenge we must never lose sight of.

We have so far discussed two value elicitation methods, swing weights and pair wise ordinal comparisons of criteria. Both methods are somewhat deficient regarding potential for eliciting emotions. The reason is that they require the decision-maker to consider the whole range of the criteria scales. In the oil-pollution example, for instance, the decision-maker must consider scenarios where as many as 1 million birds die and 1 million recreation days are lost. Since these are endpoints of the scales, they are extremely rare and correspondingly difficult to imagine, and consequently unlikely to elicit appropriate emotions in the decision-maker.

We therefore need preference elicitation methods that concentrate on more commonplace scenarios in the middle region of the ranges. They have better potential for eliciting appropriate emotions. Such methods require significant computation power in the form of decision support software.

4.2 Software Based Decision Support Tools

Pair wise trade-off between criteria is a commonly used method that resembles pair wise ordinal comparisons of criteria discussed above. The difference is that the trade-off method requires information on how much of one criterion will compensate for a certain amount of the other. With software support, it is not required to consider extreme values. Several software packages have been developed which include this method. A particularly simple one is Pro&Con,[6] which is freely available from the Internet. A demonstration of how the trade-off is done is presented below. Other software programs that use multi-attribute functions are Criterium Decision Plus,[7] DataScope,[8] Decision Explorer,[9] Decision Lab 2000,[10] EQUITY,[11] High Priority,[12] HIVIEW,[13] Logical Decisions,[14] On Balance[15] and VISA.[16]

4.3 Value Tradeoff with Pro&Con

In the following, we present the dialog of the software program Pro&Con applied to the oil pollution decision problem.

4.3.1 The Consequence Table

Pro&Con's consequence table is an Excel table that looks exactly like Table 7-1, but the user must also state whether the criteria represent benefits or costs. All criteria in Table 7-1 are cost criteria in the sense that the smaller the scores, the better.

4.3.2 Utility Functions

When the oil spill project described in chapter 3 was actually carried out, all utility functions – except for dead birds – were considered to be risk averse. Pro&Con has a simple provision for creating non-linear utility functions with negative exponential form. The decision-maker needs only state one extra point for each non-linear function. The default is linear. The utility function of *Recreation* is shown in Figure 7-2, where the decision-maker has indicated that the *certainty equivalent* is 800, meaning that the increase in utility of 0.5 from 200 to 800, equals the increase in utility from 800 to 1000.

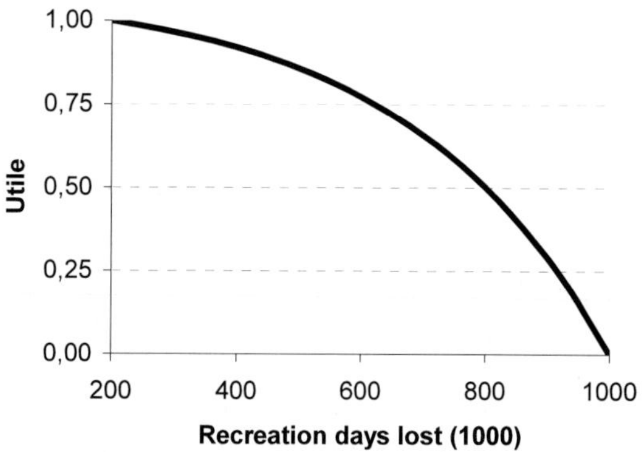

Figure 7-2. Pro&Con: The elicited utility function of the number of people that get sight of the oil spill. This function is very risk-averse.

4.4 Value Trade-Off

Value trade-off with Pro&Con is performed by comparing all criteria separately with the first one on the list in the consequence table. The decision-maker is presented with an axis diagram with two randomly placed points, and asked to move the points with the mouse until they are equally preferred. A trade-off between *industrial loss* and *recreation* is shown in Figure 7-3. The idea is that for most people, numbers have little emotional power; it is more suggestive to look at a graph when you compare consequences.

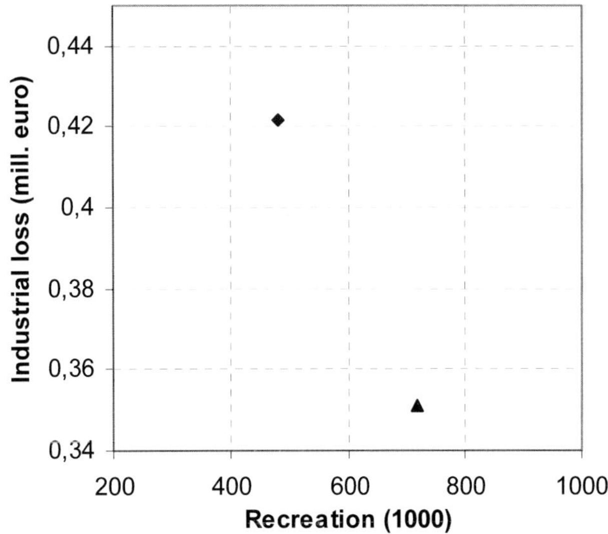

Figure 7-3. Value trade-off with Pro&Con. The user is supposed to move the two points until they are equally preferred. This establishes the relative weights between the two criteria.

Decision-makers who have real world experienced with the problems at hand may benefit a good deal from decision support such as this. But for most decision-makers, much more effort will be needed to create scenarios that are vivid enough to elicit adequate emotions. One possibility is to create pictures or videos of what alternative futures would look like.[17]

---

[1] Cowen (1993)
[2] Belton and Stewart (2002)
[3] Guilford (1954) p. 169
[4] Siegel and Castellan (1988) p. 277
[5] Damasio (1994)
[6] http://wwwe.bi.no/users/fag87027/
[7] http://www.Infoharvest.com
[8] http://www.cygron.com
[9] http://www.banxia.com
[10] http://www.visualdecision.com
[11] http://www.lamsade.dauphine.fr
[12] http://www.enterprise-lse.co.uk
[13] http://www.krysalis.co.uk
[14] http://www.logicaldecisions.com
[15] http://www.krysalis.co.uk
[16] http://www.SIMUL8.com/visa.htm
[17] Wenstøp and Carlsen (1998)

Chapter 8

WILLINGNESS TO PAY

The value of environmental resources and services

How can we find out how much people are willing to pay for environmental goods and services? Such values are commonly called "Willingness-to-pay". There are three main methods:

- The "real market pricing" method based on the assumption that the prices of environmental goods are embedded real prices.
- "Surrogate market pricing" methods based on the assumption that the value of the environment is related to the cost of either preventing – or mitigating – the effects of damages to the environment.
- The "contingent market pricing" methods based on the assumption that an artificial market can be created which is sufficiently similar to the real market so that it is possible to value the environment.

●

"It is in this manner that money has become in all civilized nations the universal instrument of commerce, by the intervention of which goods of all kinds are bought and sold, or exchanged for one another." *Adam Smith (1776)*

1. INTRODUCTION

We return in this chapter to the economist's paradigm, which assumes that the population is the locus of value – projects or policies are supposed to be evaluated according to the population's willingness to pay (WTP) for the environmental amenities involved. Economists prefer to do the evaluation in monetary terms, which makes it easy and convenient to balance benefits against costs. Information about WTP is also useful as input to politicians in their decision-making, to derive compensation payments for environmental damages already incurred, and as a part of a welfare index that includes quality of life, as well as Gross National Product.[1]

Our main goal in this chapter is to discuss how this important piece of information can be acquired. What methods do we have for estimating the population's willingness to pay for environmental amenities?

In a perfect market where Adam Smith's invisible hand operates, every good and resource has an owner and a price. Environmental resources and ecosystem services usually have no market value, however. The vast public land areas in USA are not for sale;[2] the world's largest common resource – the air – is not bought or sold. On the other hand, our next most important common resource – fresh water – is in many cases bought and sold.

On a local scale, environmental resources are often impoverished without compensation. An industry that pollutes a river so that it becomes less attractive for recreation, expends common amenities without giving compensation. According to the "polluter pays" principle, they should compensate – but how much? This is the kind of question we address in this chapter.

2. VALUE DIMENSIONS

Environmental amenities represent value to us in different ways. We value the use of a beach and breathing clean air. It is also of value to have the option to go to a beach, even though we do not plan to go there right now. Even the knowledge that a clean beach exists is of value to people, and so is the knowledge that polar bears or tigers exist even if we do not want to meet them. These value dimensions are called *use value*, *option value* and *existence value*. Some even reckon with *quasi-optional value*. All these values may be attached to the same object, and the *total value*, P_T, of this object is then the sum of all these values:

$$P_T = P_U + P_O + P_E + P_{QO} \tag{8-1}$$

- *Use value* = P_U, the user's value of the good when it is actually used.
- *Option value* = P_O, the value it may have for future use.
- *Existence value* = P_E, the value it has by knowing that it exists.
- *Quasi-optional value* = P_{QO}, the future value the resource has if new technology becomes available minus the current value. This is equal to the value of making a decision later rather than now. You can defer buying tigers for your zoological garden until it is better known how many visitors it will have.

We use the symbol P for such values. P stands for price, and corresponds to the price the population is willing to pay for it. *Use value* is probably the largest in most cases, but studies suggest that other values may be almost as large.[3] The option value of a grizzly bear has been found to be US$21.8 and the existence value of an additional US$24. For bighorn sheep the option value was US$23 and the existence value US$7.4.[4]

3. FACTORS AFFECTING WTP

The beach you visit on Sunday does not only provide recreational services. The environment regenerates nutrients, treats waste by degrading it, regulates water and supports food production. All such services are valuable, although they are normally not bought or sold at a price. However, if you terminate the services by removing plants or letting soil erode, some other means have to be invented to replace them. Valuation of environmental goods depends upon many factors, such as quality, quantity, ethical attitudes, and whether it is a question of paying for improvement or being compensated for losing them. Valuation of risk depends on whether it is voluntary or imposed. The influence of wealth of the assessor is discussed in Chapter 2, section 4.2.

3.1 Quality

Willingness to pay for an environmental amenity depends on its quality. You may appreciate immensely hiking in a pristine environment with few traces of man. If you suddenly catch a whiff of the exhaust from a car at a distance, your WTP suddenly drops sharply. But then, as the traffic increases further, your WTP stays more or less constant for a while since an extra car does not matter much. This continues until you start to become nauseated by the smells; then your WTP drops faster and faster with increasing pollution. A typical pattern is illustrated in Figure 8-1.

3.2 Quantity

In Chapter 5, we discussed demand functions for goods and services. We assumed that the price you are willing to pay for one more unit is a decreasing function of how much you already have consumed. It is reasonable to assume that the same is the case for environmental goods. It is nice to have a large area of protected wilderness, but the price you would be willing to pay for an additional square kilometer – the marginal demand curve – is probably a decreasing function of how large the area already is, even if some predators require very large tracts of land to survive.

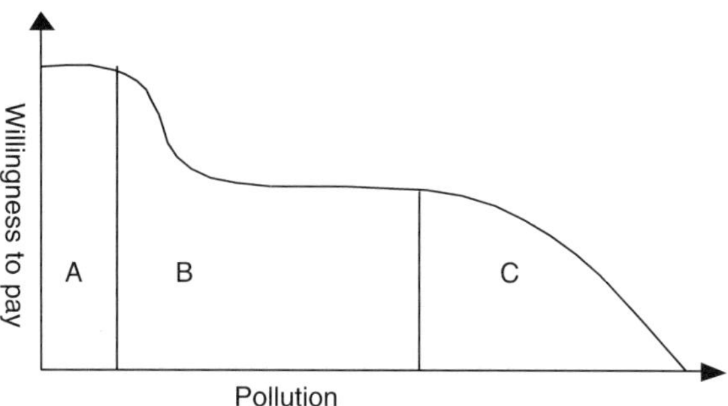

Figure 8-1. The willingness to pay for one unit of an environmental good depends on the quality. A: Pristine condition. B: Acceptable pollution. C: Serious pollution.

3.3 Consumers versus Citizens

People who participate in valuation studies are of different kinds. On the one hand we have the hedonistic *consumer* who only thinks about maximizing his own desires; on the other hand we have the idealistic *citizen* who is concerned about everybody's welfare, even including animals.

The **consumer** sets prices according to the laws of supply and demand. Some people are willing to pay more than others, because the good has greater value to them. If you buy a house, you may be willing to pay more if your mother lives in the neighborhood, or perhaps conversely. The price may also depend upon the environment, such as air quality and the amount of traffic noise in the area.

The **citizen** relates to the interest of the public, rather than his or her own self. In an interview situation, citizens typically think about their choices as input to decision-making politicians. They usually do not depend on details such as whether it is a question of the value of habitat, or the value of ecological forms, like lichens flora or bottom fauna in the water. Sometimes they will answer that the resource in question is not for sale at any price. This is **lexicographic preference,**[5] no tradeoff is admitted – the resource has a value that is unlimited compared to any other use it might be turned into. Economic theory considers such attitudes irrational, but we shall return to this question under the chapter on ethics. **Unidirectional pricing** is a weaker form: a pristine resource is not for sale, but if it already is destroyed, it has a compensation value which goes to the "offended".

3.4 Willingness to Pay and Willingness to Accept

Valuation of environmental amenities will depend upon whether it is a question of obtaining them or compensating for their loss. Suppose you are willing to pay a certain price for a house on the beach. Now, if you already owned the house and the authorities wanted to turn the beach into public land, how much would they have to compensate you to be willing to accept it (WTA) and evacuate the house? Studies show that WTA ranges from about 10 times higher than WTP for public and non-market goods to three times WTP for ordinary private goods. This means that the discrepancy between WTA and WTP is greatly reduced by market experience.[6] On the average, WTA is seven times higher than WTP. According to traditional economic theory, there is no particular reason WTP should be different from WTA. However, it is not hard to find psychological explanations: Part of the extra compensation may be claimed for uncertainty and frustration; people are not as accustomed to selling as to purchasing,[7] people have enhanced aversion against losses compared to desire for gains.[8]

3.5 Free Will versus Imposed Risks

Most people know that smoking causes deteriorated health and shorter life expectancy. The probability of acquiring smoke-related cancer during the lifetime is 50% for smokers. Still, many people accept such self-imposed risks. However, if the government for some reason imposed a similar risk, few people would accept it, and would require that something be done. By and large there is a ratio of 10:1 between WTA self-imposed risks and WTA imposed risks, but ratios as high as 100:1 have been observed.[9,10]

4. ESTIMATION METHODS

The most profitable strategy for harvesting whales is to catch them, sell them, and put the money in the bank as quickly as possible. This is because a savings account gives higher interest than the interest generated by whales' reproduction. Maximizing profits from a slow-growing species by hunting leads to its extinction.[11,12] But whales are not only meat; they are key species of the ocean ecosystem, acting in concert with multitudes of other marine species. Some whales – particularly those outside Nantucket Island – attract tourists. How should their value be measured?

There are two main principles for estimating WTP. To some degree, people *reveal preferences* through actions in the market. When possible, we may therefore observe *real market behavior* and make inferences from that.

When we cannot observe, we may conduct surveys, asking people to *state their preferences*. There are two principal stated preference methods, *contingent valuation* and *conjoint analysis*, but they do not necessarily give comparable results. A study of these two methods and a multi-attribute approach with decision panels have concluded that contingent valuation gives significantly lower WTPs than conjoint evaluation and decision panels.[13]

4.1 Revealed Preference

4.1.1 Travel Cost Method

Even though most ecosystem services are not traded in the market, people often have to travel to enjoy them; and how much they are willing to pay to travel, reveals to some degree the use value. The price of a trip to a beach in Phuket in Thailand or the boat ticket to see whales at Nantucket in USA is the result of market forces. The demand functions reflect the WTP for the experience.

The travel cost method tries to capture part of the total value of recreational amenities. The method can in general be used for resources you cannot enjoy unless you travel there, but it gives the use value, not the option value nor the existence value, and therefore underestimates the total value.

Suppose you want to estimate the use value of a recreational site. First, you have to find out where the visitors come from. You ask a random selection of visitors about that and how much they paid to get to the site. Then you group the places of origin into regions with approximately uniform travel costs. Figure 8-2 shows an example from Bretagne. You then calculate the average cost from each region, and based on that you can find a weighted average and an average per person. It is important to bear in mind that WTP will depend on season, weather and other factors, and this must be taken into account.

If the number of people interviewed is not very large, people arriving from large distances with associated large travel costs may be missed from the sample. Such visitors may still be included without actually extending the sample size by constructing and extrapolating the demand curve such as in Figure 8-3. The area under the demand curve is an estimate of the total use value of the recreational site. Travel costs are often calculated as direct travel expenses plus time costs computed at normal wage rate.[14]

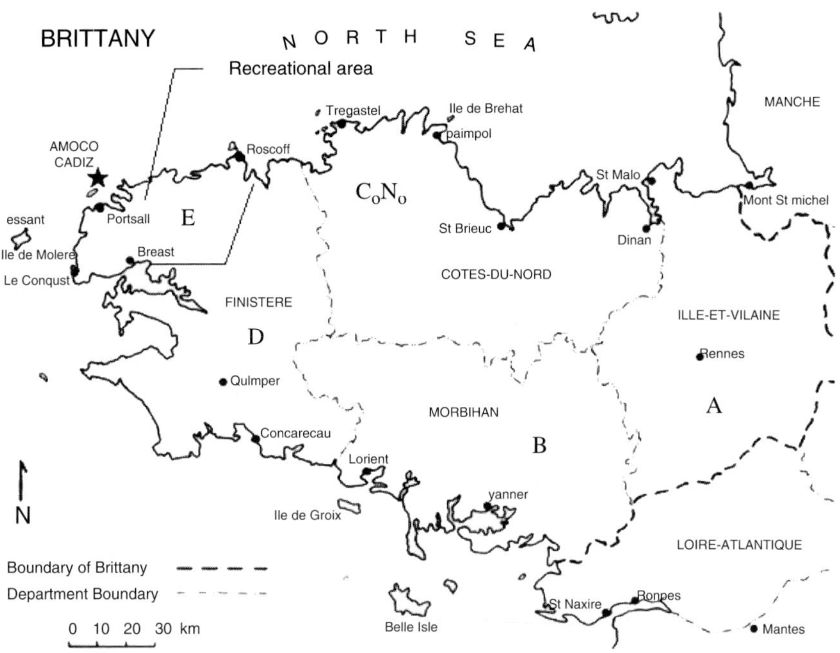

Figure 8-2. Travel cost method: The WTP to visit a recreational area near the Amoco Cadiz accident spot is estimated by dividing Bretagne into regions with uniform travel costs.

Amoco Cadiz, France 1978: About 2100 local time on 16 March 1978 the fully loaded supertanker *Amoco Cadiz* drifted onto submerged rocks 1.5 km off the coast of Brittany in France and caused a considerable oil spill near the small village of Portsall. Brown et al.[15] have estimated the changed recreation value of the beach with the travel cost method. The travel cost to the beach was approximately 1 franc per kilometer. By observing how many people came from different places just after the oil spill and one year later, they could draw demand curves as shown in Figure 8-3. The curve of 1979 is higher than of 1978, meaning that more people visited one year later than just after the oil spill. The difference represents the lost recreational value for the consumers because of the oil spill. The aggregate gain over all consumers and all regions was estimated at 6 million francs or 3 francs per visit. (1 franc = €0.12.)

The travel cost method has some obvious limitations. First of all, one can only capture use value. Then, it is not suitable for recreational areas that are used only by the local population, since they have no travel costs. There are also difficult problems to entangle if there are alternative recreational sites in the vicinity.

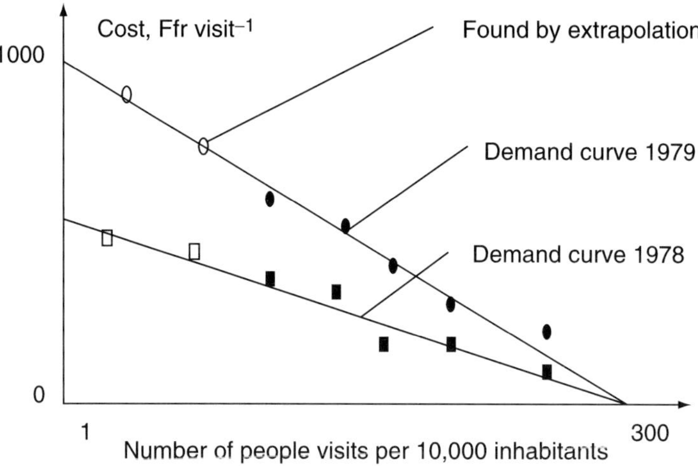

Figure 8-3. Demand curves for beach recreation in Brittany, France just after the Amoco Cadiz oil spill 1978 and one year after.[16]

4.1.2 The Hedonic Method

This method can be used if you want to estimate the use value of environmental amenities that influence observable prices of a consumer good such as housing. If the price of houses is influenced by air quality and noise level as well as other factors, you can estimate the value of the environmental goods through regression analysis. What you need is a sample of observations units (houses) that have different values of the variables. The estimated regression coefficients correspond to the use values of the independent variables.

WTP to avoid noise in Osaka, Japan: Morioka et al. have estimated the WTP of avoiding noise from road traffic in Japan.[17] Close to a ten-lane highway they observed land prices (yen/m2), noise levels dB(A) and distances to the business district as well as other factors. They found that the distance to the central business district was not significant, but noise was significant at the 5% level.

4.1.3 Surrogate Market Pricing Methods

It is often possible to observe market prices or costs that are not valid estimates of WTP, but at least they bear some relation to what we are interested in. Since they are not the real things, we call them surrogates. Here are some examples:

Damage cost: The damage cost method sums economic costs of the damage to the environment, like lost profits from wind-fallen trees after a storm. However, this does not capture other ecosystem services and some damages may not be identifiable before long after the event. The damage cost method therefore easily undervalues real damage. To compensate for this it is customary to use the upper 95% confidence bound for the mean as a substitute for the mean.[18]

Environmental resource replacement: If a non-harvestable resource such as a bird population is damaged, one can estimate the cost of replacing the population with birds from other areas. If a shore is ravaged, one can calculate the cost of replanting shore vegetation.[19] The replacement method is not a valid estimate of WTP, however, since it tends to overstate.[20] If replacement is next to impossible – such as cleaning up after Chernobyl – the cost could be almost infinite and has no connection with WTP.

Damage prevention cost: This method has been used to estimate damage from leakage from landfills,[21] but is not a valid method either, and for the same reason.

Yield loss: A change in environmental condition is translated into a change in a certain yield. The yield change shifts the supply curve for a commodity, and thereby also its consumer and producer surplus. The damage is equal to the loss in yield,[22] corresponding to use value.

Averting behavior: When people purchase equipment to avert damage, the cost of purchase is a lower limit of the WTP to avoid the damage. People buy alternative cooking equipment to avert the inconvenience of electricity outages caused by the water hyacinth infestation of Lake Victoria.[23]

Political pricing: When authorities make decisions that have impacts on the environment, one may sometimes calculate implicit WTP. The reasoning is that in a representative democracy, governmental decisions may be expected to try to reflect the WTP of the population, where politicians are seen as actors in a surrogate market. An example is the Norwegian Parliament, which ranked 540 hydropower development projects according to economy and environmental damage. The process consisted partly of political negotiations in contact with the constituencies, partly on systematic data connection, but there were no explicit notion of how much people were willing to pay to save the rivers from development. However, a post-factual study showed that the implicit WTP for environmental damages was 30% of development costs.[24]

Hypothetical Insurance Purchase: People buy insurance to obtain certainty against potential loss. Since the insurance company has to make an expected profit out of it, the insurance costs more than the expected loss, and the difference is called the **risk premium**. Suppose you were to buy insurance against theft of your car, which is worth €30,000 and with a probability of theft equal to 1%. How much would you be willing to pay for the insurance? Let us say that you are willing to pay €400, but not more. We call the €400 the *certainty equivalent* in Chapter 2, and the *risk premium* is the difference between the certainty equivalent and the expected loss 1%×30,000 + 99%×0 = €300. Your risk premium is in other words equal to €100, since you are willing to pay €400 to avoid an uncertain situation with an expected loss of €300. Such very common behavior can be explained by risk-averse utility functions, as we did in Chapter 2. Therefore, to put a monetary value on how much damage a pollution accident might cause, one could ask people how much they would pay in insurance premium to be compensated for an accidental pollution where they live. The method does not work well, however, if the risk is perceived as very small. People are bad at understanding very small probabilities, and tend not to be willing to pay a risk premium even if the potential consequences are severe.

4.2 Stated Preference

The reliability is higher if we can *observe* preferences through actual behavior – as in section 4.1, than if we ask people what they *would have done* if they had been in that situation. People are usually very uncertain about hypothetical situations. Still, this is what we have to do if more reliable information is unavailable.

4.2.1 Contingent Valuation

The word *contingent* means that the respondent is presented with a hypothetical scenario as a background for the valuation. For example, to value a recreational beach, the beach is described together with other relevant information. It is important that the description is sufficient so that additional information would not change the attitudes of the respondents. Too much information, on the other hand, will overwhelm and confuse.

There are several versions of the method. The two main versions are **open ended** (OE) valuation where the respondents are simply asked to state their maximum WTP or WTA, and **dichotomous choice** (DC) where people are asked whether or not they are willing to pay a certain amount. There are also variants of these methods, such as iterative bidding games, ranking and the use of payment cards (PC).[25] In the latter, respondents are asked to choose among amounts in an increasing or decreasing series, like €10, €20, €40, €80, etc. Sums are often doubled at each step, rather than being at equal intervals, because people often respond logarithmically.

One advantage of DC is that it mimics the decision people are making daily – they are confronted with a price and must take it or leave it.

Furthermore, it resembles a referendum situation. DC tends to give higher WTPs than OE questions. One reason is that people tend to state what they believe things *cost* rather than what they are *worth* and therefore state a *reasonable* WTP rather than their *maximum* WTP. An American expert panel headed by Kenneth Arrow has suggested rules for CVM surveys.[26] They recommend DC in the format of a referendum, and explicitly advise that a "don't know" option is included in addition to "yes" and "no" to avoid too much randomness.

Almost all CV surveys encounter the problems of zero WTP and excessive WTP:

- *Zero WTP*: In practice many people respond that they are not willing to pay anything. These **protest bidders** signal important information. Perhaps they want the proposed measures to be carried out, but protest against the idea of their paying for it, as it is seen as the obligation of the polluter or the government. This is a political signal, which should be addressed as such.
- *Excessive WTP*: High WTP-responses are another problem. Bids higher than 10% of annual income may indicate an unlimited price, meaning that they regard the good as not being for sale. "All species of wildlife have a right to exist independently of any benefit or harm to people".[27] This is an ethical signal, and an important piece of information as well.

Since zeros and excessive bids lie outside the economic WTP paradigm, they must be handled in an adequate way in the data analysis. To just include them as ordinary responses, will distort the picture of those who are receptive to the WTP idea. Usually one therefore deletes extreme values in both ends of the data set. But even then the data distribution will probably be quite skewed, resembling a geometric distribution. The average will therefore be much higher than the median. Whether to use the average or the median in a subsequent analysis depends on its purpose. Average WTP should be used if one wants to collect money for a nature information center, since both rich and poor people will supposedly contribute to its construction. However, to decide on the entrance fee, one should use median WTP to make it affordable for most people to visit it.

There is no universal agreement on how to actually carry out a contingent evaluation process, but here are some recommendations:

- Decide what type of environmental resource you want to value: benign uses like bird watching or sustainable fishing, non-sustainable activity like destruction of habitats, destruction of endangered species, climate damages.

- Decide what kind of valuation you want, "consumer", "citizens" or "offended", and whether it is WTP or WTA.
- Decide the way prices are presented to the respondent and the way in which the payment is to be done. For example, the respondent may be asked to pay an annual extra tax for ten years, or to pay a lump sum once, say, to a donation.
- Perform a pilot test to check the clarity of the questions. For example, if asked about the value of improved visibility, some respondents will assume that there is also improvement in health because of less air pollution. If aesthetics is the only intended issue, it must be stated clearly.
- Be sure that age, income and educational level of respondents are not too different from that of the intended population. If it is, correct the results with regressions on the socio-economic variables.
- Test the results for consistency.
- Interpret the results.

4.2.2 Conjoint Analysis

In conjoint analysis (CA) respondents are asked to rank a number of options where each is described by several attributes. A statistical analysis of the rankings of many respondents provides an estimate of the average value of each attribute. Table 8-1 shows an example of valuation of ill health and variety of bird species in connection with the use of pesticides. The respondents were asked to rank the four options according to preference, and then the ranks were used to estimate the underlying WTPs.

Conjoint analysis resembles the decision making approach of Chapter 3 in that both use a consequence table such as Table 8-1. However, the purposes and methods are different. In decision analysis, the purpose is to help the decision maker rank the options, and the method is to weight the attributes (criteria) by eliciting their relative importance. In conjoint analysis, the purpose is to weight the attributes, and the method is to elicit a ranking of the options. To achieve that, the options must be familiar and easy to rank. CA was first used in marketing, where consumer preferences for product traits – like sweetness of soft drinks – where elicited by asking them to rank real products.

Table 8-1. Conjoint analysis: The social cost of pesticide use in wheat cultivation, UK. The respondents are asked to rank the options. Option A was the prevailing situation.[28]

Attribute	Option A	Option B	Option C	Option D
Price of bread per loaf (pence)	60	85	85	115
Health effects, cases of ill health	100	40	40	60
Effects on birds, # of bird species in decline	9	2	5	2
Ranking				

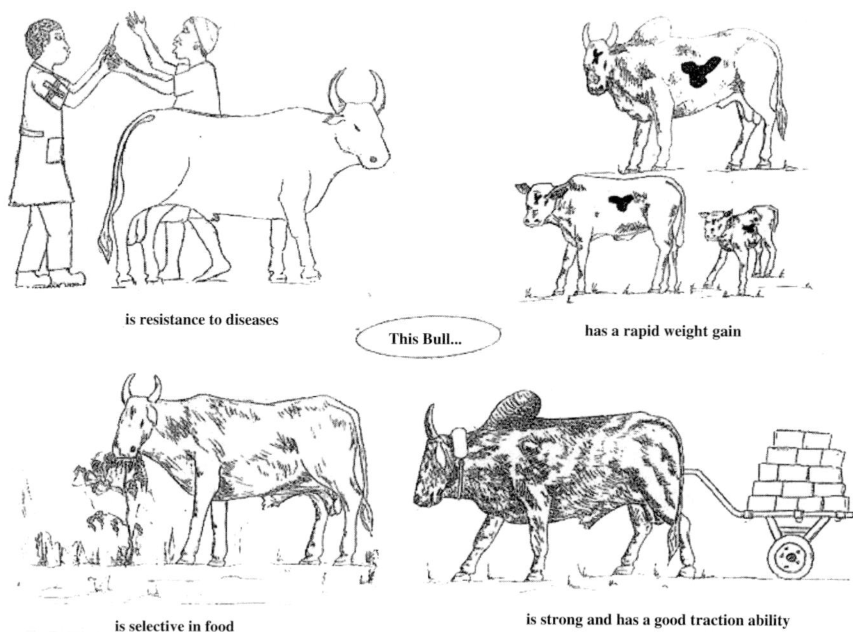
is resistance to diseases
This Bull...
has a rapid weight gain
is selective in food
is strong and has a good traction ability

Figure 8-4. Drawings used to solicit preferences for cattle traits in West Africa. Cattle perform multiple functions in this region. The most important trait turned out to be disease resistance, fitness for traction and reproductive performance. Beef and milk production was less important.[29]

Table 8-1 and Figure 8-4 give an indication of the kinds of "products" that are used in environmental management. You may try to do the ranking yourself for Table 8-1. To estimate farmers' preferences for cattle traits in West Africa, the alternatives were presented visually to the respondents, like the drawings in Figure 8-4.

4.2.3 Methodological Issues

Stated preference methods suffer from methodological problems. The two major ones being that some people respond arbitrarily without a link to real WTP, while others behave strategically.

If people do not elicit feelings for the real value of the environmental amenities in question, they answer out of the blue, or use some kind of surrogate method to provide answers. **Random answerers** can be eliminated if they betray themselves through **logical inconsistency**. They may fail to prefer dominating alternatives, or change their minds with regard to rank order, or violate transitivity requirements by preferring A to B, B to C and C to A. Nearly half of the respondents to Table 8-1 failed occasionally to provide consistent responses, and a small minority was systematically inconsistent. Respondents who are systematically inconsistent must be deleted from the sample. A weaker form of inconsistency is called **embedding**: One often observes conspicuously similar WTPs for quite different environmental resources and ecosystem services. Studies show that households were willing to pay $95 per year to prevent deterioration of visibility in the Grand Canyon, while they were willing to pay only $22 when this was valued at the same time as visibility improvements in Chicago (the site of the interviews) and throughout the East Coast of the United States.[30] Some possible explanations of embedding are:

- *Availability and misinterpretation*: The respondents are generally unaware of alternatives that have not been explicitly formulated, but which they would have assigned a value if they had been present in their mind.
- *Mental accounting*: The respondents have a mental account of their economy, with a fixed amount of money for "good causes" like environmental improvements. Thus all environmental amenities are allotted the same amount of money.
- *Moral satisfaction*: The respondent gives an amount of money that gives them the moral satisfaction from contributing to a good cause, thus they are not really assessing the real worth of what they are supposed to value.

The problem with **strategic bias** is harder to discover and control. People with environmental-political agendas may respond strategically to influence the analysis in a particular direction. Greens have an incentive to overstate their values when they do not have to pay more than others in taxes. However, if the payment were proportional to stated WTP, everybody would have an incentive to act like a free rider and understate. Such problems appear to be larger in CVM than in CA. The reason is that since CA is more complex, it is less obvious how to answer strategically.

Some researchers think that that CVM is so fraught with problems that it should not be used.[31] For instance, DC is reported to give WTP estimates that are three or four times higher than payment cards.[32] They question whether it is reasonable to use figures that depend so much on method. Other researchers agree that the methods have flaws, but know no better alternative. We believe that the most important question is how easy it is to understand what one is supposed to value. For complex issues like oil pollution, it is probably better to apply a decision-making approach like the one described in Chapter 3. For simpler issues CVM is probably useful.[33]

5. OBSERVED WTP

In this section we quote WTP measurements related to air, water, terrestrial environment, pollutants and recreation. We also examine people's propensity to pay for the future, and for statistical lives. The intention is to give the reader an idea of the kind of amenities that researchers have valued as well as typical WTPs. For example, carbon dioxide has been valued from €0.15 to €5 per ton, phosphorus at €100 per kg and lead at €500 per kg.

5.1 Environments

Wild species: An adult wild pig (*Sus scrofa*) in Kudremukha and Nagaraholé in India may sell for US$55, which corresponds to about 50% of the average annual per capita income for its hunters.[34]

Water: Estimates are reported from $0.06 to $30 per person per day to increase the water quality in American lakes from potable to fishable.[35]

Terrestrial: The bear (*Ursus arctos*), wolf (*Canis lupus*) and wolverine (*Gulo gulo*) are rare species in many regions, but important for biological diversity. The WTP to preserve these species in Norway has been measured as €600 million, or €150 per person.[36] In USA the WTP per person per year for a bald eagle has been valued at $12.4, a grizzly bear at $18.5, a blue whale at $9.3 and a California sea otter at $8.1.[37]

5.2 Pollutants

By definition, pollutants do damage, and people are therefore willing to pay to reduce polluting emissions. WTP is often expressed in euro per emission unit, but it is not obvious how to define one unit of emission. A common measure is average rate of emission in a region, in terms of kg per m^2 and year. Once WTP is estimated, it may be used in damage functions as described in Chapter 5.

However, WTP per emission unit is a rather crude measure for several reasons, and should be used with care. It does not take into consideration that damage may increase more than proportionally with emissions; 30 kg of a pollutant may be more than three times as dangerous as 10 kg. Neither does it consider that geographic averages may underestimate local damages, such as if other pollutants interact synergistically to increase toxicity.

5.3 Recreation and Ecotourism

- *Backcountry recreation*: It is estimated that 50% of US citizens and 84% of Canadians spend time and money on bird watching, wildlife photography and other outdoor recreational activities that involve wildlife without killing it. WTP per day has been estimated at Canadian $60 per day at prevailing backcountry crowding levels, but people were willing to pay more if the crowding were less.[38] For comparison, the existing daily backcountry fee was $4.25 per day.

- *Water quality*: A change in water quality along the Chesapeake Bay (Baltimore, Annapolis) measured as an increase in fecal coliform counts with 100 counts per 1000 mL produced about 1.5% drop in property prices.[39] This is a substantial change since mean coliform reading is 103 counts per 1000 mL. The drop corresponds to US$2 per m^2 property.
- *Noise*: In a residential district in Osaka, Japan, use of the hedonic method showed that from the average noise level of 58dB(A), an increase in the noise level of 1dB(A) decreased property values by 2.4%, corresponding to US$50 m^{-2}. An increase in 10 dB is perceived as a doubling of noise.[40]

5.4 Statistical Lives

If the life of a child is in danger because it is trapped under the ruins of an earthquake stricken house, society is usually willing to incur great expenses to rescue it. In most societies, it is considered immoral to set a price on the life of an *identified* individual – everybody has the right to live, and no effort should be spared, which in a theoretical sense implies that the value of a given person's life is infinite. In practice, of course, this is impossible – there are always limits. Decision problems in environmental management contexts are a bit simpler, however. Here, decisions have impacts only in a statistical sense, as a change in the *average probability* of people dying. For instance, cleaner air will improve the *average survival rate* in a population, since fewer people will suffer from lung-related ailments. To evaluate projects that have impact on the average survival rate, we need to balance costs against benefits, and then we need to know the value of a **statistical life**

5.4.1 The Concept of the Value of a Statistical Life

People are willing to pay something to reduce the risk of dying – we install safety equipment in cars for instance. If we add the WTPs of all persons in a population, we obtain an estimate of the value of a statistical life (VSL). Rather than the value of any particular person's life, VSL represents what the whole group is willing to pay for reducing each member's risk by a small amount. For example, suppose that each member of a population of one million persons is willing to pay 20 euro per year to reduce the death risk in the population from five persons to three persons per year. The total

WTP is 20 million euro, and since two lives out of one million are saved, VSL is one million euro.

To estimate the value of a statistical life year, VSLY, one assumes that people value reductions in current mortality risk in proportion to the expected number of years left to live, *T*. More precisely, *T* is the expected number of years left to live, averaged over all people in the society. Thus $VSLY = VSL/T$.

5.4.2 Estimation Methods

Since we have defined VSL in terms of death risk, we can use the term interchangeably with the *value of change in mortality risk*.[41] We can therefore measure WTP for a statistical life through both revealed and stated preference methods.

Revealed preference studies assume that people reveal their true preferences through market behavior. The most common method is *wage-risk analysis*, where one observes the wages of persons that have jobs with different death risks, and tries to deduce a relationship between wage and death risk through regression analysis. Another method is *consumer-market studies*. One may for instance observe peoples' willingness to pay for smoke detectors to be installed in homes, and the added safety they give.

Stated preference methods typically ask questions such as "if you are going to travel by air to another city, how much more would you be willing to pay in air-fare if the risk of death were reduced from 2 in a million to 1 in a million?"

5.4.3 Estimates

Literature surveys of revealed preference studies from several countries including US, UK, Canada and Japan indicate *VSL* estimates that are clustered in the range from four to nine million US$ (1999). Results from consumer-market studies tend to fall between one and five million US$.[42,43]

How much people are willing to pay for safety obviously depends on how much they *can* pay, which depends on their income. This would imply that rich people are worth more than poor; and if we continue this line of reasoning to nations, it means that inhabitants in rich countries are worth more than those in poor countries. In fact, we might even consider calculating VSL in one country from a VSL estimate in another country, by just correcting for GDP per person. From an ethical perspective, this does not sound right, however, since another line of thinking tells us that all men (and women) are worth the same.

Some observers find reported *VSL*s implausibly large. For example, a *VSL* of five million US$ implies that a family of four would annually be willing to spend about US$ 2100 of its US$ 40,000 income to reduce the mortality risk of each member by 10%. On the other hand, this is a large risk reduction, roughly equivalent to eliminating the adult motor vehicle fatality risk and reducing the children's risk by half.[44]

VSL can also be defined for age classes. *VSL* peaks near age 40 and is less than half as large at ages 20 and 65. It is also interesting to see how people discount the value of future life years. While discount rates between 11% and 17% over a five-year horizon have been observed for American and Norwegian citizens, Ethiopian citizens appear to discount with a 40% rate – thus putting much less emphasis on what will happen in the future. In other less developed countries no value is attributed to lives saved more than five years into the future.[45]

A central issue in health policy is to determine whether operations that save or improve lives are cost-efficient. The reasoning is that a life year with illness is worth less than one in perfect health. To adjust for that, one has introduced the concept of **quality adjusted life year**, QUALY. There is a general consensus – at least in the USA – that society would benefit from medical interventions that costs less than $30,000 per QUALY.

5.4.4 Exercise: Cost-Benefit and the Value of a Statistical Life

Assume that it has been established that the *VSL* for a worker with a life expectancy of 40 years in New York is $5 million. The per capita GDP in US is $26,037 and in Indonesia $1019. Life expectancies at birth are 76 years and 63 years respectively, and median discount rate for future lives years is 57% in Indonesia, but unknown in the USA (and for the New York worker).[46]

Required:

1. What questions could you ask to elicit the NY worker's discount rate?
2. What is her *VSLY* without and with discounting the value of future life years?
3. Would you recommend a medical operation on the NY worker if it costs $70,000 and keeps him in perfect health for exactly a year?
4. What if the worker were from Indonesia?

Solution:

1. Ask the person to evaluate two programs: Program A would save 100 lives this year. Program B would save *x* lives next year. Elicit the value

of x that makes her indifferent to A and B. Supposing that x is 110 lives, we obtain the discount factor f = 100/110 = 0.91. This gives a discount rate $f = 0.91 = 1/(1+r)^2$. This gives $r = 0.05$ or 5%.
2. *VSLY* is $5,000,000/40 = $125,000 without discounting. With 5% discount rate, you must discount a series of 40 future life years – that is a series of 40 figure ones: 1 1 1 1 1 … 1 1 1. If you do that, you will find that the present value of 40 years life is about 17,3 years. (See Chapter 5, section 6.) Thus her *VSLY* is $5,000,000/17,3 = $290,000.
3. You would recommend the operation both with the undiscounted and the discounted VSLY.
4. Compensating for different GDP per capita, *VSL* for the Indonesian worker would be $5,000,000 × 1019/26,037 = $200,000 and *VSLY* = 200,000/40 = $11,000 Thus without discounting you would not recommend the operation. However, the discounted *VSLY* is 200,000/2.75 = 72,700 just sufficient to recommend the operation.

As you probably have noticed, there are an abundance of ethical questions related to the use of VSL and discount rates. Of course, you would never apply a questionnaire to measure the value of an identified person, but is it fair to extend your WTP for your own life to other persons?

Try to find your own discount rate by asking a friend to give you the choices of program A and B and thereafter compare your response to the heuristics for discount rate choices: Wealthy, healthy, young and better educated people tend to have low discount rates because their future is likely to be longer. Impatient, risk-taking people, and people living in a risky environment tend to have high discount rates.

6. APPLICATION: SETTING TARGETS FOR EMISSION REDUCTIONS IN THE NETHERLANDS[47]

There are several ways to set targets for a country's environmental efforts. The Netherlands has set targets for emissions of six polluting substances, based on beliefs of relationships between emissions and end impacts. The time horizon corresponds to time frames in national agreements, approximately ten years. In the introduction of Chapter 4 we introduced the concept of unit prices of pollutants. These prices are gross averages, and developed to be able to handle complex decisions on a national or super national level. They are based on damage and willingness to pay estimates, and carry with them all the uncertainties associated with such estimates.

6.1 Decision Analysis

Below, we outline a decision-analytic framework for this policy question.

Decision maker and stakeholders: The decision maker is the government of the Netherlands. Among the stakeholders is the general public who will have to live with the consequences, consumer associations, green NGOs, industrial research firms, and industry that produces and sells pollution reduction equipment. See Chapter 3 section 2.

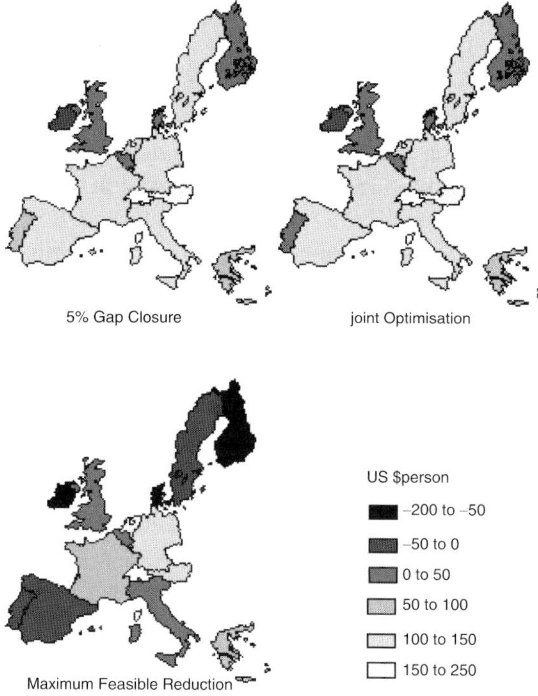

Figure 8-5. Annual net-benefits in EU member states. "50% Gap closure" is reaching 50% of the target levels; "Joint optimization" explores the interaction of ozone and acidification related activities.[48]

Objectives: The overall goal is efficient reduction of pollutant emission in the Netherlands. The objectives can be organized in classes of end-impacts: environmental damage, social damage, and cost.

1. *Minimize pollution effects*
   - Minimize greenhouse effects
   - Minimize acidification of waters and forests
   - Minimize eutrophication

   - Minimize photochemical oxidation
   - Minimize discharge of toxic substances
   - Minimize waste disposal
   - Maximize compliance with signed international agreements
2. *Minimize social costs*
3. *Minimize monetary costs*

Alternatives: The alternatives are different choices of pollution reduction targets. A *target* may be to reduce a pollutant load with 50% of the difference between the present load and a final goal corresponding to the maximum feasible reduction. Such a target is called a *50% Gap closure*. In the present case, the abatement measures could be implemented according to two schemes. With the first scheme, the benefits of emission reductions were calculated separately for each type of damage, and abatement measures implemented according to the cost benefit ratio of each measure. With the second scheme the benefits of reductions were calculated according to the improvement in both ozone related damages and acidification, and the abatement measures implemented according to this cost benefit ratio. This is called joint optimization Figure 8-5.

Consequence analysis: For this particular problem one would need to calculate the consequences in the Netherlands of reducing, say CO_2 with 1 kg (Chapter 4 on unit prices and Chapter 22 on climate effects). One option is to use published "unit prices", in monetary terms, for each substance one wants to reduce. The "unit pricess" are often calculated as the average damage caused by the pollution in a certain region. For example, the effects of SO_2 and other pollutants to human health effects are discussed in Chapter 14 on toxins, the effects on agriculture in Chapter 20, and effects on building materials in Chapter 21. One has found that the average damage cost in 15 EU countries is six USD per kgSO_2. There are many uncertainties in this estimate, however, due to quantification of emissions, air quality modeling, dose-effect modeling, and valuation.[49] "Unit prices" for a selection of substances are shown in Chapter 21, section 2.3 and can be used to give rough estimates of pollution damages. To estimate the utility of complying with international agreements, is difficult, and might preferably be scored directly by the decision maker on a subjective scale. The social costs are also difficult to estimate, including threats to employment in factories that may close down if they cannot afford the cost of pollution abatement. Monetary costs could be represented by the control costs summed over all sources that emit the pollutant.

6.2 What Happened?

The Netherlands' 1999 to 2000 reduction targets for the six pollutants ranged between 10% and 70%, with 10% for CO_2. The estimated control cost in 2000 was between 70 and 560 euro per person per year in the Netherlands, or on the average about 12% of each person's contribution to GDP.[50] Balancing the gains of pollution control with the control costs, the Netherlands would avoid damages to health, crops, and materials corresponding to a value of 2770 million euro, and obtain a net benefit of 100 to 150 euro per person by reducing SO_2, NO_x, NH_3 and VOC with on the average about 50% of the maximum feasible reduction. Reducing emission down to the feasible minimum would give a net cost for many of the EU-countries as shown in Figure 8-5.[51]

---

[1] Blamey et al. (1995) p. 267, 265, Constanza et al. (1997) p. 259, Max-Neef (1995)
[2] Vatn and Bromley (1994)
[3] Pearce and Turner (1990)
[4] Brookshire and al (1983)
[5] Lockwood (1999) p. 107
[6] Horowitz and McConnell (2002)
[7] Cropper and Oates (1992) p. 711
[8] Mansfield (1999)
[9] Cummings et al. (1984)
[10] OECD (1989) p. 34
[11] Clark (1990)
[12] Pimm (1997)
[13] Halvorsen et al. (1998)
[14] OECD (1989) p. 44
[15] Brown (1983)
[16] Brown (1983)
[17] Morioka et al. (1996)
[18] Ginevan and Splitstone (1997)
[19] Seip (1984)
[20] Mattson (1979)
[21] Tellus (1991)
[22] Cropper and Oates (1992) p. 720
[23] Kateregga (2004)
[24] Carlsen et al. (1993)
[25] Frykblom (1997)
[26] Arrowet et al. (1993)
[27] Blamey et al. (1995) p. 264
[28] Foster and Mourato (2002)
[29] Tano et al. (2003)
[30] Cropper and Oates (1992)
[31] Vatn and Bromley (1994)
[32] Cameron et al. (2002) p. 394.
[33] Lumley (1997) p. 79

[34] Madhusudan and Karanth (2002)
[35] Cropper and Oates (1992) p. 717
[36] Navrud (1992)
[37] Pearce, Cline et al. (1996) p. 200
[38] Rollins (2001)
[39] Legget and Bocksteal (2000)
[40] Morioka et al. (1996)
[41] Blomquist (2001)
[42] Blomquist (2001), Costa and Kahn (2004), Viscusi et al. (2003)
[43]Viscusi et al. (2003)
[44] Hammit (2000)
[45] Poulos and Whittington (2000) p. 1450
[46] The problem is based on Poulos and Whittington (2000) p. 1450 and Hammit (2000)
[47] Krewitt, et al. (1999), footprint after Wackernagel and Rees (1995)
[48] After Krewitt et al. (1999)
[49] Krewitt, et al. (1998) p. 181
[50] Sas et al. (1997)
[51] Krewitt, et al. (1999)

Part 2

THE SUBJECTIVE EXPERIENCE

Overview

Chapters

9. ETHICS: Ethical principles for environmental management.
10. QUALITY OF LIFE: Recreation, cultural experience and life quality standards.
11. AESTHETICS: The aesthetical value of landscape and cityscape.

> "...the fundamental principles of ecology govern our lives wherever we live, and ... we must wake up to this fact or be lost." Karin Sheldon (b. c. 1945), U.S. lawyer specializing in environmental protection. As quoted in Ms. Magazine, September 1973

According to the quote, the principles of ecology act as a template for our lives, and many people will agree to that. However, it could also be argued that the principles of human culture are the opposite of those of ecology. Whereas ecological systems oscillate, humans try to smooth them. In ecological systems "the fittest survive". Human systems tend to be democratic and have instruments that check those that dominate. So, there are different views. But although subjectivity steers human activity, the three chapters in this part put emphasis on how people perceive the environment with their own values as the point of departure.

In each chapter, we first give background information, then we venture into the question of quality standards, although such standards are not common, neither in ethics nor in aesthetics. Finally, we present algorithms that capture essential relations among central variables.

Chapter 9

ETHICS

Ethical principles for environmental management

Environmental management is in a deep sense a question of ethics. The reason many issues are controversial is not only conflict of interests, but conflict between ethical mindsets. We therefore start with an exposition of different mindsets and follow up by adressing some central ethical questions:

- What are the central classes of ethical mindsets?
- Is rationality ethical?
- Is rationality and deep ecology compatible?
- What is valuable and who should do the valuation?
- Why is ethics important in private business decision making?
- Who should decide and whose values should be applied in public policy-making?
- Why is the precautionary principle important?

"The well-being and flourishing of human and nonhuman life on Earth have value in themselves. These values are independent of the usefulness of the nonhuman world for human purposes." *The Deep Ecology Platform*, Arne Næss and George Sessions

1. INTRODUCTION: WHAT IS ETHICS ABOUT?

Ethics is about decision-making. It separates situations we aim for and situations we seek to avoid. This book takes the decision-making approach to environmental management, but up to now we have mostly discussed methods and techniques, and only briefly touched deeper questions such as:

1. *Ethical principles for choice*: What basic principles should guide us when we make choices? Ethical theory describes three different decision-maker *mindsets*. They emphasize virtue, duty or consequence. We have hitherto tacitly assumed that decision-makers want to optimize

consequences, but this need not be what actually happens – and it is not even obvious that this is what *should* happen. We shall therefore have a closer look at it.

2. *The anthropocentric question*: This book assumes an anthropocentric perspective, meaning that humans are the masters on earth. When we make decisions about the environment, we apply our own values; neither animals, nor plants nor mountains are sacred in and by themselves, but may be attributed a status of sacredness from humans. Many people – such as those involved in the deep ecology movement – consider such attitudes arrogant, unethical and even dangerous; we need to have a closer look at our assumptions and their consequences.
3. *The source of value*: Once we have adopted an anthropocentric perspective, considering human beings as the source of value, should we then apply the values of the population or those of experts, or should the decision-makers use their own values? So far, we have adopted a pragmatic attitude by recommending use of popular values when available and assuming decision-maker values when necessary. But the underlying question is in reality one of ethics and needs to be considered further.
4. *Formation of values*. The term "value" has several meanings. Plato distinguished between "value in use" and "value in exchange". One definition is that "value" is the capacity of a limited good, or a limited service, to provide a benefit. In an ethical context the benefit is that the people in the society feel good.[1] People derive values from religious or philosophical texts, from parents and teachers, through actual behavior, or by observing the market and society at large.

2. ETHICAL PRINCIPLES FOR CHOICE

2.1 Mindsets

Public discussions of environmental policy tend to be emotional, and it is useful to understand why. Take the whaling question, for instance. Japan and Norway want to hunt whales for economic reasons. Whales could be an important source of food, if they were caught in a sustainable way. The rest of the world seems to be against it, however, with Americans as particularly outspoken antagonists. The conflict does not seem to be about limited resources, but something more basic like ethics. Below, we have phrased the whaling question in three different ways, each representing a particular ethical mindset.

"Is it OK to kill whales?" Interpretation: The question reveals an ethical stance; it may be taken to be a question of whether a whale-killer lacks *virtue* since he kills a being that is almost as advanced as humans, and moreover kills it in a bestial way.

"Should whale-hunting be allowed?" Interpretation: The underlying moral issue is whether we have the *right* to hunt an endangered species.

"Should the whale hunting quota this year be set at zero, 100 or 200 whales?" Interpretation: This is a question about which level will have the best *consequences*, when objectives like economic profit, ecological balance, and the future stock of whales are considered.

The point is that decision makers have different *mindsets*. One decision-maker may tacitly apply principles for choice that are fundamentally non-commensurable from those of another. Some are preoccupied with the act itself, some are concerned about rights and duties, and some are concerned about consequences and are willing to trade off benefits against costs. The principles are known as different kinds of ethics, which often lie at the roots of the most emotional conflicts that arise in environmental management. Ethics separate situations we aim for and situations we seek to avoid.[2] Virtue ethics, duty ethics and consequentialism are classical principles to guide decision-making.

2.2 Virtue Ethics

If we are primarily concerned with displaying the right attitude when we act, we are guided by virtue ethics. Aristotle (384-322 BC) is a well-known classical authority. Forgiveness, gratitude, regret, remorse, loyalty, humility, compassion, courage, prudence and loyalty are examples of Aristotelian virtues. The focus is on acts and attitudes – or moral character, not on consequences. Courage is important, not survival. Virtue promotes human flourishing, according to Aristotle – and in saying so; he gives a consequentialistic argument for virtue ethics as a principle of decision-making. The important point, however, is that of mindsets – the way the decision-maker thinks or feels. The mind of a virtuous person is set on attitude, not consequence.

Today we see an emphasis on virtues on the web pages of organizations, displaying for the particular organization an intended moral character as a list of core values. Integrity, honesty, equality, impartiality, loyalty, respect, prudence and tolerance are currently on the top of the list of corporate virtues. This reflects a felt need for more virtuous behavior, and respect and prudence are important virtues in questions of environmental management. And so is compassion – a whale-killer appears to lack a compassionate attitude towards the suffering of others, regardless of other characteristics of a virtue ethics mindset.

2.3 Duty Ethics

Duty ethics – or deontological ethics – classify actions according to whether they are right or wrong with respect to a system of rules. The Ten Commandments is a duty based system where actions are judged according to whether the commandments are obeyed or not, not because of the consequences of the actions.

Immanuel Kant (1724-1804) is the most important contributor to a deontological system. He could not accept that arbitrary personal sentiment should be the basis for moral choice, and tried to formulate a principle whereby rules for ethical choice could be deduced through introspective reasoning. He insisted that our actions possess moral worth only when we do our duty. Consequences do not matter. His principle for determining what is our duty is known as the categorical imperative:

> "Act only on that maxim through which you can at the same time will that it should become a moral law!"

The United Nations' Universal Declaration of Human Rights of 1948 is a notable example of a system of Kantian rules for public policy, where the rights are considered absolute regardless of their consequences. The objective of a Kantian system of moral laws is to create a good society. Thus, the underlying motivation is consequentialistic – and again, the important point is that the mindset of a Kantian decision maker is not concerned with those ultimate ends, but with whether actions are right or wrong, judged according to rules only.

In the previous chapter, we discussed the value of a human life and noted that while we put value on statistical lives, we cannot for moral reasons do the same when it is an identified person. Kant assumed that human beings have a special value – a **dignity,** which prevents us from doing that.

2.4 Consequentialism

For a consequentialist, the value of an action derives entirely from the value of its consequences, and this contrasts both with virtue ethics and duty ethics.[3] Consequentialism has broader applicability than virtue or duty ethics. In principle, it can be applied to any decision problem where consequences can be identified. The mindset of a consequentialist is forward looking, and the basic method is to separate facts from values. One of David Hume's (1711-76) important contributions was to describe the apparent "gulf" between belief and value. He used the term "**belief**" to denote conception of facts – what "is". Beliefs are, according to Hume, vivid or lively ideas regarding matters of fact, and beliefs about consequences require cause-effect reasoning. He noted that reason can show us the best way to achieve our ends, but it cannot determine our ultimate desires:

> "Tis not contrary to reason to choose my total ruin, to prevent the least uneasiness of an Indian or a person wholly unknown to me".

In other words, he has such preferences because he has certain values and feelings; they are not at all matters of reason, and we have no right to call his action irrational as long as he acts according to his feelings.[4] In environmental management we need to develop well-founded beliefs about ecological facts. We need for instance to learn more about the ecological importance of species, how species depend on each other as threads in a web. But we also need to balance the costs of making beliefs well founded against the benefits. It is for instance not obvious how much effort should be invested in preparing environmental assessment statements, and estimating the effects of proposed projects.

3. RATIONAL DECISION-MAKING

Can we accommodate all three mindsets in the rational decision process we developed in Chapter 3? Most individuals harbor all three. In routine type decision situations, we stick to rules rather than considering consequences; and when no rules apply or the consequences appear counterproductive, we may switch to a different mindset and use better judgment. Virtues are omnipresent; we do not litter, even if nobody watches. Nevertheless, people are different and under dominion of the mindsets to different degrees.

Perhaps virtues and duties could be incorporated in the rational decision process, simply by regarding them as consequences and adding appropriate attributes to the list of decision criteria? The problem is that it does not work that way; virtuousness and rule following are different from ordinary consequences in that a gain in one dimension does not compensate for a loss in another. Fraudulent money does not compensate for virtue lost by cheating.

An alternative to subsuming all decision problems under the sway of consequentialism is to try to deal with virtue and consequence separately. Rules that are not themselves in question can sometimes be handled as constraints in an optimizing problem. Virtue is much more problematic, and the reason is that virtuous behavior is controlled by very strong emotions.

Think of shame and guilt. Those are feelings we go at great lengths to avoid. They are much stronger than the rather feeble emotions we try to elicit when we are asked to choose between consequences in a multi criteria decision problem, or between options in a WTP scenario.

Strong emotions make us apply lexicographic preferences, where one criterion gets an infinitely higher weight than the others: "Whales must not be killed no matter how good the benefits might be". A similar rule is **categorical exclusion** that applies to the items on UNESCO's list of world heritage objects. The problem with lexicographic decision rules and similar rules is that they break down if the consequences are overwhelming enough.[5] Most people *will* kill one person to save thousands. When politicians say that they will restore a polluted landscape no matter the costs, it is simply not true – costs always matter.

3.1 Recommendation

Values are sometimes classified according to priority. One scheme that shows increasing orders of priority looks like this.[6]

1. Instrumental value
2. Intrinsic value ≈ innate value ≈ inherent worth
3. Moral value ≈ moral concern ≈ moral standing ≈ with rights
4. Legal standing, legal rights

Here, the law goes first (duty ethics). Then follows moral values and rights, which are protected by duty ethics, but to a lesser degree. Intrinsic values are on the next level. They are subject to trade-off. On the lowest level are instrumental values, which are not in themselves valuable – money is an example, money is not edible.

Although persuasive, we find such propriety lists too simplistic, however. One cannot for instance categorically decide that the law comes first. One must always judge the whole situation.

Our conclusion is that we recommend consequentialistic ethics: Actions should be judged according to consequences, with rules and regulations functioning as restrictions on the set of viable options. We recognize that emotions are important in all value judgments, but the strong emotions associated with virtue should be tempered by coaxing decision-makers or respondents to look forward and consider consequences. Adam Smith put it like this: "Virtue is more to be feared than vice, because its excesses are not subject to the regulation of conscience".

As for the whaling question, this recommendation would imply that we set a quota for whale hunting that maximizes long-term profit and minimizes suffering of whales that are killed. This would be an example of consequentialism, with the restriction that no quota may jeopardize the survival of the species, which is a rule-based restriction. This integrates soft values as well as our duty not to endanger species. But it will not satisfy those who strongly feel that whale-killers lack virtue – that it is not OK to kill whales, no matter what. Thus, while the conflict may be eased, it is not resolved.

4. THE SOURCE AND LOCUS OF VALUE

We shall find it useful to distinguish between source and locus of value. The valuator is the source of value, and the locus of value is what is valuable – clean air, a beautiful scenery etc.[7] The ethical approach we recommended above makes human beings the source of value – this is what we call **the anthropocentric principle**.

Once the anthropocentric principle is accepted, however, the next question begs itself: Should public servants or the general population – or perhaps a business manager – be the source of value in environmental decision-making? We have described two different approaches to valuation – multi-criteria utility analysis in Chapter 7, and measurement of willingness-to-pay (WTP) in Chapter 8. Decision makers were the source of value in the first method, and the population in the second. The important point now, however, is not which value-elicitation scheme is best, but who should be entitled to be the source of value. That answer depends on whether we are considering a private or public project. We shall discuss this in sections 5 and 6.

4.1 The Anthropocentric Principle and Deep Ecology

The anthropocentric principle may seem obvious; who other than human beings could be the source of value? The Deep Ecology movement (DEM), with the philosophers Arne Næss and George Sessions as spokesmen, is deeply concerned about the environment. They have posted eight statements about ethical conduct. Let us have a look at what they say. Statement 1 says:

> "The well-being and flourishing of human and nonhuman life on Earth have value in themselves. These values are independent of the usefulness of the nonhuman world for human purposes."

This means that DEM attributes existence value to all forms of life on earth and that the existence value is independent of *use value*. It does not preclude that existence value can be traded off against other values as we did in the oil pollution case in Chapter 3, and does therefore not appear to be in conflict with consequential ethics. Statement 2 says:

> "Richness and diversity of life forms contribute to the realization of these values and are also values in themselves."

This simply states that richness and diversity are values. Such values can without objection be taken into account in a consequentialistic analysis. Statement 3 says:

> "Humans have no right to reduce this richness and diversity except to satisfy vital needs."

Although close to a duty-based formulation, the statement may be read as an allowance for trade-off between richness and diversity in one scale and vital needs in the other. It requires judgment of vitality of needs. Statement 4:

> "Present human interference with the nonhuman world is excessive, and the situation is rapidly worsening."

This is a valuc judgmcnt: the current importance weight attributed to nature is to small. It is also a statement of belief: they believe that the current situation is unsustainable. Statement 5 reads:

> "The flourishing of human life and cultures is compatible with a substantial decrease of the human population. The flourishing of nonhuman life requires such a decrease."

This is an interesting value judgment concerning population size, life quality and the "flourishing" of nature is too small. Statement 6 concludes:

> "Policies must therefore be changed. The changes in policies affect basic economic, technological structures. The resulting state of affairs will be deeply different from the present."

This conclusion is based on the values and beliefs above. Statement 7:

> "The ideological change is mainly that of appreciating life quality (dwelling in situations of inherent worth) rather than adhering to an increasingly higher standard of living. There will be a profound awareness of the difference between big and great."

Here, they encourage a revision of values, which will be more wellfounded if there is higher emphasis on nature and lower emphasis on consumption. Statement 8:

> "Those who subscribe to the foregoing points have an obligation directly or indirectly to participate in the attempt to implement the necessary changes."

The final statement is a meta-statement phrased in the language of duty ethics. This exercise serves to illustrate that in spite of the duty ethics character of DEM, it is actually compatible with consequential ethics. What it advocates is basically a revision towards more nature oriented values and a corresponding change of policy. This is in accordance with the rational approach.

5. PRIVATE BUSINESS DECISION MAKING

5.1 The Traditional Economic View

According to traditional market economic thinking, the society is best served if everybody maximizes their own economic gain within existing rules and regulations. Thus, as we discussed in Chapter 5, it is the task of the authorities to provide the right economic incentives to make profit-maximizing business behave in an optimal way for the society. Friedman (1970) is quite clear:

> "The leader of a company has a direct responsibility to the firm's owners, who employs him. This responsibility consists of running the enterprise in accordance with their wishes, and that generally implies that one should earn as much money as possible, while at the same time following the rules of society."

In this case, value maximization is one-dimensional and profit oriented. The owners of the firm are both source and locus of value. The concept of corporate social responsibility provides an alternative view, which we shall discuss below, but it is important to be aware that the traditional economic view is by no means abandoned. Influential economists still advocate strict profit maximization, and maintain that anything else leads to destruction of value.[8]

5.2 Corporate Social Responsibility

Although lawful profit maximizing may be a good strategy in the short run, many maintain that to maximize long-run profit, one should show **corporate social responsibility**[9] (CRS). According to CSR, it is not enough to act according to laws and regulations. Something more is required: One should include the interests of the stakeholders in the goal hierarchy, and ensure that they receive due weight.[10] Stakeholders are anybody with an interest in the operations of the firm, such as employees, owners, costumers, suppliers and the local community – and the local community has an interest in the environment. If this is not enough, some also list the environment itself as a stakeholder.

This latter means that when projects are evaluated for long-term profit, the managers should list such things as clean air, clean water, absence of noise, recreational areas, wildlife etc. among the decision criteria. How should these criteria be weighted? Since the overall goal is to maximize long-term profit, these criteria are not considered end impacts, but instruments. Therefore, the management should weight them according to what impact they believe the stakeholders will have on long-term profit. This is a *moderate form of CSR*, where the managers are the source of value, and the stakeholders the locus of value.

Not everybody thinks it is ethical to consider stakeholders as instruments for the firm, they would rather see the firm as an instrument for the stakeholders: stakeholder interests should be maximized, not stockholders'.[11] This is a *radical form of CRS* and implies that the managers should attribute weights according to the values of the stakeholders. The stakeholders are sources of value, and the locus is with whatever they think valuable. For the company to make the right choices, it would be necessary to elicit the stakeholders' preferences. Some companies – like oil companies – that are especially vulnerable to public opinion – and indirectly to political pressure – try to acquire this information anyway to have a firmer basis for their decisions.

The concept of corporate social responsibility is today well integrated in the operations and image building of proactive companies. They establish rules and procedures, with external auditing to ensure that they are followed. The problem is however, that the ethical emphasis tends to be on virtue, not on consequence. Even though a company *looks* good, it does not necessarily *do* good.

5.3 The Global Compact

At the World Economic Forum meeting in 1999, the UN General Secretary challenged international corporations to contribute to a stable and just society in a globalized world by subscribing to nine principles regarding human rights, workers' rights and environmental protection. The nine principles are:

Human rights:
1. Support and respect the international human rights within their area
2. Make sure their own operations do not conflict with human rights

Workers' rights:
3. Support freedom of organization and the right to collective negotiations
4. Eliminate all forms of forced labour

5. Effective elimination of child labour
6. No discrimination regarding employment and work

Environment:

7. Support the "Safety First" principle in environmental questions
8. Endorse initiatives that contribute to environmental responsibility
9. Encourage development of environmentally friendly technologies

We see that the nine principles for the most part are based on virtue ethics. They focus on actions, not on consequences. Although laudable, one must also bear in mind that virtue sometimes can be detrimental. One controversial example is child labour, which under some circumstances may be necessary for survival of the family. The global compact obviously touches on hard ethical problems: which principles should govern a comparison of values in rich and poor countries? What right have rich people to acquire poor peoples' commons by buying land and using it for recreation?

6. PUBLIC PLANNING AND POLICY-MAKING

Economists generally agree on the ethical principle that the population should be the source of value in public planning and policy-making. This is in sharp contrast to views held by many philosophers, sociologists and politicians, who find reliance on WTP/WTA uncompelling as an overall theory of the good.[12] The economist view is based on the presumption that the market knows best – there is no natural higher authority, and market externalities, such as environmental impacts, can be internalized through measurement of WTP. Individuals are after all likely to be the best judges of what will satisfy their intrinsic desires.

There are at least three basic problems with the economist view; one is practical and two are ethical.

6.1 Low Validity and Reliability of WTP Estimates

We have discussed some of the difficulties in Chapter 8. To these we may add that a meta-analysis of estimates between 1967 and 1988 of the value of air quality based on land property value produced such unreliable results to indicate that markets are unable to value air quality.[13] Even more devastating perhaps, was Kahneman et al. who in 1992 argued that stated preference methods are virtually worthless since people do not actually state their WTP, but instead quote a convenient number to purchase social responsibility.[14] Another investigation underscores this, where they found no statistically significant differences in WTP to prevent 2000, 20,000 or 200,000 migratory waterfowl deaths, all numbers less than 2% of the waterfowl population.[15]

6.2 Imperfect Information

When WTP/WTA surveys are carried out, the respondents are presented with scenarios describing the consequences of different projects. But it is well established that humans have severe cognitive limitations that prevent us from forming correct beliefs.[16] To name a few; we systematically overestimate small probabilities, we imagine patterns that are not there, and we are prone to framing effects. It is virtually impossible to describe scenarios in an objective way – that is, exhaustively, so that additional information would not change the attitude of the respondent. How could we in a WTP survey, for instance, describe the effects of global warming in an exhaustive way to make the respondent develop appropriate preferences? This is clearly extremely difficult, and it does limit the applicability of the WTP principle.

6.3 WTP and Fairness

6.3.1 The Rich and the Poor

The WTP criterion skews the analysis in favor of rich people.[17] As a simple example, let us assume that the authorities consider turning an open field into a golf course. The open field now serves as a recreation area for local families, who are generally poor. According to the WTP criterion, one should add together the individual WTPs for recreation and do the same for golf. Whichever sum is largest wins. Since golfers are probably richer, they might win even though they are fewer. The question is: Is this an ethically good decision? The same dilemma applies, as we have seen, to global commons.

In Chapter 5 on economic incentives we showed a graph for calculating the optimum pollution level seen from the viewpoint of a polluter. In Figure 9-1, the marginal control cost curve describes the cost of pollution abatement. The rising curve A shows marginal damage caused by increasing pollution. The optimum pollution level Q_A is where the marginal cost curve and the marginal damage curve intersect. However, who is to estimate the correct damage costs? Poor people – here called the losers – may be manipulated and convinced by the winners that “pollution seems palatable and worth the cost”.[18]

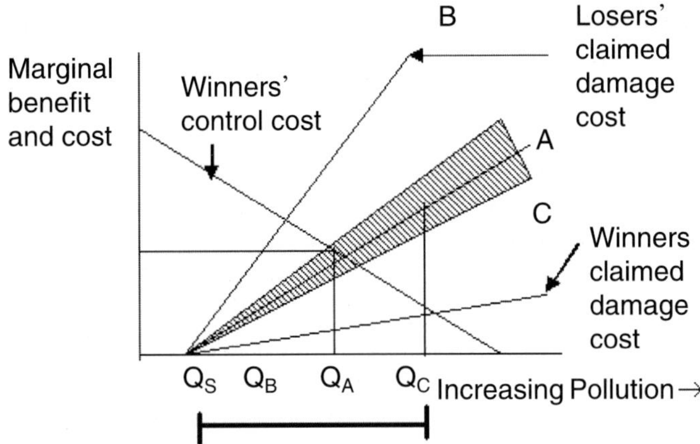

Figure 9-1. The optimum level of environmental pollution, as seen from the perspectives of rich (winners) and poor (losers) people. Q_S represents pollution with no observable effect. Q_A is the optimal level where marginal control costs equals pollution damage costs. Q_B and Q_C are deviating cases discussed in the text. The line from Q_S to Q_C symbolizes *ethics deficit* if the "winners" claimed damage cost C is used. Around line A there is a fan of conceivable lines that can be defended by "reasonable" persons,[19] thus a solution may be "weakly" right.[20]

6.3.2 Rich and Poor Countries

People in poor countries have little money to spare and will consequently have a smaller WTP for environmental goods and services than people in rich countries – and even for human lives. This distorts comparison of WTPs, which should be adjusted according to **Purchasing Power Parity**, PPP.[21] The PPP exchange rate is calculated from the relative value of a currency based on the amount of a basket of goods the currency will buy in its nation of usage. PPP will often differ from the relative values of the Gross Domestic Product in the two countries.

6.4 The Population does not Always Know Best[22]

A well-known problem with WTP/WTA is that preferences may be different before and after a project is carried out – this is called **the problem with endogenous preferences**. When a dam is built for hydroelectric purposes, there is almost always opposition, notably from environmentalists. But if the dam is built anyway, people will adapt – starting with using the lake for new purposes, like fishing and recreation for example – and actually enjoy the lake. If the authorities for some reason propose to restore the valley to its original form, one may very well see reversed preferences: Now people want to keep the lake.

Another problem concerns the **welfare of future generations**. We have already discussed the fact that in WTP-elicitation surveys, some people act like citizens and others like consumers. The "consumers" have a myopic perspective, they want to enjoy life now, and – perhaps unthinkingly – leave the problems to future generations. But even the citizens have a problem. How can they make decisions on behalf of future generations when they do not know the preferences of those generations? This is actually not only an ethical problem, but also a logical paradox. Let us look again at the example above of a hydroelectric dam, and particularly consider the building of the Three Gorges Dam on the Yangtze River in China. That dam will create a 500 km long lake, which will exist for many generations. Long stretches of river as well as whole cities will disappear, millions of people will be relocated, and the famous Three Gorges will be lost as a tourist attraction. If the dam were not built, future generations would probably continue to have strong preferences for what they now lose. On the other hand, since the dam is built, the preferences of future generations for the lost cities and the river valley with the gorges will probably weaken, giving way to stronger preferences for the new lake, for the new cities and the areas down-stream that have then become protected from flooding. This makes the WTP principle problematic to defend from an ethical point of view.

6.5 WTP and Demography

The intensity of environmental concern appears to depend on demographic variables. Examples are:

- WTP seems to be greatest at medium incomes around US$7000–US$10,000 and falling to both sides.[23]
- Small societies tend to show more concern than large societies because it is easier to take care of defectors.
- People in democracies are more concerned than under other regimes.[24]
- People living close to nature are more concerned than more urban people.
- Age, gender and education also seem to play a role.

All this means that it is quite a challenge to conduct WTP surveys in an ethically defensible way so that all these groups are heard.

6.6 Recommendation: Expert Panels as Source of Value

Our conclusion is that to rely only on WTP/WTA is inadequate as an ethical principle in public planning and policymaking. Lack of reliability and validity, limited information, the question of fairness, and the problems with embedded preferences and future generations all point in the same direction. The underlying problem with WTP/WTA surveys is that it is not really a question of measuring existing preferences. On the contrary, people do not walk around with ideas of the monetary worth of environmental amenities, just waiting for somebody to ask. Rationality requires well-founded values, and that requires information, consideration, dialogue and reflection – in other words, it takes time. WTP/WTA surveys simply do not allow for that. An alternative or supplement is to interview just a small selection of people who already are informed and have reflected on the issues.[25] These people might be public servants such as planners, or representatives of *non-governmental organizations* or NGOs. It is convenient to create groups of approximately three persons, forming **expert panels**. Each panel partakes in computer-interactive interviews, using methods described in the first two chapters of this book. This gives ample opportunity for dialogue within the panels, which helps build both well-founded beliefs about reality and well-founded preferences. Of course, there is no guarantee that these panels represent the will of the population, but they may be appointed deliberately so that they represent a range of opinions. For instance, one panel may consist of public planners, another of producers and a third of greens from NGOs. These panels should not be regarded as *decision makers*, only as *sources* of value. They provide *input* to the real decision makers, who are politicians. Presumably, a high quality interview process will mitigate many of the problems of WTP/WTA survey methods – and they are much cheaper.

7. ETHICS AND ECOLOGY

The "safety first" principle is also called the **precautionary principle**. If the consequences of a decision alternative are unknown and there are threats of serious or irreversible damage to the environment, "the burden of scientific proof" lies with those who propose the alternative. They have to demonstrate that the implementation will not result in serious or irreversible damage.[26]

Consequential ethics require rationality, and rationality requires well-founded beliefs about reality. Ethical decision making regarding the environment must be based on a minimum knowledge of ecology. Seven issues are important in this respect. Climate is a basic determinant of habitat suitability, and complete ecosystems or biomes are a requisite for the maintenance of large carnivore species. So-called "key-stone" species are found to be most important to maintain traditional ecosystem, but there are also rare ecologically unimportant species. Some species carry a higher proportion of evolutionary history than others,[27] some species spread malaria

(Mosquitoes) or are by their invading ability (the zebra mussel) endangering indigenous species, some species are endangered[28] and some species are rare locally, but have a wide distribution regionally.

8. APPLICATION: LAND REFORM AND DEFORESTATION IN THE BRAZILIAN AMAZON

Figure 9-2. Forest fires in Brazilian Amazon. "Beneficial" use often turns out to be deforestation by burning.

One goal of land reform in Brazil is to distribute land equally. The Brazilian constitutions have given squatters the opportunity to obtain ownership of up to 50 hectares of vacant public land if they develop and occupy their claims for one year, and on private land after 5 years if they occupy and develop the land without opposition from the owner. Adam Smith gave a justification for the law: "when private property of land is not expedient, it is unjust.[29]" Private owners, however, may put the land into "beneficial" use themselves to avoid occupation by squatters, and this often amounts to burning the forest, Figure 9-2.

In this section, we will apply concepts and methods described in this book to see how a policy analysis of the dilemma – which has been called "Too poor to be green" – could be carried out.

8.1 The Decision Problem

The decision problem is how to increase income and life-quality among the poor, and at the same time preserve the forest. The area in question is the Amazonian Atlantic forest, and the time frame is 200 to 500 years, more if erosion becomes significant. The decision makers are Brazilian bureaucrats reporting to various Brazilian constituencies. Among the stakeholders are the poor (the squatters), the landowners, and national and international environmental groups.

Important goals are to maximize income for the poor, to maximize equality, to maximize social order or avoid violence, to maximize stocking of CO_2 to reduce climate effects, and to maximize wilderness preservation.

There are two main alternatives: to maintain status quo and to enforce the constitutional law described above.

Consequence calculations. To estimate the land area required to support a farm family data from the section on land-use can be used (Chap. 20), and if you want to compare this area to a corresponding area for hunters and gatherers in a tropical forest, data on sustainability can be used, (Chap. 13, section 4). However, the soil in large parts of the Brazilian forests are not well suited for agriculture, (Chap. 15, section 3.8). Brazil belongs among the countries with medium Human development index (no 74, year 2000), high inequality: (poorest 20% has 2.5% share of income or consumption as compared to 9.7% for the country with the largest share. It is probably reasonable to assume that the equity criterion increases linearly with the number of squatters that obtain title to land. The Brazilian forest may stock about 200 ton CO_2 ha^{-1} (Chap. 22). To estimate species diversity in the forest several calculating models can be used, (Chap. 12, section 5). The extent of non-sustainable foresting and wild-fires put on by humans for the sake of demonstrating "beneficial" use, as well as incidences of violence, is discussed by Alston et al. (2000). A normal incidence rate for violence would be 300–600 in 100,000 and per year, (Chap. 10, section 3.3).

Preferences. Since the stakeholders are among the poor, soliciting preferences fairly is difficult and discursive formats should probably be used. The decision analysis may end up as a weak ethical decision with "disagreement among reasonable persons", Rawls (1993, p. 48) and Chap 9, section 6.3).

8.2 What Happened?

Alternative ii) was chosen by the Brazilian government. As of 2005 the "land-use title rulings" are still maintained. Conflicts that include violence are still prevalent. This example is formulated after Alston et al. (2000), but several articles by Fearnside (1993), (1996), (1999) and student projects have been used to discuss the problem.

---

[1] Based on a definition by Haksver et al. (2004)
[2] Blackburn (1998)
[3] Blackburn (1994)
[4] Hume (1988)
[5] Sen (1995)
[6] Brennan (2002)
[7] Lee (1996)
[8] Anonymous (2005)
[9] Freeman (1984)
[10] Jensen (2001)
[11] Freeman (1984)
[12] Cowen (1993)
[13] Smith et al. (1995)
[14] Kahneman and Knetsch (1992)
[15] Boyle (1994)
[16] Kahneman and Tversky (1979), Kahneman and Tversky (1996)
[17] Hubin (1994)
[18] Galbraith (1973)
[19] Rawls (1993)
[20] Adapted partly after Boyce (1994)
[21] The UN has a list of PPP for all countries in the world on their home page
[22] Cowen (1993)
[23] Andreoni et al. (2001), Ezzati et al. (2001), Gangadharan et al. (2001), and Bruvoll et al. (2003)
[24] Equity: Boyce (1994); democracies: Neumayer (2002)
[25] Wenstøp and Seip (2001)
[26] Barbier et al. (1994) p. 172, Palmini (1999)
[27] Nee and May (1997), Purvis et al. (2000)
[28] Possingham (2000)
[29] Mill 1848 (1965) 230; III-4.15

Chapter 10

QUALITY OF LIFE

Recreation, cultural experience and life quality standards

We address the fundamental question about the main sources of life quality and look at how life quality standards are composed.

●

"Culture is the process by which a person becomes all that they were created capable of being." Thomas Carlyle (1795-1881) Scottish essayist and philosopher

Figure 10-1. The Mesa Verde dwellings   Climbing up to the dwellings of Mesa Verde

A great concentration of Anasazi Indian dwellings, built from the 6th to the 12th century A.D. are located at the Mesa Verde plateau in southwest Colorado. To protect the structures from wear and tear, tourists are not allowed to climb them. Photo: 1977.

1. INTRODUCTION

We discussed ethical principles for choice in Chapter 9 and suggested an anthropocentric approach. Anthropocentrism means that human beings are the source of value, and that we are free to value anything we deem valuable. Valuables give life quality, and this chapter focuses on efforts to systematize and measure values. In this chapter, we will discuss what gives life quality, with an emphasis on environmentally relevant themes like recreation and tourism. We will also discuss ways to measure *quality of life*. Gross Domestic Product (GDP) per inhabitant is the traditional measure of the *living standard* in a country. But the problem with GDP is that it does not take into account externalities – things that are not priced in the market, such as ecosystem services. Thus, many of those qualities that add value to life are not included and we need more comprehensive measures. Several indices have been developed, and they usually include sustainability, the natural environment, human rights and equity. You will find more information about this in a United Nations Human Development Report.[1]

People find it valuable to do such things as to stroll along a shoreline and experience the landscape or townscape; to participate in commercially arranged recreation by tourist bureaus; to stay overnight in hotels or camping grounds; to visit historical parks and monuments.

There are for example in the European Union 11,435 publicly approved seaside beaches and 4376 freshwater areas for swimming. This amounts to one beach per 20,000 people.

By the term *recreation* we will mean an activity – which need not be very active – for which the environment is used to increase quality of life, without the users needing to pay for it. *Tourism*, on the other hand will mean that there are commercial actors who facilitate recreation activities and are paid for it, like hotel owners. Finally, *cultural experience* can be free or not, but the object of the experience is man-made: parks, townscapes or monuments.

1.1 Recreation and Tourism

There are at least five types of tourism, which are different with regard to environmental impact:

- *Ethnic tourism*: traveling for the purpose of observing cultural expressions and life-styles of indigenous people; also called "event" tourism. It easily conflicts with the right to privacy. For example, to ensure privacy for aborigines, visitors to Kakadu National park about 250 km east of Darwin, Australia, are not allowed access to highly sacred sites.[2]

- *Historical tourism*: townscape, museum- cathedral tours to see the life and sometimes glory of the past.
- *Environmental tourism* often draws tourists to distant places with environmental attractions, rather than ethnic ones. It often conflicts with fragile landscapes and cultural artifacts. Hotels close to the beach may conflict with local fishery and the shore ecosystem.
- *Recreational tourism* emphasizes sports, walking and sun bathing.
- *Business tourism* centers on conventions or conferences and often emphasizes the themes taken up by the conference.

Vacation patterns will change over time, but some typical and some extreme values may give guidelines for first estimates of volumes. In the US there are 58 million international tourist departures each year, which amounts to 22% of the population (but some individuals are responsible for several departures.). In Norway, a small country, the corresponding number is 3.1 million departures or 73% of the population. Probably due to Norway's size, location, and long dark winters, Norwegians spend more holidays abroad than the average citizen in countries where the climate is more favorable. Holiday trips may typically last for one or two weeks. The most active holiday travelers are the young and the middle-aged, students, white-collar employees, the rich and people living in cities.

1.2 Culture

Human cultures create artifacts that remain for future generations, like buildings, monuments or particular landscape features. The concept of cultural landscape was probably first used by the German geographer F. Ratzel in 1895 to refer to landscapes formed and influenced by human activity.[3] It may help to protect such artifacts if they are assigned a value; or alternatively, exempted from modification or destruction by categorical exclusion rules, for instance by being included on the UN list of cultural heritage objects.[4]

1.3 Value and Externalities of Recreation and Tourism

The tourist industry is a substantial value creator in many countries. Such values can be estimated with methods described in Chapter 8. To paint a picture of the magnitudes involved, an estimate of the global value of recreation and culture is shown in Table 10-1. Based on forecasts of domestic and international travel, the World Travel and Tourism Council, WTTC, has estimated that global tourism generated US$3.4 trillion ($10^{12}$) per annum in gross output for the year 2000.[5]

Table 10-1. Estimated global values of recreation and cultural experience.[6]

Ecosystem service	Examples	Values 10^9US$ per year
Recreation	Eco-tourism, sport fishing, and outdoor recreational activities	815
Cultural experience	Aesthetics, artistic, educational, spiritual, and or scientific values	3015

Ecotourism is a growing industry, with an estimated value of US$500 billion ($5.10^{9,}$ 1997).[7] The recreation and tourist industry is a viable option for many developing countries, with activities like camera safaris and whale watching becoming increasingly more popular.

Adverse effects: Generally, service industries like hostelling, trucking and courier services produce per monetary unit about a half to a third of the emissions, wastes and energy consumption generated by the manufacturing industry.[8] There are other costs that are not easily measured, like damage to the environment and historic monuments in areas frequently visited. Other adverse effects are interference with local living conditions, and commercialization of indigenous culture, but few cost estimates are available.

2. QUALITY STANDARDS

Quality standards for goods or services that enhance life quality are not easy to identify or agree on. However, for some goods or services – such as tourism – the market imposes standards. In other cases, such as cultural experience, authorities set standards explicitly or implicitly, for example by subsidizing theatre tickets.

2.1 Recreation and Tourism

There are few administratively established quality standards for recreation and tourism. Instead, market forces direct people to places that offer the greatest benefit for the money, and commercial standards are developed that can be inferred from tourist and travel guides that categorize tourist attractions into quality classes identified by a certain number of stars – the more stars the better. If you prefer ethnic tourism, you would probably prefer the *Rough guide* that emphasizes ethnic particularities to the *Lonely Planet*, which tells you about cultural artifacts.

There are guideline standards for recreation in water, however. The Clean Water Act in the USA aims to make all navigable water bodies fishable and swimmable, and water quality measurements decide when beaches have to close because the waters become dangerous to public health.[9]

2.2 Cultural Heritage

The second half of the 19th century witnessed a major change in the focus on cultural heritage objects. Before that time only monumental buildings were regarded as worthy of protection, but now buildings and remains from the life of common people were considered worthy of protection as well. In addition, whereas the focus previously was on single buildings, whole areas with buildings and surroundings are now often protected as a unit.

UNESCO defined cultural and natural heritage in 1975.[10] Article 1 concerns cultural heritage:

- Monuments; architectural works, works of monumental sculpture and paintings, elements of structures of an archeological nature, inscriptions, cave dwellings and combinations of features that are of outstanding universal value from the point of view of history, art or science.
- Groups of buildings: groups of separate or connected buildings which, because of their architecture, their homogeneity or their place in the landscape are of outstanding universal value from the point of view of history, art or science.
- Sites: works of man or the combined works of nature and man, and areas including archeological sites, which are of outstanding universal value from the historical, aesthetic, ethnological or anthropological points of view.

Article 2 deals with natural heritages:

- Natural features consisting of physical and biological formations or groups of such formations, which are of outstanding universal value from the aesthetic or scientific point of view.
- Geological and physiographical formations and precisely delineated areas which constitute habitats of threatened species of animals and plants of outstanding universal value from the point of view of science or conservation.
- Natural sites precisely delineated and of outstanding universal value from the point of view of science, conservation or natural beauty.

The criteria for being listed in the National Register in the USA are:

- Being associated with events that have made a significant contribution to broad patterns of history.
- Or being associated with the lives of persons of significance in our past.
- Or embody distinctive characteristics of type, period or method of construction.

- Or represent the work of a master.
- Or possesses artistic values, or represents a significant and distinguishable entity whose component may lack individual distinction.
- Or may be likely to yield important prehistory or history information.

3. LIFE QUALITY

Several attempts have been made since the mid-1950s at constructing indices that supplement GNP in expressing quality of life and also address environmental sustainability.[11] An index for quality of life, QL, should arguably have the following components:

- Income or purchasing power.
- Governmental non-monetary support.
- Leisure activities.
- Working environment.
- Cultural identity.
- Environmental consciousness.
- Pollution.
- Crime.
- Exposure to accidents.

Table 10-2. United Nation's Human Development Indices[12] The GEM includes female empowerment issues like women in government, and the year of the first female elected to parliament (not listed below).

Index	Longevity	Knowledge	Decent standard of living
HDI	Life expectancy at birth	Adult literacy rate Combined enrolment ratio	Adjusted per capita income in purchasing power parity (PPP) US$
GDI	Female and male life expectancy at birth	Female and male adult literacy rate Combined enrolment ratio	Female and male per capita income in PPP US$ based on female and male income shares
HPI-1 for developing countries	Probability at birth of not surviving to the age of 40	Adult literacy rate	Deprivation in economic provision: 1. % people without access to safe water, 2. % people without access to health services, 3.% children under 5 that are underweight.
HPI-2 for industrialized countries	Probability at birth of not surviving to the age of 60	Adult functional literacy	% people living below the income line (50% of median disposable household income). Long term (>12 month) unemployment rate.

Since not all these aspects are equally important for all citizens, sub-indices are created that capture themes that relate to people with different life styles or values.

We show three examples, the UN Human Development Index, the Swedish "Level of Living Index", and the British Headline Indicators.

3.1 United Nation Human Development Index

The United Nations human development index consists of four companion indices that provide summary information about human development in a country, see Table 10-2.

While HDI expresses human development in general, GDI is a composite measure reflecting gender inequalities in human development. A gender empowerment measure, GEM, measures gender inequality in terms of economical and political opportunities. The Human poverty index, HPI, measures deprivation. It has one version for developing countries and another version for industrialized countries.

3.2 Level of Living

A Swedish index for the "Level of living" is shown in Table 10-3.

Table 10-3. Components of the Level of Living Index.[13] The numbers in parentheses in the first column refer to a comparable listing in the British Headline Indicators shown below.

Components	Indicators
1. (5) Health and access to care	Ability to walk 100 meters, various symptoms of illness, contact with doctors and nurses
2. Employment and working condition	Unemployment experiences, physical demands of work, possibilities to leave the place of work during working hours.
3. (1) Economic resources	Income and wealth, ability to cover unforeseen expenses up to US$ 1000 within a week
4. (4) Education, skills	Years of education, level of education reached
5. Family and social integration	Marital status, contact with friends and relatives
6. Housing	Number of persons per room, amenities
7. Security of life and property	Exposure to violence and thefts
8. Recreation and culture	Leisure time pursuits, vacation trips
9. Political resources	Voting in elections, memberships in unions and political parties, ability to file complaints

3.3 The British Headline Indicators

The British government has defined headline indicators to help society move in the right direction or to prevent a reversal if a target has been reached. The system is shown in Table 10-4 and consists of 14 indicators that include economic conditions (1, 2, 3), quality of life indices (5, 6, 7), direct pollution levels (9, 11, 12), and sustainability indices (8, 13 and 14).

Table 10-4. British Headline Indicators.

Indicator	Value about 1980	Value 1996	Highest value reached
1. Total economic output, GDP yr^{-1}	120	180	180
2. Investment in public, business and private assets, % of GDP	18	18	23
3. Working men/women, %	78/56	80/70	84/65
4. Qualifications at age 19, level 2, %	45	70	70
5. Expected years of healthy life, men/women	71/76	74/78	74/78
6. Homes judged unfit to live in, %	8	6	8
7. Burglary/violence per 100,000 and year	500/250	800/600	1200/600
8. Emission of greenhouse gases, 10^6 ton yr^{-1}		180	180
9. Days when air pollution is moderate or high urban/rural, %		40/40	60/50
10. Road traffic, 10^9 vehicle miles	150	280	280
11. Rivers of good and fair quality, %	94	94	94
12. Population of wild birds 1970 = 100	115	110	118
13. New homes built on previously developed land	50	50	50
14. Waste arising and management, 10^6 ton yr^{-1}		145	

For some indicators there are clear targets that are either possible to achieve as high level (1, 5) or that should reach zero or 100% (6, 7, 8, 9, 11, 13, 14). For some indicators the targets are fuzzy (3, 4, 10). The indicator for road traffic contributes to the quality of life as a car makes life easier for most people, but it also contributes to air and noise pollution, and for non-electric cars, to exhaustion of natural resources.[14]

The British index includes indicators that cannot be related directly to life quality and where there is little consensus on targets. There are no rules for combining sub-indices into one overall index that expresses the "quality of life". The crime sub-index can be used as an example of the difficulty at defining pain from crime; there is no disutility meter available that would describe, say three "thefts from a car" as equal to one "home burglary". None of the indices above include criteria that measure cultural, national, regional or ethnic identity although these are valued by people and can be sources for conflicts.

3.4 Nuisances

Life quality also depends upon lesser things that affect us, called nuisances. Among these are odor and noise, the latter probably the most important factor that reduces feelings of well-being among people. A common problem in project planning is to take into account that people often are subject to different levels of nuisance from the project. Noise from road traffic and windmills are typical examples.

Assume that there are two alternative trajectories for a new road. One decision criterion is to minimize the noise impact on the population, and the question is how to calculate the overall noise load. For that purpose a Sound Weighted Population, *SWP*, index has been defined.

$$SWP = \sum N_i \times W_i \qquad (10\text{-}1)$$

N_i is the number of people exposed to the i-th decibel interval, and W_i the weight attributed to the i-th decibel level. *SWP* is by convention normalized to a sound level of 73 dB, which therefore has weight 1.0. A level of 73 dB is higher than in a business office, but less than in a street with traffic. A set of reasonable weights is shown in Table 10-5.

Table 10-5. Sound Weighting noise levels. Db is day-night average noise level, *W* is weight.[15]

Db	35	40	45	50	55	60	65	70	75	80	85	90
W	.006	.013	.029	.061	.124	.235	.412	.664	1	1.43	1.93	2.65

4. DOSE-RESPONSE MODELS

A dose-response model is a functional relationship between a cause and an effect. The basic idea with applying the dose-response concept to quality of life, is that quality of life generally increases as a function of the availability of environmental resources. Closer inspection reveals, however, that the relationships can be quite complicated. We start with an examination of how monetary income increases quality of life – and there is no proportionality.

4.1 Income

Income in a country is measured in monetary units, but the satisfaction derived from the income depends on other factors as well, such as family size, climate, income history, and expectations about future income. Studies show that poor citizens believe that almost everybody is fairly rich, while rich citizens believe that almost everyone is poor. People shift norms with income, that is, their actual incomes "anchor" their preferences. The most pessimistic case is that a 10% increase in income causes a 10% increase in the expected standard and no increase in satisfaction. Observations show that the income effect on satisfaction is about 60%. That is, if the income is doubled, the satisfaction increases by 60%.[16] Studies of family size shows that a 10% increase in household size requires 2.5% higher income to reach the same satisfaction level. In some countries, like Pakistan, higher household size in terms of more children is an asset.

4.2 Tourism's Chain Impact

Money left by tourists provides income to the resort that produces a chain of expenditure-income-expenditures. The details depend on how much of the money left is spent on goods and services by the initial receivers, and how much is saved. The more that is spent locally, the more it stimulates the local economy. Economists use the term *marginal propensity to consume*, *MPC*, which is the fraction of additional available income that is spent on consumption. The opposite is the marginal propensity to save, *MPS*, which is assumed not to stimulate the local economy. To calculate the effect of tourist expenditure on the economy, one sometimes uses a multiplier, *M*:

$$M = 1/(1 - MPC) \tag{10-2}$$

With $MPC = 0.5$, US$ 1000 spent becomes a US$ 2000 benefit to the local community. The value of multipliers depends upon many factors, of which the most important is the size of the leakages. This is smaller for large geographic areas, and for areas at a distance from neighboring economies, like Australia. For services in the tourist sector, a multiplier of 1.37 has been proposed, and for manufacturing and agriculture values of 1.91 and 1.79, respectively.[17]

4.3 Climate

A climate index *C* that includes average annual temperature *TEMP*, humidity *HUM*, and precipitation *PREC*, has been estimated:[18]

$$-\ln(C) = 0.15 \ln TEMP + 0.40 \ln HUM + 0.10 \ln PREC \tag{10-3}$$

With Paris set at 1.00, Berlin scores 1.11, Copenhagen 1.10, London 1.08, Rome 0.95, Nice 0.91 and the Channel islands 0.87. The numbers mean that one needs 11% higher income in Berlin than in Paris to reach the same satisfaction.

The US Agriculture department has performed a similar study focusing on counties in the US.[19] They identified six measures of living quality with emphasis on climate: January temperature, January sunshine, temperature difference between July and January (less is better), July humidity, water area, and topographic variation. They then created a *Natural Amenities Index* (*NAI*) based on these measures and rated each county in the contiguous 48 states, producing a *NAI*-map of the US which shows that the western states and Florida are on top, while the upper mid-west – from Ohio through the Dakotas – shows a relative lack of natural amenities. Interestingly, it also turned out that this pattern correlates strongly with population change. Counties with very high *NAI* typically doubled in population from 1970 to 1996, while those scoring low barely held even. In a dose-response framework, *NAI* may be seen as the dose, and population movements as the response.[20]

4.4 Cultural Identity

We are not aware of any studies of the relationship between factors that increase cultural identity and quality of life. However, artifacts that signal cultural identity appear to be often valued out of proportion to their potential market value if culture were not part of it. Since a nation ought to have an opera house and a building for a symphony orchestra that signal cultural identity, such buildings, and the performances held in them are often heavily subsidized. Parliament buildings – potentially forceful cultural signals – are usually more costly than their functional equivalent.

5. APPLICATION: CULTURAL HERITAGE MONUMENTS

Heritage monuments is a finite resource

Over the years, cultural, economical and social factors have led to the development of distinct landscapes and building complexes all over the world. One such complex is the Al^{c}kas settlement in the highlands of southwestern Saudi Arabia. To assess the value of Al^{c}kas, Saleh (2000) described the settlement in terms of its social structure, and the visual characteristics of the cultured landscape that consist of natural, agricultural, and built landscape elements.

Figure 10-2. 1 Alckas settlement, Saudi Arabia. A new water tower adds a new physical feature to the Alckas skyline. Photo: (Saleh 2000).

5.1 Decision Analysis: The Case of the Al^{c}kas Settlement, Saudi Arabia, and the Mesa Verde Dwellings, Colorado, USA.

The decision problem for Al^{c}kas is to find a way to supply water to the settlement, and one question is whether the planned water tower enhances the visual experience as illustrated in Figure 10-2 – or if not, is the benefit greater than the disadvantage?

For the ancient dwellings of Mesa Verde, USA, the management problem is different. Here the problem is that tourists visiting the site cause wear on the monument, and slowly change its character. How then should cultural artifacts be managed, both those that belong to the world heritage, and those that contribute to local cultural identity?

Decision makers: The ultimate decision makers for Al^{c}kas settlement and the Mesa Verde dwellings are the central governments of Saudi Arabia and the USA, respectively.

Stakeholders: Important *stakeholders* are the people of the Al^{c}kas settlement, and in both cases local and central government offices responsible for the preservation of cultural artifacts. There are also citizen groups that want to protect cultural artifacts, and among them both users and non-users. There are also international organizations that take care of human cultural heritage objects, such as UNESCO (this chapter, section 3)

Identification of decision makers and stakeholders are discussed in Chapter 3, section 2.3.

Decision objectives: The *goal hierarchies* are similar. The main goal is to preserve the cultural sites at the least social and economic costs. For the settlement, the major conflict is between supplying water to the settlement and preservation of the skyline. In Chapter 11, section 2, view elements, including "misfits" are discussed. In the Mesa Verde case the major conflict is between visible impacts on the rock foundation of the dwelling set by the tourists climbing the rock, and the enhanced ability of people to identify with their past by viewing the site at close distance and climbing around the dwellings. Economy is a central criterion in the Al^{c}kas case, since water probably could be supplied by other means than by a water tower.

Decision alternatives: Relevant alternatives for the Al^{c}kas settlement are not to build the water tower, or to build it at the proposed site. In the Mesa Verde case the alternatives were to let people freely climb the buildings, or to restrict access to the buildings. How to structure objectives and goals is discussed in Chapter 3, section 3.

Consequence analysis: The construction in Al^{c}kas could be made temporary, whereas the effect of wear for Mesa Verde is permanent. The benefit of the water tower in Al^{c}kas may be large for the people living there, and contribute to preservation of the settlement, because it may make the settlement into a mixed-use site, preserved, used, and probably taken care of. The preservation value of the sites may be examined by

- Studying written and oral legends to determine if the sites are referred to in that material, and thereby contributing to cultural identity (this chapter, section 3).
- Applying willingness to pay, WTP, studies like the travel cost method or the contingent evaluation method described in Chapter 8.
- Trying to get the sites included among the UNESCO world heritage objects. If the latter effort is successful, the value of the heritage site may be beyond economic terms (Al^{c}kas is not, Mesa Verde is, Fig. 10-1 and Fig. 10-2).

Utility analysis: Some of the primary utility functions will probably be judged as risk-averse, and for some criteria one will probably require categorical exclusions, see this Chapter, section 3 on Cultural Heritage standards. Primary utility functions are discussed in Chapter 2, section 2. The analysis should result in a consequence and a utility table like Tables 3-5 and 3-6, and primary utility graphs, like those in Figure 3-3.

Preferences: Since the Mesa Verde dwellings are included in the World Heritage list, their value extends to mankind and should not be subject to

WTP studies, and the preferences of the local community should probably not be given high weight. In the Al^{c}kas case, it may be adequate to elicit WTP-data by consulting the local community. Preferences may be solicited by one of the methods described in Chapter 7, section 2. The results can be used as a background for determining importance weights as in Table 3-8.

5.2 How Did It Go?

- *The Al^{c}kas settlement*: In Figure 10-2 we see that the tower actually was built.
- *The Mesa Verde dwellings*: People have now only restricted access to the dwellings.

---

[1] http://www.undp.org/hdr2000/english/HDR2000.html
[2] Mercer (1994) p. 136
[3] Saleh (2000) p. 62
[4] UNESCO(1972), see also :http://www.unesco.org/whc/nwhc/pages/doc/dc_f2.htm
[5] McIntosh et al. (1995) p. 472
[6] Constanza et al. (1997)
[7] Pimentel et al. (1997)
[8] Rosenblum et al. (2000)
[9] Cropper and Oats (1992)
[10] Search for UNESCO cultural heritage
[11] Erikson (1993), UN (2000) UK (2000)
[12] UN (2000)
[13] Ericson (1993) p. 68
[14] MacLean and Lave (1998)
[15] After von Gierke (1977)
[16] van Praag (1993) p. 373
[17] James (1994), McIntosh et al. (1995) p. 326
[18] van Praag (1993) p. 376
[19] McGranahan (1999)
[20] Doyle (2005)

Chapter 11

AESTHETICS

The aesthetical value of landscape and cityscape

Aesthetics is an important concern when projects have impact on landscape or cityscape. This chapter adresses some of the questions that arise when one endavours to take aestethic values into account. In particular:

- Why do we find some environments aesthetically pleasing and some not?
- How can we describe aesthetic elements of landscapes and cityscapes?
- What are the determinants of aesthetic preference?
- How does aesthetic preference vary with time and experience?
- Is the concept of visual quality standards viable?
- How can aesthetic preference be predicted?

"What is the use of aesthetics if they can neither teach how to produce beauty nor how to appreciate it in good taste? It exists because it behooves rational human beings to provide reasons for their actions and assessments. Even if aesthetics are not the mathematics of beauty, they are the proof of the calculation." Franz Grillparzer (1791–1872), Austrian author. Notebooks and Diaries (1820)

1. INTRODUCTION

This chapter deals with the aesthetics of landscapes and cityscapes. Aesthetic considerations are often part of environmental impact studies since most projects have effect on visual quality. But visual quality is notoriously hard to describe – it is in the eye of the beholder. Also, aesthetic preference changes across geographic areas and historical eras, as well as with age and gender. Like art, public awareness of visual landscape quality seems to be cyclic – waxing in one period, then waning in the next.

In a personal account of environmental aesthetics, Porteus (1996)[1] coined his own terms for ugly landscapes or cityscapes:

- "America the ugly": A generic term for cityscape with myriads of street signs, cars and electric wires crossing the sky view.
- "Blandscape": A uniform suburban landscape with no trees or chimneys, but with pretentious door-cases and windows.
- "Mindscapes" from Japan.
- "Cityness" the image of skyscrapers.
- "Identykit": A set of modern building towers.
- "Anywhere": A cityscape without character.

In spite of differences among cultures, some vistas appear to be admired almost universally. Most people express admiration for the Great Wall in China, the Fjords of Norway, and the colors of flowers on Mount Rainier.

Aesthetics is probably the single attribute of people's quality perception that is easiest to confound with other factors such as habitat quality, ability to support emotional experience, health, environmental concern, and largeness and vastness. Landscape painters have been observed to prefer sunset to sunrise (35 against 11), but this preference may be confounded with the artists' tendency to get up late.[2] Aesthetic values are manifest not only in landscapes, but in poetry, visual arts, gardens and buildings as well. Many authors have tried to examine the relationships between landscapes, poetry and painting to find similarities in expressions that are characteristic for certain periods or certain cultures.[3]

A central question for a decision maker in a project is: How do I score my alternatives according to aesthetic value? Aesthetic quality measures are not easy to define, and there are probably no generic quality measures that are valid in all cultures. Even so, we will present three so-called *actuarial methods*, where the landscape is broken down in elements and each element assigned a quality score. The idea is that the weighted sum of scores provides an overall aesthetic score of the landscape. Actuarial methods have the advantage of reducing the possibility of bias, but the danger is that synthesizing properties of the whole from its parts may appear like a bad caricature of holistic processes.[4] It is possible to control the validity, however, by checking consistency with expert ratings or popular vote.[5]

2. BACKGROUND INFORMATION

2.1 Values and Externalities of Aesthetics

Aesthetics has value when it gives personal or social satisfaction. A walk in a natural environment can give physical and mental restoration.[6] When we erect buildings or make roads or other objects that are visible parts of the landscape or cityscape, we usually try to improve how it looks. The cost is an indication of implied willingness-to-pay for aesthetics, although it sometimes may be difficult to distinguish aesthetics from technical or functional features. In Norway, the budget for artistic decorations of public buildings is typically 0.5–2.0% of total cost. Altogether, the public sector in Norway uses some €3 million per year on buildings and a comparable sum on support for enhancement of built land.

Other activities to improve aesthetics are enhancement of dams, erection of monuments, use of architects in design of buildings and built land, buying art from art-galleries, visiting exhibitions, and running and attending art schools. A crude estimate gives €5 to €10 per capita and year for aesthetic improvements in a developed country, or about 0.02% of GNP per capita. The compensation given for loss of aesthetic experiences caused by reduced flow in a major waterfall in Norway was estimated at €0.017 per kWh.[7]

Ugly or disturbing monuments or buildings may cause negative aesthetic experience, but we have not seen studies of the willingness to pay to avoid such experiences, although there must certainly be a positive WTP to avoid tagging.

2.2 Landscape and Cityscape

The discipline of aesthetics deals with the questions of what sort of landscapes people actually prefer and why. The managerial objective is to preserve landscape quality as well as possible in the face of project development. To understand aesthetics, it is useful to start with history, and then gardens are important. Old gardens give us an understanding of what was considered beautiful when they were originally conceived, and gardens are metaphors of what the world's landscapes will look like in the future.[8]

Landscapes and cityscapes have two major aesthetic aspects, *scenery* and *corridor*. **Scenery** is landscape viewed from a distance. It is a two-dimensional image, which roadmaps of "scenic routes" tell you to take pictures of. We have other senses than vision, however. If you move through a scene, or smell the scent of a nearby shoreline, or feel the discomfort on a steep cliff, you are said to have the aesthetic experience of a scenic *corridor*, which is a three-dimensional experience. Landscapes are best seen while the eye moves, spending more time on complex and "mystery" features than on easily interpretable ones. Man-made landmarks are sometimes erected expressly to contrast with, or enhance, the natural landscape – painters now and then add old ruins to landscape paintings to improve their beauty.[9]

2.3 Aesthetic Theory of Landscape and Cityscape

Why is a landscape considered beautiful? Here are four aesthetic theories that seek to provide answers and predict whether a particular landscape will be perceived as a positive or negative contribution to the human environment. The theories are not necessarily in conflict, all may contribute to a full conceptual understanding of human responses to landscape features. They differ in terms of what response components they focus on. Some deal with the evolutionary basis for responses, others deal with responses that are molded by individual cultural experience. Nobody doubts that culture and experience influence aesthetic judgments, but whether emotional responses are strictly determined by experience, or experience primarily modifies genetically based responses is among the questions current research attempts to find out.

2.3.1 Habitat Theory

According to the habitat theory, man's landscape preferences have been molded by biological needs. Aesthetic preferences are rather based on a "live-in" than a "look-at" experience. Even if we control the environment so that it is safe and can sustain life, basic or genetically determined responses influence preferences for habitat-like landscapes.[10] The habitat theory can be discussed in the context of predator (hunter) and prey theory. The hunter needs the ability to see the prey and therefore needs an open landscape, a "prospect". The prey seeks to hide, and thus needs a refuge. The same person may be both a hunter and a prey, and thus prefer landscapes that contain elements of both prospects and refuge.

2.3.2 Control versus Mystery

This theory is also called the "tranquility" hypothesis.[11] It holds that some people appreciate tranquil landscapes with few surprises and ample time for reflection, in contrast to others who prefer landscapes that promise new discoveries.[12] Such landscapes have attributes like complexity and mystery. The opposite is coherence.

2.3.3 Compatibility or "Fittingness"

This theory holds that people perceive landscapes as beautiful if landscape use and landscape components "fit together". This approach is sometimes called the fine arts/design approach.[13] The compatibility theory postulates why components "fit together". Turner (1775–1851) considered

Scotland a more picturesque country than Wales, as the lines of mountains are finer, and its rocks are more massive.[14]

2.3.4 Contemporary Fine Arts

The contemporary fine arts theory holds that a landscape is perceived as an object of art, like a painting, a sculpture, a performance, or a music video. This means that colors do not have to be natural, the sequence in which the eyes search the landscape has no relation to physical experiences, and outsized objects are not bothersome. This theory differs from the compatibility theory in that there is no requirement for compatibility *per se*, but only that the landscape has attributes in common with contemporary art expressions.

2.4 Landscapes and Cityscape Types

There are two fundamental visual quality dimensions: *openness* and *complexity*. The landscape forms *prospect* and *refuge* represent maximum and minimum openness. *Coherent* landscapes are simple, while *mystic* landscapes are complex. This gives rise to four archetypical scenic forms, which are illustrated in Figures 11-1 and 11-2. The pictures show real scenes, therefore the forms are not pure, but also contain elements of other types. Figures 11-3 and 11-4 show corresponding cityscapes.

Figure 11-1. Prospect landscape types. Left: Coherent wine fields north of Lyon, France. Right: Complex and mystic winter forest.

Figure 11-2. Refuge landscape types. Left: Coherent. Right: Complex and mystic.

Figure 11-3. Prospect cityscape types. Left: Coherent. Right: Complex and mystic.

Figure 11-4. Refuge cityscape types. Left: Coherent. Right: Complex and mystic.

2.5 Landscape Descriptions

Landscape description requires identification of elements of the landscape and patterns that connect them. The **fractal theory** of landscape forms suggests that landscapes can be described as sequences of line-segments connected according to rules from dynamic chaos theory. Such rules can be very simple, but still produce results that appear very complex. Figure 11-5 shows an example of a leaf. If you look at the details, you will find the patterns repeating themselves. The results can become both unexpected and beautiful.

Figure 11-5. Fractals. Self-similarity in Nature. Patterns on one scale (the leaf right) consists of an assembly of smaller leaflets with similar patterns (left).[15]

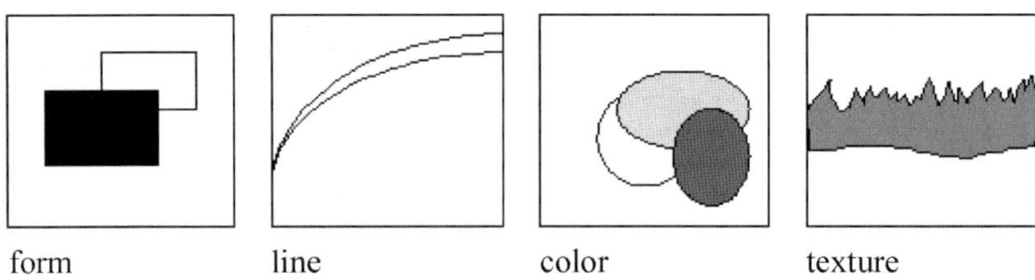

Figure 11-6. Basic elements.

The elements of landscape forms can be sorted in four classes:

- Basic elements.
- View elements.
- Compound elements.
- Vegetation and use-elements.

2.5.1 Basic Elements

Line, *color*, *texture* and *form* are four categories of basic elements that have been used to describe landscapes and cityscapes as predictor variables for how people will assess their aesthetic qualities. See Figure 11-6.

- *Form*: the perceived aggregate of elements, a consciousness of the distinction between the whole and its parts, resembling the concept of assembly rules.
- *Line*: very thin threadlike mark.
- *Color*: the color of the view.
- *Texture*: the arrangement of the particles or small parts of any material.

Associated with these elements are rules for assembling elements that would allow some combinations, but not others, to form a particular type of landscape or cityscape. For example, since few vertical lines are found in most natural landscapes, a high occurrence of vertical lines would indicate a certain type of landscape. In theory, it could be possible to quote elements and a set of combination rules which would elicit ideas of what the scene would look like, much like the term "renaissance garden" would give a knowledgeable person a feeling of what the garden would look like.[16]

2.5.2 View Elements

View elements influence landscape preferences strongly, positive or negatively Figure 11-7.

- *Relief*: steepness of the landscape.
- *Symmetry*: arrangement of objects in the view.

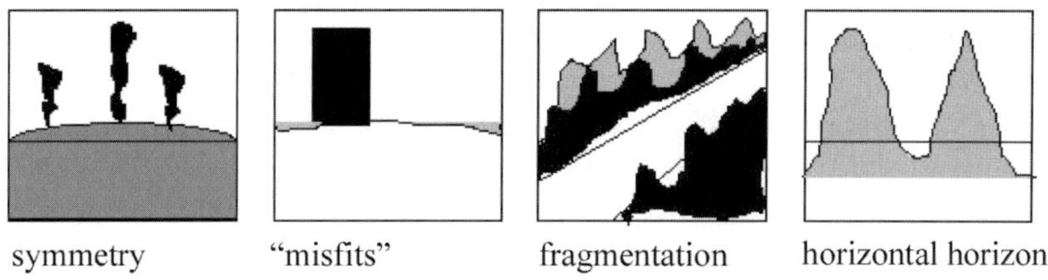

Figure 11-7. View elements.

- *Misfits*: visual detractions, like a disproportionately tall building, or electric transmission towers.
- *Fragmentation*: separation of elements which are usually perceived as belonging together, like the separation of a forest by a highway.
- *Horizontal line-breakage*: elements that break floating horizontal lines in the landscape.

"Misfits" has negative associations, but misfits may also contribute positively to a landscape or cityscape experience, like the State of Liberty in New York harbor, or the mast tree in Figure 11-8, which is left standing for ecological reasons.

2.5.3 Compound Elements

- *Complexity*: the number and arrangement of the forms of the landscape. It is sometimes also called "ambiguity".
- *Mystery*: elements suggesting unknown and surprising elements that may hide in the landscape.

Figure 11-8. Mast tree left in a rolling landscape.

- *Coherence*: The ability to rapidly organize the scenic elements into a coherent and well-understood whole, in contrast to the mystery element
- *Identifiability*: also called legibility.
- *Spaciousness*.

2.5.4 Vegetation and Use-Elements

Vegetation and use-elements are used in "actuarial" studies of landscape aesthetics. These elements describe vegetation, topography and man-made constructions. Some examples are shown in Table 11-1.

Table 11-1. Examples of vegetation and use elements. Landscape elements used to describe whole landscapes. Characteristics starting with bold letters were significantly (positively or negatively) included in regression equations for aesthetically pleasing English landscapes. After Briggs and France (1981).

Vegetation	Topography	Built up land
Deciduous woodland(%)	**S**lope(m km^{-1})	**Wa**steland (%)
Coniferous woodland (%)	**L**akes and Reservoirs (%)	**Res**idential land (%)
Permanent pastures (%)	Stream length (km.km^{-2})	**I**ndustrial land (%)
Cultivated land (%)		**Ra**ilways (%)
Parkland (%)		**Ro**adways (%)
Hedgerows (km.km^{-2})		

2.6 Aesthetic Preferences Across Landscape Types

Several studies of landscape preferences have been undertaken in the UK and the USA. They are based on presentation of landscape scenes to a representative selection of "judges" and thereafter analyzed according to vegetation and use-elements in the scenes. Uplands with lakes and waterfalls, headlands and bay coastline are found most attractive while rolling woodlands, sandy beaches backed by cliffs come in second in popularity. Least attractive are city centers, slagheaps, and polluted areas.

Table 11-2. Correlation between landscape components and aesthetic preferences. The number of plusses or minuses indicates the number of studies with this finding. Summary of 4 studies reported by Briggs and Courtney (1994).

Landscape type	Positive/negative correlations with aesthetic preference.
Relief	+++
Deciduous woodland	++
Unimproved pasture, bare rocks, coastline, streams	+
Residential land	-
Moorland, wasteland, railways	–
Industrial land	—

Topography, water, evidence of Man's influence, seasonality, foreground vegetation and cloud formation are factors that influence visual assessments.[17]

One American study of preferences for tree forms, shows that broken branches, deformed tree trunks and highly asymmetric canopies were least attractive.[18] The finding of a selection of studies is summarized in Table 11-2. The studies reported were restricted to England and the USA. Corresponding studies in other countries and cultures may give different results.

2.7 Aesthetic Preferences Across Social Characteristics

Aesthetic quality preferences tend to vary with:

- Gender and age.
- Familiarity with landscape.
- Occupation and interests.
- Culture traits.
- Time.

Generally, the overall aesthetic preferences for most landscapes or cityscapes – whether they are natural or built – are similar across demographic characteristics. Avant-garde architecture is an exception.[19]

2.7.1 Gender and Age

Landscape preferences seem to be established at the age of 18, but familiarity may be a confounding factor.[20] Age-related changes in preferences may be related to habitat, since people of different ages have been molded by different experiences throughout their lifetime. One study showed that almost half of women's paintings had strong refuge symbolism compared with only 25% of men's paintings (*chi square* = 6.89, $p = 0.03$). Further, the horizon covered at least half the width in 58% of the landscapes painted by men, whereas 75% of women's paintings had no horizon or only a peephole.[21] This may reflect differences in sociological adaptation or inherent traits between the genders.

Another study found that women tended to adopt a systematic exploration strategy for landscapes more often than men.[22]

2.7.2 Familiarity with Landscape

People are familiar with the landscape forms where they live – or within their *home range* to use an ecological metaphor. The familiar landscape has little mystery in it, and people may express preferences for non-familiar landscapes irrespective of their home range landscape forms. The opposite view may also be argued, but studies of single cultures indicate that landscape familiarity has little influence on landscape preferences. For example, respondents expressed similar preferences for a mixed set of mountain slides independently of whether they were familiar with the landscapes of Virginia or Utah.[23]

2.7.3 Culture Traits

Porteus (1996) claims that there are substantial differences between English and American (US) tastes.[24] The former is characterized as "elegantly tamed" whereas the latter is preference for the vast, limitless, and huge. On the other hand, a meta-analysis of demographic effects on environmental aesthetics showed that Australian and Italian students agreed on landscape preferences ($r = 0.92$).[25] Also, a quote from the Chinese writer Shih-Táo (c. 1642-1718) suggests that western and Chinese aesthetics have similarities: "One must avoid laborious details, flatness, of falling into a set of patterns, being woolly, lacking coherence or going against the inner nature of things."[26]

2.7.4 Occupation and Interest

Particular occupations or interests often cause prejudice and bias. Planners often claim that aesthetic training makes them better judges of aesthetic qualities than the general public. Farmers who depend on agriculture for a living prefer rolling hills and park-like landscapes. Environmentalists prefer landscapes in their present form, whatever that may be. Designers have a more positive preference for avant-garde architecture than lay people.[27]

3. SPACE, TIME AND EMOTION

A particularly interesting theory is that aesthetic preferences change with time and experience. This is illustrated in Figure 11-9, which shows how a person may walk through a space with shifting complexity and coherence. Possible paths are shown as zigzag lines where people move through land- and cityscapes with time. The x-axis shows emotional preference, ranging from need for coherence and legibility to need for complexity and mystery according to the "control versus mystery" theory (section 2.3). The y-axis represents the degree of physical complexity, which increases from the midline and up as rural complexity increases, and from the midline and down as urban complexity increases. A flat or rolling rural landscape is close to the midline. A flat suburban countryside, or a well-regulated modern city, would also score close to the midline, but on the downside. A rugged wilderness area would score on the top of the upper half-plane.

There are two zigzag paths, showing how two imaginary groups of people move with time according to their preference development. The theory is that people do not visit each position in the graph in a random fashion. Several testable hypotheses can be generated from the graph. For example:

- The residence time of an average person is longer in the middle of the graph than in the circumferences.
- If a person has stayed at position A2 for a while, she will change preference and move in direction D1 with a probability of more than 50%.

Staats and coworkers in the Netherlands have in fact made somewhat similar experiments. They had students make a simulated one hour walk in natural and urban environments by showing them a series of slides. The students experienced natural environments as more attractive and pleasant than the urban ones, and being fatigued before the walk enhanced the strength of this feeling.[28]

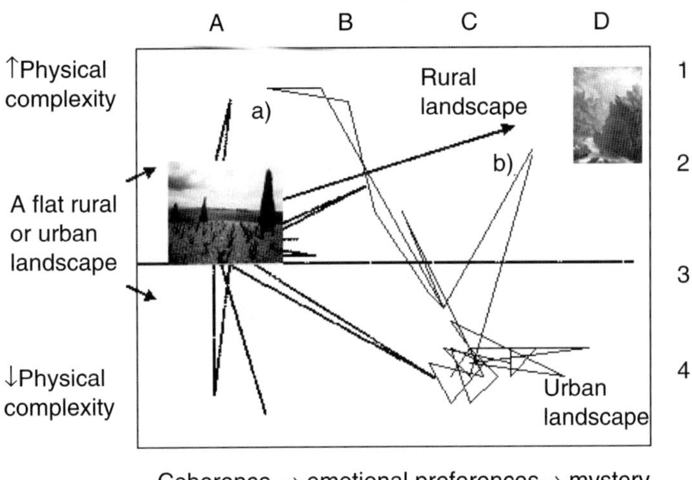

Figure 11-9. Preference for coherence and complexity change with experience as people steer their course through shifting landscapes. Trajectory a) shows the path of a hypothetical farmer from a rural community, trajectory b) the path of a city dweller.

4. VISUAL QUALITY STANDARDS

From a managerial point of view, it would be nice to have standards for visual quality. Then one could determine whether proposals for new projects that change the landscape or cityscape – like skyscrapers, dams and windmills – meet the standards or not, and act accordingly. Alas, this seems to be a forever-elusive dream. There are no lethal thresholds for visual quality, as there are for noxious gas. On the contrary, aesthetics views change with time, experience and culture – sometimes in cyclic patterns where old fashions return and become modern again. Nevertheless, municipalities often employ building codes, and landmark buildings are erected according to architecture competitions where proposals are ranked or rejected. Thus, there are obviously tacit quality standards for aesthetics at work, but they are much less tangible than within other areas of the human environment. Wc shall therefore not try to describe aesthetic quality standards here, but at least we can approach the theme by exploring how statements about visual quality are produced. We have the following chain of events:

1. *Sensory input*: Vision of scene that is steep, flat, tall, voluminous, etc. In some instances even scent and humidity may play a role.
2. *Emotional response*: The vision causes emotional reactions, which can be labeled according to increasing arousal: interest → excitement; enjoyment → joy; surprise → startle; distress → anguish; fear → terror; contempt → disgust; anger → rage.[29]
3. *Verbal response*: Utterances that characterize the scene in terms of aesthetic quality. Three main dimensions have been proposed for verbal expression of aesthetic quality. The dimensions are independent in the sense that the same scene may score well in one dimension and poorly in another:

   - *Beautiful*: Landscapes are nice, pastoral landscapes included[30]
   - *Picturesque*: Landscapes lend themselves to picture representation. The landscapes may include animals (in groups of three), and ruins.[31]
   - *Sublime*: Landscapes are limitless, endless, vast, and huge.[32]

Landscape characterizations sometimes confound features – like steepncss – with the feeling you get from the experience – fear. For example, visual quality has been characterized as "amusing" (feeling), "noble formation" (description) and "benign" (I do not feel threatened).[33]

It is possible to use verbal or numerical quality gradation to indicate degree of *beauty*, *picture-likeness* and *sublimity*. Verbal assessment can be made operational through answers to questions like "How well do you like the scene?" The answers may range from "very well", to "very bad", and "don't know".[34] Although on the curious side, one may also say that a view is "breathtaking", and measure the frequency and duration of non-breathing in people exposed to that view. Some views even enhance morphine production in the body as Leon Battista Alberti (1406-1472) wrote:

> "Those who suffer from fever are offered much relief by the sight of painted fountains, rivers and running brooks, a fact which anyone can put to test; for if by chance he lies in bed .. (then) turns his imagination on limpid waters and fountains, .. or perhaps some water .. sleep will come upon him as the sweetest of slumbers"

There is an extensive literature on how to measure electronically people's emotions and facial expressions (frowning, smiling) when looking at pictures that elicit positive or negative responses.[35] A numerical scale from –10 to +10 has been used to fix the extreme values when assessors were asked to imagine the worst and best landscapes in England.[36]

An important question is who is entitled to – or even qualified to – determine landscape standards? Planners and architects may claim that it is their duty to guide public taste. Others hold that visual attractiveness of landscape is ultimately a product of the aggregated opinions of all concerned.[37] If aesthetic value were decided by a representative selection of a population as is done in political polls, how should one construct an average value, and does this average express any type of meaningful consensus? How does one handle "don't knows" and outliers? One option is that experts suggest contemporary qualities, but that their assessments may be overruled by popular votes at a later stage.

5. NATURAL AND MAN-MADE IMPACTS ON ENVIRONMENTAL AESTHETICS

Nature as well as humans influence the environment, yet the impacts that cause greatest concern are those made by humans.

Natural impacts: Wildfires, volcanic eruptions, and landslides change landscapes, while grazing wild herds change the vegetation. Recurrent fires caused deforestation and prairie type landscapes in North America.

Man-made impacts: Man changes the controlled landscape by erecting buildings, roads, bridges, transmission towers, windmills, and dams for hydropower and irrigation. Where construction codes apply, they usually reflect visual preferences. The most common restrictions are signal restrictions (88%), height restrictions (84%), general compatibility of buildings (81%), off-road parking and loading zones (78%), conservation of existing vegetation (75%), specific landscape elements requirements (70%), and percentage of open space requirements (69%).[38] Human activities require space, which is sometimes called human "footprints". For example, to produce 1 MW of electricity by hydropower, it is necessary to change 7 km^2 land into a water reservoir.[39] The numbers quoted for human footprints in Chapter 13 on harvesting suggest the extent of aesthetic changes of the environment.

6. DOSE-RESPONSE FUNCTIONS

There are three major types of aesthetic dose-response models.[40] The models do not relate directly to any of the theories for aesthetic evaluations described above, although results obtained with the models may strengthen some of the theories and weaken others.

- Intuitive or mental models based on an appreciation of the whole landscape.
- Statistical models based on elements of landscapes or cityscapes. The method has to be calibrated with an intuitive classification. The statistical techniques remove some of the subjective bias when predictions are repeated.
- "Actuarial" techniques that include questionnaires.

The landscapes studied are often restricted to certain landscapes with elements of the landscape within a reasonable range. *To inform about the landscapes* for aesthetic studies one use *in-situ* observations, slides or computer animations. 35-mm color slides have been shown to elicit preference responses comparable to those expressed on site.[41]

Preferences can be elicited in different ways. The three dose-response models described below assume that it is possible to assess the visual quality of the whole from the visual quality of its parts. The *response* of aesthetic dose-response functions is the net visual quality, *NVQ*, of a landscape or a townscape, viewed as a whole. *NVQ* is expressed by descriptors like "beautiful" or "breathtaking" or scaled from –10 to +10. The *dose* is the parts of the landscape, with or without a new project, which may be a building, a road, or a dam. *NVQ* can be formulated as follows:

$$NVQ = f(X_1, X_2 .. X_n) \quad (11\text{-}1)$$

The parts X_1, X_2 .. are landscape characteristics in one form or another.

6.1 Benchmark Models

For aesthetic values there are no fixed benchmarks such as background concentrations or "No observable effect level" for toxicity models, but intuitive evaluation of landscapes as bench-marks against which they compare other landscape models.[42]

6.2 Statistical Landscape Models

Model 1: Briggs and France (1981) used statistics to predict landscape quality from landscape elements. They established a multivariate model by asking 10 persons to assess the landscape quality, *LQ*, in 47 grid squares from –10 (worst aesthetic quality) to +10 (best aesthetic quality). Multiple regression of landscape quality, *LQ*, against the components gave the following model.

$$LQ = 3.14 + 0.12\ Lakes + 0.05\ Deciduous\ forest - 0.23\ Railways - 0.22\ Hedgerows - 0.14\ Roads - 0.06\ Wasteland - 0.06\ Residential - 0.02\ Slope;\ r^2 = 0.891 \qquad (11\text{-}2)$$

All elements were significant at the 5% level. The authors suggest that the parameters "slope" and "hedgerows" gave negative contributions because the landscape studied was flat (South Yorkshire), and because hedgerows acted as a surrogate for other negative variables. Incidentally, the length of hedgerows has been used as a measure of favorable habitats for birds. The 10 observers gave very similar scores, except on landscapes on the urban-rural fringe.

Model 2: *Visual absorption capability* is the capacity of the landscape to screen or accommodate proposed projects and still maintain its inherent character. A study of the visual impact of electricity transmission towers showed that the distance to the tower was the most important factor and the effect of the transmission towers on a landscape faded away at about 1 km viewing distance.[43] These results were corroborated by other results showing that a landscape's attractiveness is very much oriented towards the near and middle view distance (within a radius of 0.5-1 km). At distances between 3 and 5 km colors become indistinguishable. Landscape type did not have a significant impact on the scenic quality, yet there was a tendency for the largest impacts to occur in rural, rolling landscapes, while medium impacts occurred in residential townscapes, and the least impacts occurred in flat landscapes.[44] Although there were no positive distance effects,[45] if the constructions are far enough away, they may still contribute positively if we are to believe the English painter William Gilpin (1762-1843):

"When all these regular forms are softened by distance.. when hedge-row trees begin to unite, and lengthen into streaks along the horizon – when farm houses, and ordinary buildings loose all their vulgarity of shape, and are scattered about – it is inconceivable what richness, and beauty, this mass of deformity, when melted together, adds to a landscape."

6.3 Questionnaire Type Visual Models

Questionnaire models will be explained through an example that shows how to score the capacity of a landscape to screen or accommodate a project and still maintain its character. This is called Visual Absorption Capacity, *VAC*. The model consists of six factors, with *slope* being the most critical since the steeper the slope, the less vulnerable is the landscape to visual impacts. Slope is therefore a multiplier in formulae 11-3. The factors and their scores are explained in Table 11-3:

$$VAC = S \times (E + R + D + C + V) \qquad (11\text{-}3)$$

Table 11-3. Visual absorption capability. After Yeomans (1988).

Factor	Characteristics	Factor score
S; Slope	Steep (+55% slope)	1
	Moderately steep (25%–55%)	2
	Flat (0–25% slope)	3
D; Vegetation diversity	Barren, grass, brush	1
	Conifer, hardwood, cultivated	2
	Diversified (mixed open and woodland)	3
E; Soil stability and erosion potential	High erosion hazard or poor regeneration potential	1
	Moderate erosion hazard or moderate regeneration potential	2
	Low erosion potential or good regeneration potential	3
V, Soil/vegetation contrast	High contrast between exposed soil and adjacent vegetation	1
	Moderate contrast	2
	Low contrast	3
R, Vegetation regeneration potential	Low regeneration potential	1
	Moderate	2
	High	3
C, Soil and rock color contrast	High contrast	1
	Moderate	2
	Low	3

7. APPLICATION: VALUATION OF AESTHETIC QUALITIES

If a region or a landscape is a candidate for inclusion under some scheme of landscape protection, such as being designated as "Area of outstanding Natural Beauty",[46] the decision problem is whether to include the land or not. John Rawls' famous phrase "Disagreement among reasonable people"[47] may apply to many disputes, but is probably particularly relevant when aesthetics is discussed. Here is an outline of how a decision analysis might be structured:

7.1 Decision Analysis

Decision maker: In such cases, it is usually the government or the parliament that decides.

Stakeholders: Among the *stakeholders* are landowners, or if the land is "common" land, commoners with particular use rights, or those owning or leasing such rights. Environmental groups will probably be involved, as well as people using – or intending to use – the land for recreation, and people that want to protect natural land for other reasons. See Chapter 3, section 2.3.

Decision objectives: *The goal* is to maximize use of land. One kind of use is protection of worthy land as "Outstanding Natural Beauty". The goals in this case should reflect that the value of natural amenities can be divided into use value and existence value as well as other components, Chapter 8, section 2. For use value one draws on other valuing techniques than for existence value, Chapter 8, section 4. Examples of sub-goals are:

- Maximize use of the land for grazing.
- Maximize extractive use (sand, gravel, firewood and turf/peat).
- Maximize use of the land for recreation and tourist industry.
- Maximize wildlife preservation.
- Maintain the beauty of the land.
- Minimize costs.

Consequence analysis: Here we will only deal with one criterion: the aesthetic value of the land. We use a methodology adapted from Porteus (1996). The method elicits different aspects of the aesthetic value if one includes respondents that are, and respondents that are not, planning to visit the landscapes:

- *Study task.* The task is to assess if the scenery in the candidate landscape is suitable for inclusion among areas with "Outstanding Natural Beauty" (ONB). One may select slides, photographs or computer presentations from landscapes that already are included among the ONB and mix them with comparable scenes from the candidate landscape. The number of pictures may range from 10 to 50. Respondents have a tendency to be fatigued by too many pictures since the number of comparisons can be large for sets greater than 10. (For the paired comparison technique, N = n (n-1)/2 or 45 comparisons for ten pictures). Separate the slides into a "learning set" and a "test set", or use cross validation. In order to reduce the number of comparisons, see Chapter 7, section 2.2.
- *Select respondents* according to the scope of the study, e.g., according to gender, age, occupation, "home-range" characteristics, etc. (Chapter 8, section 4.2).
- *Show the slides to the respondents.* They can rate their preferences on a semantic or numeric scale. In the paired comparison technique, responses are 1 for the better and 0 for the worse of two pictures respectively. If no ranking can be given, both pictures obtain a score of 0.5 (see Chapter 7, section 2). Each view may be shown for five seconds.
- *Rating.* The paired comparison technique outlined in Chapter 7 on preferences can be used to assess people's assessment of visual quality. One must be sure to use control respondents or control views when relevant. The order of the slides has to be balanced so that each landscape appears equally often in each pair, and each landscape type should be proportionally distributed in front and back order.
- *Conclusion.* If all scenes from the candidate landscape are ranked lower than all scenes from landscapes belonging to the ONB, the landscape should probably not be included among the ONBs. Use techniques from pair wise comparison to examine the significance of the results in terms of consistency and judges' agreement. If one also wants to examine which view elements were responsible for the conclusion, one may proceed as follows:
- *Context domain.* Scenes are grouped into context domains by scaling or cluster analysis.
- *Statistical analysis.* The responses are analyzed in terms of predictor variables like the view elements discussed above. With a regression analysis landscape quality, *LQ*, is related to the view-elements or landscape components such as percentage relief >60%; percentage rolling hills, etc.: $X_1, X_2, \ldots, X_n$.

$$LQ = a + b_1X_1 + b_2X_2 + \ldots + b_nX_n; \tag{11-4}$$

This method is similar to the Hedonic method described in Chapter 8, section 4.1.

Preferences: The consequence analysis we have outlined above may use experts as respondents, or local citizens from the actual area. The resulting aesthetic value of the land should be converted to an importance weights for the aesthetic value criterion, and the weights for the other criteria (costs, grazing value, etc.) should be determined and entered as well in Table 3-8. Then the decision maker makes the final assessment.

7.2 What Happens?

We cite some results from *The case of the Yorkshire Dales National Park*, which demonstrates that agreement among people about aesthetics are difficult to achieve, and that aesthetic views change with time.

The Yorkshire Dales is a National Park in England. The philosophers O´Neill and Walsh (2000) quote three contrasting statements about this landscape: First from about 1700:"

> "Nor were these hills high and formidable only, but they had a kind of an unhospitable terror in them…(they were) ..., but all barren and wild, of no use or advantage either to man or beast."

Second from about 1990:

> "This is a tough landscape shaped by tough people. It is unique to the British rural scene and impressively distinctive"

and the third from about 1995:

> "Most of the district still remains overrun by, and severely eroded by, sheep which none but a subsidized and distorted market system would support. Many recreationalists appear to prefer impoverished grasslands, treelessness. (instead of ecological restoration)"

The aesthetic questions are: what is aesthetically pleasing: useful landscapes, unique and distinctive landscapes, natural ecological landscapes, or some other options? Who should determine what is aesthetically pleasing?

---

[1] Porteus (1996)

[2] Heerwagen and Orians (1993)

[3] e.g., Fitter (1995), Porteus (1996), Andrews (1989)

[4] Cold et al. (1998)

[5] Briggs and France (1981) argue for the second option
[6] Staats et al. (2003)
[7] Navrud (2001)
[8] Janzen (1998)
[9] Porteus (1996)
[10] Appleton (1975), Heerwagen and Orians (1993)
[11] Porteus (1996) p. 136
[12] Kaplan and Kaplan (1983)
[13] Swithart and Petric (1988)
[14] Andrews (1989) p. 199
[15] Picture from Hastings and Sugihara (1993) p. 2–3
[16] Assembly rules in ecology
[17] Wellman and Buhyoff (1980) p. 106
[18] Heerwagen and Orians (1993) p. 159
[19] Stamps III (2004))
[20] Porteus (1996) p. 140:140
[21] Heerwagen and Orians (1993)
[22] De Lucio et al. (1994)
[23] Wellman and Buhyoff (1980)
[24] Porteus (1996)
[25] Stamps III (2004)
[26] Shih-táo (1997) p. 70
[27] Stamps III (2004)
[28] Staats et al. (2003)
[29] Tomkins (1991) p. 18
[30] Andrews (1989) p. 90
[31] Porteus (1996) p. 61–64
[32] Porteus (1996) p. 103:103
[33] Andrews (1989) p. 80
[34] von Winterfeldt and Edwards (1986) p. 99
[35] Bradley et al. (2001) Vrana and Gross (2004)
[36] Briggs and France (1981)
[37] Briggs and France (1981)
[38] Preiser and Rohane (1988) p. 264
[39] Chapter 19. On terrestrial environment
[40] Porteus (1996) p. 121, Marlatt et al. (1993), Briggs and France (1981)
[41] Wellman and Buhyoff (1980)
[42] Briggs and France (1981)
[43] Hull and Bishop (1988)
[44] Briggs and France (1981)
[45] Hull and Bishop (1988)
[46] Short (2000)
[47] A headline in J. Rawls (1993): Political liberalism

Part 3

ECOLOGICAL PRINCIPLES AND PROCESSES

Overview

Chapters

12. ECOLOGY: Ecological concepts.
13. HARVESTING: Sustainability, footprints, and harvesting economics.
14. TOXINS: Toxicity of pollutants for humans and ecosystems.
15. SOIL AND SEDIMENT: Physics and chemistry of soils and sediments – soil loss and formation.
16. HYDROLOGY: Rain, runoff, evaporation and storage.

A fundamental question about ecology is this; "Is ecology a predictive science?" The late Robert Henry Peters claimed in his book "A critique for ecology" in 1991 that the science of ecology lacks scientific rigour, has weak predictive capability, and fails to harness modern technology. Others opposed that view. A related key question is therefore what rigour we require for sound and wise environmental management, and what the alternatives are for making consequence assessments.

This part deals with ecological principles and processes that are common for all environments; the fourth part describes the individual environments.

The chapters are organized as follows. We give first a short introduction where we discuss typical conflicts with human uses of the environment. The next sections give background information, including an overview of the services supplied by various environments. Finally, we suggest simple models that may help us make predictions. However, there are more complete models available for almost all issues, and they should be consulted once screening surveys have been completed.

Chapter 12

ECOLOGY

Ecological concepts

This chapter outlines basic concepts of ecology with relevance to environmental management. We discuss optimal and acceptable ecosystem standards, and present the basic equation for biological growth, including abiotic and biotic factors that affect the abundance and distribution of species. We show models for estimating the carrying capacity of a habitat, and models for estimating species diversity. We also discuss models for estimating the size of populations that are large enough to persist for 100 years, the minimum viable population.

●

"Die Natur weiss allein, was sie will." (Nature alone knows what she wants.) Goethe.

1. INTRODUCTION

Ecology is the science of the relationship between organisms and their surroundings. Of particular importance is the *abundance and distribution of species*, which is affected by physical, chemical and biological factors. Figure 12-1 shows how abundance and distribution can be illustrated graphically in the form of contour maps.

Environmental management requires that we include ecological concerns in our decision processes, and to do that we need to know how human activity influences the ecosystem, how such influences can be detected, and how they can be mitigated. A major ecological objective is to maintain habitat and species diversity. This objective translates into maintaining natural habitats and solving conflicts with other land uses.

Figure 12-1. Distribution and abundance of Puffins Fratercula arctica (L.) in Scandinavia Black circles show densities from 5,000,000 to 1 million pairs. The bird's distribution is strongly associated with food availability in the sea and temperature. Note that there is a grand distribution scale (arctic versus non-artic) and a smaller scale for the colonies.[1]

Basic questions regarding a given ecosystem are:

- What is its natural state?
- How large are the natural fluctuations?
- How do human impacts affect it?
- How do we measure human impacts?
- Are some parts of the system more important than others?
- What is the restitution time? That is, if disturbed, how long will it take before it returns close to its "natural" state, if it ever does?
- Why do we want to keep ecosystems in the natural state?

Humans are part of ecosystems and thus influence it. The most severe influences come through urbanization and agriculture, and also introduction of nonnative species.[2] Common human activities that have detrimental effects are biological pesticide control, harvesting, establishment of monoculture fields, fragmentation of land and pollution. However, most human activities can probably technically be performed in a sustainable way, or the ecosystem can be brought back to pristine state within a limited time, such as 30 years. But technical progress is one thing; the real issue is to address social and economic challenges that hinder or help ecological practice. We must consider natural events, like volcanic eruptions, hurricanes and tsunamis,[3] which by themselves or in combination with human interference may profoundly change ecosystems.

Table 12-1. Relationships between consequences of human activity (column 1), the pollutant carrier (column 2), end-impacts (column 3) and how damages to ecosystems and humans can be measured (column 4). PEASL = people exposed to concentrations above safety levels.

Resource	Carrier	End-impact	Measurement
Energy			
Fossil fuel use, coal, oil, gas	Air	Exhausted resources, air pollution	% Remaining life time for fossil fuel sources
Renewable fuel, wood, waterpower, wind, light	Soil, water	Land-use, impacted natural areas, socio-economy	Km^2 vegetation, wildlife in less than pristine state Lost recreation days
Global pollution			
Carbon dioxide, CO_2	Air	Global climate	Global warming potential, GWP
Methane, CH_4	Air	Global climate	GWP
Ozone, O_3 (atmospheric)	Air	Global climate, health	Human mortality, sub-lethal effects, PEASL
Air pollution			
Ozone, close to earth surface (urban ozone)	Air	Health, vegetation	PEASL, Vegetation (km^2), Agricultural loss (ton ha^{-1})
Carbon monoxide, CO	Air	Health	PEASL, Angina attacks
Nitrogenoxides, NO_x	Air	Health, corrosion, enrichment	PEASL, number of monuments damaged, area damaged (km^2)
Sulphur dioxide, SO_2	Air	Health, air and water pollution	PEASL, area damaged (km^2)
Suspended particulate matter, SPS	Air	health	PEASL
Water pollution			
Total nitrogen, TN	Soil, water	Eutrophication, nuisance algal blooms, recreation	Area (km^2); lost recreational visit days
Chemical oxygen demand, COD	Water	Eutrophication; algal blooms, fish kills, recreation	Area (km^2); lost recreational visit-days; dead fish ha^{-1}
Others			
Noise	Air	Nuisance	People exposed above nuisance levels
Barriers	Land	Wildlife, plants	Area damaged (km^2), nuisances

This chapter focuses on discharges and emissions that threaten ecosystems. Other factors will be discussed in connection with chapters that focus on particular ecosystems, such as soil, lakes and rivers.

Table 12-1 relates resource use, emissions and discharges to end-impacts to ecosystems and humans. The last column suggests how impacts can be measured. Ecosystem damages are often formulated as areas not in pristine state. Damages to humans are measured in terms of morbidity and mortality. In addition to the compounds listed in the table, other chemicals as well occur in toxic concentrations, such as lead, cadmium, mercury, copper, dioxins, flour, PAH, PCB and tin organic compounds.

2. BACKGROUND INFORMATION

This section gives information on ecosystem services. Data are continually updated internationally; the numbers provided here are therefore only meant to indicate the magnitude of what we are talking about.

2.1 The Value of Ecosystem Services

What kinds of services do ecosystems actually provide that we find valuable? Here is an example of services from a mangrove forest.[4]

- Habitat, nutrients, and breeding of shrimp, crustaceans and mollusks.
- Spawning and breeding of fish species.
- Provision of construction materials and fuel wood.
- Source of salt, plants, herbs and small game used locally.
- Controls of storms, floods and erosion.
- Sequestration of toxins, carbon, nitrogen, phosphorus and other nutrients.
- Protection of other marine and coastal ecosystems such as coral reefs.
- Habitat for birds – amenity value of bird watching.
- Option values related to future uses.
- Existence value.

Particularly valuable ecosystems are often protected from human impact in the form of national parks, wildlife reserves or other sorts of nature reservation areas. Within countries, the protected areas range from an average of 2.6% in Europe to 11.3% in Latin America and the Caribbean.

Ecosystem services have been valuated by willingness-to-pay (WTP) methods. For example, WTP to preserve visibility in the Grand Canyon and nature reserves in the United Kingdom have been estimated at US$ 27 and US$ 40 per person and year in 1990, respectively.[5] Environmental management is of course also practiced without explicit information about WTP. In those cases, its cost sometimes can give an indication of underlying WTP that is not measured, but still reflect the will of the population. We should be careful with circular arguments, however: We measure WTP to find out how much to spend on environmental management. To infer the other way and assume that management without WTP information, still reflect true WTP, is a rather bold assumption. Anyway, let us look at some examples: the Directorate for the Environment in England manages and gives advice on conservation, enhancement of the natural beauty and amenity of the countryside with its flora and fauna, and on provision of public access and open-air recreation. It also manages national parks, public rights and commons. Its annual expenditures are about £12 million, which amounts to €250 per capita and year, or 2% of GDP per capita. See Table 12-2.

Table 12-2. Expenditures on nature conservation. We have assumed 1€ = 1US$.[6]

Expenditures	Per capita globally	Per capita in England	Per km^2 protected land (2000)	Per km^2 (goal)	Per red listed species in US
Euro yr^{-1}	1	250	450	470	4.3×10^6

Globally, 13×10^6 km^2 of land is presently protected, with corresponding expenditures amounting to $\$6 \times 10^9$ per year. With 6×10^9 people on earth this amounts to about €1 yr^{-1} $capita^{-1}$, as shown in Table 12-2. Only \$93 per km^2 is spent in tropical countries.[7] Protection measures in Madagascar's Mantadia national park caused an annual loss in income around US\$40 for households that based their subsistence income on agriculture and forest product collection. To manage a broadly representative system of nature reserves covering nearly 15% of global land area would cost annually about US$\$10 \times 10^9$.[8] The annual 1996 expenditures in the US are estimated to be $\$3.9 \times 10^9$. Since there are currently 911 species, subspecies and populations on the **Endangered species act** list in the US, this amounts to US\$ 4.3×10^6 per endangered species per year.

The annual value of the services offered by the species is between \$100 and \$200 per hectare,[9] which can be compared to the annual expenditures on protected land, which is about \$5 per hectare.

2.2 Ecosystem Resources

The **biosphere** contains the ecosystems of the world. We divide it in three spheres: The **hydrosphere** harbors the ecosystems of the water environments, including salt and fresh water. The **lithosphere** harbors the terrestrial ecosystems. The **atmosphere** reaches 250 km up in the air and harbors the air born ecosystems. Biological material circulates between the spheres, which we for scientific purposes described in terms of carbon, nitrogen, phosphorus or energy. Organic matter is 50% carbon.

From a man-ecosystem perspective, the most important objective is to preserve a high degree of species and habitat diversity. With a conservative estimate, there are about seven to 14 million species of eukaryotes, which include all plant- and animal-like organisms, but exclude bacteria.[10] Each species has on the average about 220 populations. With 14 million species, this gives 1.1 to 6.6×10^9 populations globally. A **population** is defined as a geographical entity within a species, distinguished genetically from other populations because of limited gene flow.[11] The number of populations per 10,000 km^2 ranges between 0.01 for reptiles to 316 for land mollusks. There are for example 1.7 plants, 0.06 birds, and 5.6 mammals per 10,000 km^2, and on the average 1.2 generic populations per 10,000 km^2.

Species diversity is often measured by observing a subset of indicator species, like vascular plants, mammals or birds. However, lower taxa do not necessarily overlap spatially with higher taxa; thus, protection of higher taxa does not guarantee protection of species at lower levels, and vice versa.[12] The number of species varies among taxa and with the region. The number of tree-, vertebrate-, and bird species are all about 200 in North America. Bird species diversity in Central America reaches 600 species, however. Table 12-3 shows the number of threatened or endangered species in selected countries and in the world. In the US about 30% of land mammal species are in this category. Endangered species are said to be on the **Red list**, that is, a list that formally declares a species as endangered or needing protection.[13] On a world basis 20% of mammal species are threatened, 31% of the amphibians, 12% of the birds, 4% of the reptiles and 3% of the fishes. Of the plant species, 3% are threatened, with gymnosperms as the most vulnerable.

Species are not equal in value. Some species are very efficient colonizers, and are able to "hold the land" for unforeseeable duration. Such species are actually unwanted and have thus a negative value, like the zebra mussel with 40 000 individuals per m^2, depleting water of oxygen. Other examples are the European house sparrow, the Argentine ant and the Australian rabbit. These species dominate nature everywhere like a global McEcosystem and are called "injurious species" under the federal Lacey Act.[14]

Biodiversity hotspots are areas where exceptional concentrations of endemic species are suffering large losses of habitats. **Endemic species** are species with restricted geographical distribution. As many as 44% of all species of vascular plants and 35% of all species in four vertebrate groups are confined to 25 hotspots comprising only 1.4% of the land surface area of the world.[15] The land area in the US containing endangered species is small compared to its total land area.

Table 12-3. Endangered species in selected countries, regions, and the world.[16] The total known number of species in the world is 1–1.5 million, and an estimated 6–50 million unknown species (eukaryotes). The figures will change as species are listed or removed. World figures are from 2004.

	Plants	Invertebrates	Land mammals	Birds	Reptiles	Amphibians	Fish	Total
Europe	2676	66	–	396	16	15	48	
Antarctica	0	0	0	1	0	0	0	1
Norway	13	3		8	0	0	1	23
United States	503	–	58	72	43	57	107	924
S. America		4	29	66	4	9	8	
Madagascar	162		50	27	18	2	13	302
China	167		76	73	31	1	33	385
Word	8321	1992	1101	1213	304	1770	800	15503

For each group of species, plants, birds, fish and mollusks, 50% of the endangered species occupy between 0.15% and 2.0% of the land area.[17] Hot spots are concentrated where the range of many endemic species overlaps with intensive urbanization or agriculture. Thus, endemic species are prone to extinction mostly because of habitat loss or habitat degradation. Another reason is introduction of exotic species that out-competes native ones.

Typical hotspots are the Brazilian's Atlantic forest with 8.7 endemic plants per 100 km^2 and 0.6 endemic invertebrates per 100 km^2, and the Caribbean with 23.5 endemic plants per 100 km^2 and 2.6 endemic invertebrates per 100 km^2. In studies of hotspots all species are counted with the same weight. However, theories that assign different importance values according to evolutionary history have been suggested.[18] With species importance as a secondary attribute, the areas defined as hotspots and therefore of particular concern, could change. Many people value apex predators and large animals, like whales, as more valuable than others. A third position might be that hotspots should be defined separately for all regions in the world. Hotspots in arctic areas (pun not intended) may be considered of particular importance because there are so few species in these regions.

3. ECOSYSTEM THEORY

From an operational point of view an **ecosystem** can be defined as a distinguishable part of the biosphere where dominant, key, or functional groups of populations have strong interactions among themselves and with their physical environment. Normally they also have weak, or a few easily identifiable, strong interactions with the outside world. Within an ecosystem, abiotic (physical and chemical) forces and biotic interactions act together to produce a distribution and abundance of species. Outside forces also act on the ecosystem, driving it transiently or cyclically, such as seasonal driving forces that are important for most ecosystems in the Northern and Southern hemispheres. Figure 12-2 shows a schematic diagram of stocks and flows in an aquatic ecosystem. The biotic component of each ecosystem is ultimately driven by the harvesting of light through photosynthesis.

3.1 Photosynthesis

Green plants are the main converters of sun-energy for ecosystems. During the first "light" stage of photosynthesis, chlorophyll-a – a green pigment in the chloroplasts – absorbs enough energy from sunlight to split electrons from water molecules, simultaneously releasing oxygen and driving the production of adenosine triphosphate (ATP), which is used in a second "dark" stage. This next step begins when RuBisCO – an enzyme with the full name ribulose-1,5-biphosphate carboxylase-oxygenase – combines with carbon dioxide to produce 3-phospho-glycerate, or PGA. Powered by energy from PGA, a series of reactions transform PGA into a host of starches, sugars, and other organic compounds. Photosynthesis as a whole is not particularly efficient, a crop plant that stores as much as 1% of the total received energy is exceptional.[19] In contrast, a photoelectric panel transforms 5–7% of the light energy into electricity. Energy is harvested from light as shown in Table 12-4. The top equation in the table is representative of the production of carbohydrates such as glucose, in which the number of carbon dioxide molecules that is fixed is equal to the number of oxygen molecules released, that is $\Delta O_2/\Delta CO_2$ is 1.0. The oxidation of one mole of glucose (180g) releases 673 kcal or 3.7 kcal g^{-1}. Because gases are often measured as volumes, and 1 mole of a gas occupies 22.4 L under standard conditions, the equivalents are often expressed as 5 kcal L^{-1} of either oxygen or carbon dioxide. Some carbohydrates are slightly higher than glucose in energy content, and the calorific value for carbohydrates is therefore often rounded to 4 kcal g^{-1} dry weight. Organic material also contains nitrogen, phosphorus and silica. Approximate ratios are as follows: C/N = 6, C/P = 40, Si/C = 0.8, C/Chl-a = 30. The equations describe how ecosystems function. If you walk along a greenish (eutrophic) lake or river you may sense a rotten smell. The reason is that organic matter uses oxygen at the bottom of the water when it degrades, degradation then continues anaerobically according to the second equation in Table 12-4, and produces the smell. You may observe a mechanical device floating in the middle of small fishponds. This is probably a pump oxygenating the water to give the fish sufficient oxygen. If you continue your walk through the forest, you may appreciate that it sequesters carbon dioxide, the major greenhouse gas. This is described by the first equation in Table 12-4. This reduces the amount of CO_2 in the air. About half the dry weight of wood is carbon, and the vegetation stores about 40% of the world's CO_2 that is not dissolved in water.

Table 12-4. Energy equations.

$12H_2O$	$+6CO_2$	+ 709 kcal	→	$C_6H_{12}O_6$	$+6O_2$	$+6H_2O$
water	Carbon dioxide	Energy from light	Chloro-phyll	Carbohydrates	To air	water
$C_6H_{12}O_6$	$+6O_2$	→	$6CO_2$	$+6H_2O$	+ energy	
Carbo-hydrates	oxygen	Metabolic enzymes	Carbon dioxide	water	energy for work and maintenance	

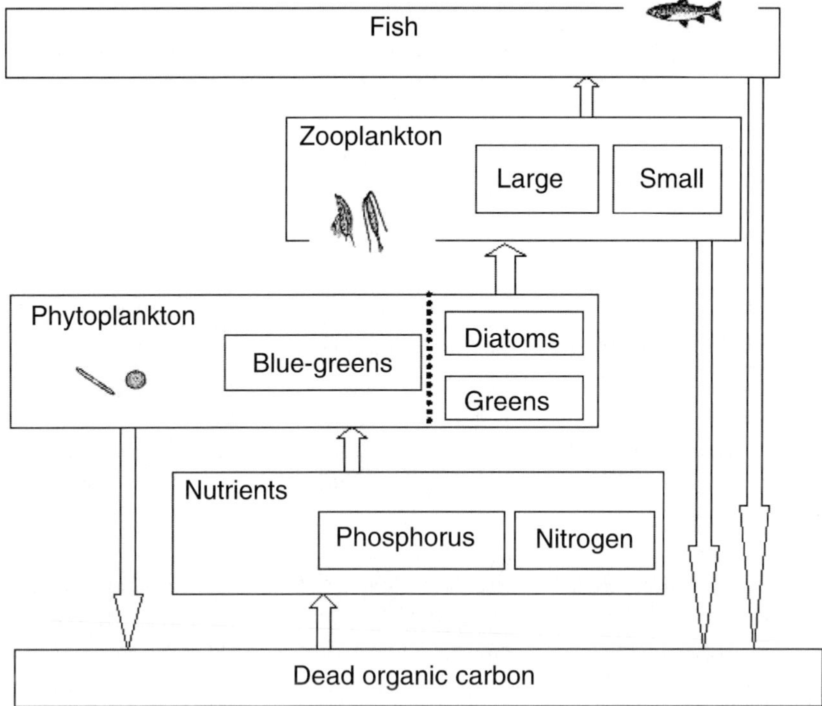

Figure 12-2. Aquatic ecosystem with standing stocks and flows. Standing stocks may be defined in terms of energy (Joule), biomass (dry weight) or similar units. Flow rates are often given in units of energy per day (J day^{-1}) and occasionally as oxygen consumption. Blue-green algae are not easily eaten by zooplankton.

3.2 Description of Ecosystems

An ecosystem is described by several attributes. *On the primary level*, the description consists of an inventory of all the individuals in the system, and all factors affecting them. *On the secondary level*, individuals are characterized on the species level, each species by abundance (number of individuals) and distribution. At this level, several indicators have been developed, such as species richness (number of species present in the ecosystem), species diversity (a compound measure of species richness and species abundance), species dominance index, and food-web linkage properties.

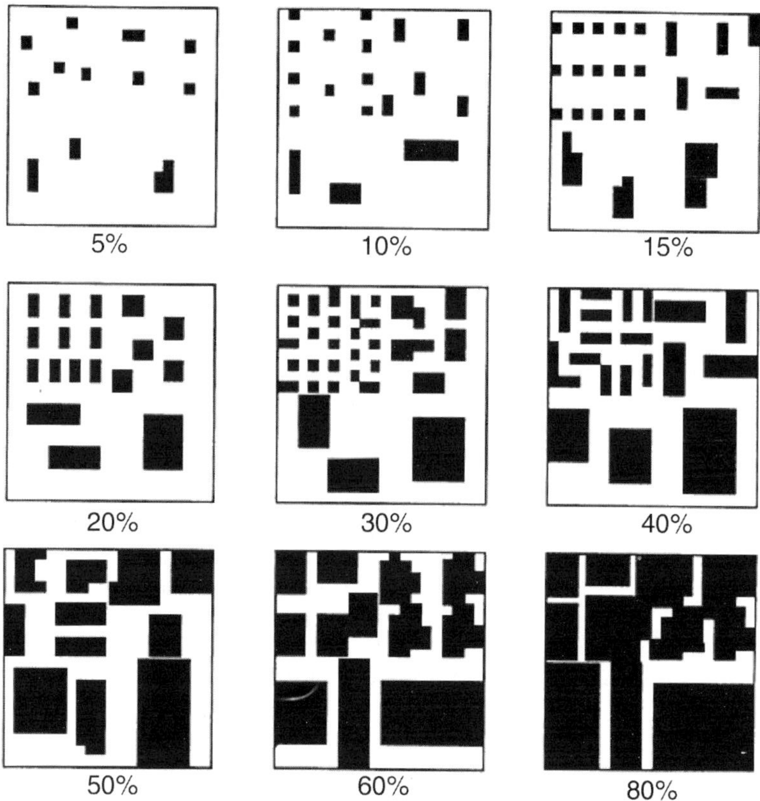

Figure 12-3. Percentage cover. Habitat fragmentation is an important process, depending upon species behaviour and size. Larger species require larger connected habitats to survive, and all species require access to neighbour populations for gene exchange After Northcote(1979). Permission of Rellin Technical Publications.

Ecosystems are also characterized by the biomass of each species, for example as kg standing biomass per m^3 or m^2. Another parameter is **production rate** measured as kg per hectare and year, corresponding to the flow of biomass or energy from one species to another, such as from prey to predator in units of kg carbon m^{-3} day^{-1}. There are also flows to microbial communities where organic molecules are broken down to inorganic compounds.

In flow diagrams like the one shown in Figure 12-2 one often adds sources and sinks for materials or energy flows. The source of energy is mainly light trapped by chlorophyll in green plants, and the sinks are energy trapped in carbohydrates, and buried permanently in soil or sediments. Standing biomasses and flows may be expressed in terms of:

- *Biomass wet weight, live weight, or dry weight.* About 2–30% of "wet" biomass is actually dry weight.
- *Carbon*: About 45–60% of biomass dry weight is carbon, and the ratio of wet weight to carbon is about 9:1.[20]
- *Limiting nutrients*: Phosphorus is about 1% of biomass dry weight and nitrogen about 10% of dry biomass weight.
- *Energy*: The energy content of various organic materials is 4–5 kcal per gram dry weight for plant materials and 1.5 kcal per gram fresh weight of meat.

For terrestrial ecosystems and for riverbeds, biomass may be expressed indirectly as percentage cover, that is, the area of the projection of the biomass on the ground, such as tree crowns, divided by the total area. See Figure 12-3. For rivers, the cover percentage may exceed 100% when there are several layers of vegetation.

On the tertiary level, species are assembled into guilds that are assemblies of species that have almost the same role in the ecosystem. In aquatic ecosystems, several species of diatoms are for example treated as a guild. Zooplankton guilds are often defined according to size classes.

3.3 Factors that Limit Species Abundance

A diagram like the one shown in Figure 12-4 is useful to understand why a species is in a certain place at a certain time, and to predict its abundance and distribution. To use the graph to find out why a species is not on a particular site, ask first if there are physical or chemical factors that prevent it. For example, is the summer temperature too cold for the species? If not, examine species interactions. Could a superior competitor have crowded it out? And so on.

The so-called **carrying capacity** of a habitat refers to a single species and often to food availability as the limiting factor. The carrying capacity is the maximum abundance that can be sustained. It can be estimated in two ways. The first is simply to observe the maximum biomass within a certain area. The second is to calculate food requirements for the species and then to establish how much food is available for the species in that habitat. When observing it, one may estimate the maximum biomass when it appears to be stationary or slightly oscillating, as in the right part of Figure 12-5. Carrying capacity also depends on habitat characteristics like the presence of shelter and breeding grounds.

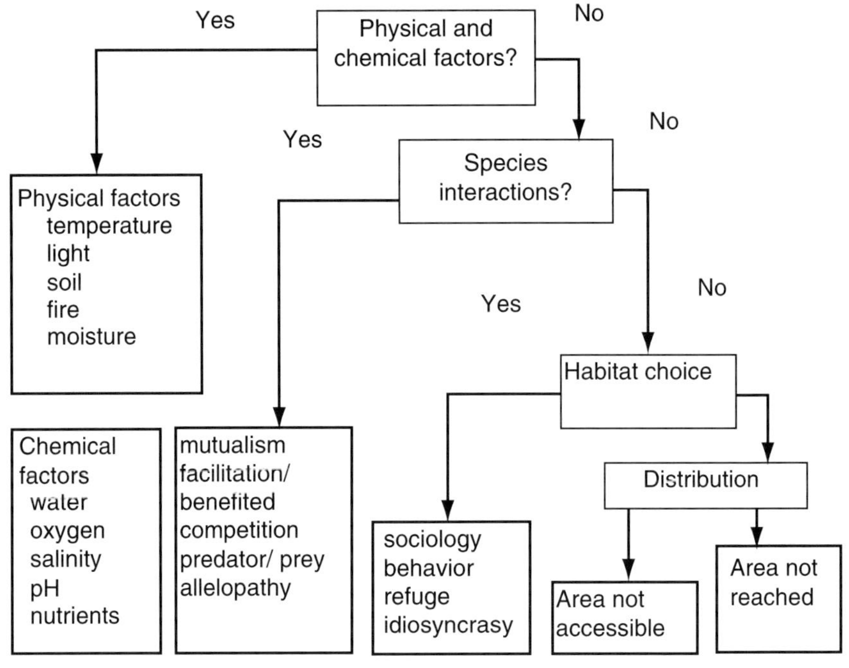

Figure 12-4. Finding out why a species is or is not in a habitat[21].

3.4 Dynamic Variation in Species Abundance

We distinguish between two basic dynamic patterns, growth and fluctuations. The initial development of any population is characterized by growth, and then normally a phase is reached where the abundance varies around the carrying capacity. This is called the climax state, which may be of limited duration. A fire may for example wipe out an aging forest.

3.4.1 Growth

If a population has unlimited resources, simple reproduction will cause it to grow with a rate that is proportional to the number of species at any time. If N is the biomass of the population (or number of individuals), and r_0 is the intrinsic growth rate, that means the change in biomass with time can be written:

$$dN / dt = r_0 N \tag{12-1}$$

The solution to (12-2) is the simple exponential growth function:

$$N(t) = N(0)e^{r_0 t} \qquad (12\text{-}2)$$

In practice, the growth will be limited by the carrying capacity *K* and approach zero as the *N* approaches *K*. A simple model of this behavior is provided by *the logistic growth equation* which turns out to fit empirical growth patterns rather well:

$$dN/dt = r_0 N(1 - \frac{N}{K}) \qquad (12\text{-}3)$$

This equation has the solution:

$$N(t) = \frac{K}{1 + ce^{-rt}} \text{ with } c = \frac{K}{N(0)} - 1 \qquad (12\text{-}4)$$

The function is shown graphically to the left in Figure 12-5. To find at which level of the biomass the growth rate is maximum we differentiate Eq. 12-3 with respect to *N*. The growth rate is largest when $N = K/2$ and at this value the growth rate is $r_0 K/4$. When the biomass is very low so that growth can be assumed to be unrestricted, the doubling time is $\ln 2/r_0$.

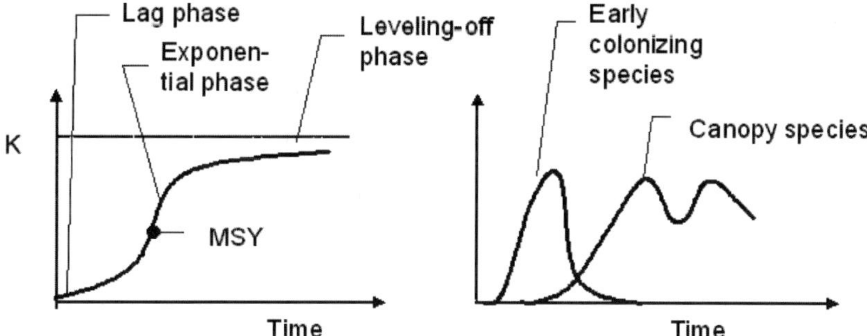

Figure 12-5. Biomass as a function of time. The left figure shows logistic growth. MSY stands for Maximum Sustainable Yield. MSY points to the time and to the biomass level where growth is largest. The right curve shows a sequence where early colonizing species first settle on a site, and thereafter give way to canopy species.

3.4.2 Fluctuations

Table 12-5 shows a typology of dynamic patterns of species abundance. An important finding is that even for single populations, the dynamics change with increasing growth rate, from a smooth growth that levels off at carrying capacity, over stable cycles, to what is called chaotic behavior, which resembles epidemic outbreaks. Metaphorically, one can say that fluctuations will be more probable if the gas power (growth) is much stronger than the braking power, or the braking power is slow to activate. It has been shown that species at higher trophic levels are more prone to dynamic chaos than species at lower trophic levels.[22] Since the growth rate is related to the amount of food resources available, epidemic thresholds are often formulated as a function of food resources. In a similar manner it appears that cycles and their amplitude increases with food resources. This phenomenon is called the paradox of enrichment.[23]

All three of the time courses in Figure 12-6 represent possible "natural" developments of a species. The first one is typical for large bodied, slow growing species (large mammals), and the last one is typical for small bodied, fast growing species (insects). The relationship between body mass and growth rate was established by Peters (1993).

Table 12-5. Population variation typology. After Sugihara and May (1990), Tømte et al. (1998), Andersen (1997).

Type of variability	Description	Diagnostics
Seasonal	Variation caused by seasonal variation in physical and chemical factors	Often sinusoidal variations in species abundance coinciding with seasonal drivers like temperature
Stochastic	Variation caused by stochastic physical and chemical processes	Large deviations from the mean. Look for concurrent deviations in climatic drivers outside of two standard deviations. Use a non-linear time series predictor and see that predictive power based on earlier biomass observations is close to nil.
Limit cycles	Medium sized variation in abundance	Look for variations in food and consumer abundance, but shifted in time. The same type of cycles occurs for competing species in a seasonal environment and for facilitator-gainer organism pairs.
Dynamic chaos	Variations caused, for example, by too small braking power in organism growth compared to growth	Difficult to separate from stochastic behavior. Use a non-linear predictor and look for exponential decay in prediction with time in contrast to instantaneous decay. Effect also identified as bifurcation.

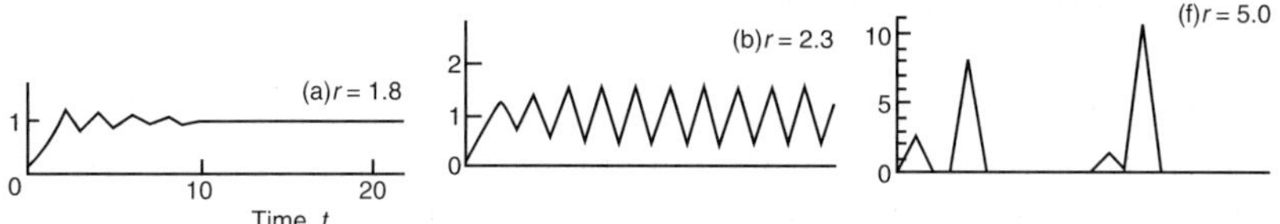

Figure 12-6. Examples of dynamic behavior of population density with increasing growth rate. After May (1976).

3.5 Abiotic Factors

Physical and chemical factors also determine ecosystem behavior. Among the most important are precipitation, soil condition, humidity, temperature, and habitat heterogeneity. Such **abiotic factors** can be classified in two groups, according to whether or not their effect depends on population density.

Finite food resources that are used up, and toxic waste product that are generated by the growing organisms, are examples of *density-dependent* factors. Such factors will inhibit growth when the biomass of the organism becomes sufficiently large. The sunlight factor is also often density dependent. If there is too much biomass, individuals will shade each other like trees competing for sunlight. This is called self-shading, and is the ultimate limiting factor in aquatic algal systems.[24]

The second category of growth retarding factors is *density-independent.* This includes factors like temperature and humidity, such as water-holding capacity of the soil.

3.5.1 Niche Space

Abiotic factors are often discussed in terms of niche space. A **nichee** is an abstract body in n-dimensional space defined as:

$$X = (x_1 - x_2, y_1 - y_2, z_1 - z_2, \ldots) \tag{12-5}$$

Each axis in the n-dimensional space, x, y, z, ... represents a required resource for the species, like nutrient and light. The **niche** of a species is defined as the volume in the n-dimensional space where the species can survive. The living conditions are optimal only in a limited area within the niche. In the sub-optimal space outside the inner core, other species with more optimal conditions may out-compete the first species.

3.6 Biotic Interactions

In this section we discuss how biotic interactions affect the distribution and abundance of species. A particularly interesting aspect is the prevalence of alternative stable states in nature. This will be discussed in the next section.

3.6.1 Organism Strategies

Biotic interactions are reciprocal in the sense that both species are profoundly affected. Species interactions can be classified in four major categories: Competition, facilitation, mutualism and prey/predation. These interactions, as well as some others, are briefly discussed below.[25]

- *Competition*: Two or more species compete for the same resource. Competition may occur by direct physical interference, or indirectly because two or more species use the same resources.
- *Facilitation*: One species facilitates the growth of another species, but is subsequently forced to die or leave the system.
- *Engineering*: Some species have the ability to structure the environment for other species, but can still be present in the system.[26]
- *Mutualism*: Both species require the presence of the other.
- *Prey/predation*: One species eats the other. The first has to be present for the latter to get food. Thus, lack of prey is as important to the predator as the presence of a predator to a prey. To avoid prey/predator cycles, the predator could switch to other prey items if the preferred food becomes scarce. The prey seeks a refuge to avoid predation, for example, a caribou moves to wolf-free high elevations.[27]

Two features are important in studying interactions: two species may show one type of interaction at one life stage and another type at another stage. Second, the interaction type one register depends upon time and space of the observations. During one season zooplankton eat phytoplankton and a regression between their abundances shows a negative association. However, zooplankton and phytoplankton abundance are positively associated if phytoplankton abundance increases over several years.

The term "trophic cascades" was developed by Shapiro et al. (1984) and Carpenter et al. (1998) and refers to the mechanism by which species at the top of the food chain determine the structure of the community at lower levels. This mechanism is often referred to as top-down control and the species exerting the control are called "drivers" or key species. The other species are sometimes called "passengers". From a managerial perspective, it may be important to maintain the "drivers" in a natural ecosystem. The opposite control mechanism is referred to as "bottom- up control". There is currently no clear diagnostic tool available to determine which of the two processes are most important for the ecosystem structure.

3.7 Ecosystem Stability

Almost all efforts in ecosystem management aim at stability. A resistant ecosystem continues to function when stressed by a disturbance, and a resilient ecosystem has the ability to recover from a disturbance. Some ecosystems, however, appear to be able to exist in alternative stable states.

3.7.1 Alternative Stable States

There are feedback mechanisms that maintain stable states, but it may still be possible under certain circumstances that the system jumps into another state and remains there for a long time. The situation can be visualized as a marble ball being restricted in a trough, but able to roll out of the trough and into a second trough at lower or higher elevation if forces on it are sufficient. See Figure 12-7. A classic example is forested areas that can remain forested as long as not all trees are logged. The standing forest will secure moisture, shade, windshield, and even protection from sheep and cattle grazing for seedlings.

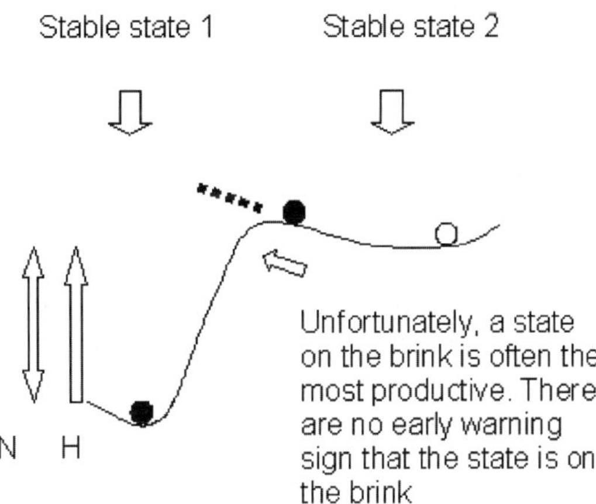
Stable state 1
Stable state 2
N
H
Unfortunately, a state on the brink is often the most productive. There are no early warning sign that the state is on the brink

Figure 12-7. Alternative stable states of an ecosystem. The marble - ball metaphor. Examples are forested (2) versus barren areas (1), Harvestable fish stocks (2) versus fishless sea (1). N is natural causes, and H is human induced causes for change. Based on Scheffer et al. (1993)

However, once the trees are removed, an alternative stable state could be treeless moorland. Similar examples are found in shallow lakes and in saline, but previously wooded areas in Western Australia.[28] To make a system transfer from a bad state to a desired state, one may take advantage of windows in the climate, like the El Nino.[29]

3.7.2 Species Succession

Species succession most often occurs when pristine ground is left open by natural biotic or abiotic factors or by man-made influence. New sand dunes or retreating glaciers can produce bare land ready for invasion by the first successional species. This is called primary succession. The first species to invade a new area is often called *r*-selected species, because they usually grow rapidly. Late successional species, *K*-selected species, are slow growing (growth rate 2% to 20% of *r*-selected species). A current issue is whether early colonizing species should get protection of their habitat.

4. ECOSYSTEM STANDARDS

Standards are benchmarks set for administrative purposes. In Chapter 5, we discussed the idea of imposing emission standards when damage costs are not known. That gives the polluters an incentive to find the cheapest way to meet the standards, with a cost-effective solution as a result. To set good standards, however, it is important to identify critical concentrations and critical loads.

4.1 Critical Concentrations and Critical Loads

Critical concentrations of chemicals refer to levels that are on the threshold of being toxic or otherwise unwanted. A critical load is the maximum emission rate that can be sustained without concentrations increasing above critical levels. An ecosystem can usually recycle, self-purify, or absorb a certain load of natural or man-made discharges and still be in balance.[30]

It is notoriously difficult to quantify critical concentrations and loads, and natural balance. The concepts are closely related to the notions of *carrying capacity* and *sustainable development*. Andersen[31] has contributed to the understanding of critical loads by showing that "excessive" loads of phosphorus to lakes[32] corresponds approximately to levels where there is a high likelihood for a bifurcation or dynamic chaos. At this level species biomass may increase very rapidly, as was illustrated in the right graph of

Figure 12-6. This means that at this level stable grazer control of algal biomass is difficult or impossible to achieve. Critical loads are defined for some substances, like nutrient loads to freshwaters. Pollution control authorities may also define critical loads for particular chemical discharges at particular sites.

5. DOSE-RESPONSE RELATIONSHIPS

There are three classes of dose-response models. Here we discuss those models that are common for all compartments of the biosphere. Other models are discussed in the relevant chapters.

- Organism growth models.
- Species diversity models.
- Species creation and extinction models.

5.1 Organism Growth Models

An organism has to satisfy its **basic metabolic needs** for essential maintenance of the tissue before it can grow. The rate of energy uptake for maintenance is related to body weight in animals:

$$E_m = a\, W^{z} \tag{12-6}$$

W is wet weight in gram, E_m is energy in kcal per day (1 cal = 4.19J), a is a coefficient (also kcal per day), z is an exponent, $z < 1.0$. Since z is smaller than 1.0, large animals require proportionally less energy than a small animal. Some coefficient values are given in Table 12-6.

To calculate the food biomass required to support a given animal biomass, one needs to know three ratios.[33]

1. c_g: Energy in the food material eaten, (E_{eaten}), versus energy available for consumption, E_{con}: The latter is typically 5–15% of the former for terrestrial herbivores and 40–50% for aquatic herbivores, partly because of table manners, partly because of digestion.
2. c_{con}: Energy consumed versus the energy assimilated, E_{ass}: This is typically 60% for terrestrial herbivores and 80% for carnivores.

Table 12-6. Organism energy requirements[34] Efficiency is the fraction of energy used for maintenance and stored as tissue to gross energy intake.

Organism	a, kcal day^{-1}	Exponent	Typical size, W, g	Efficiency
Sea urchin	0.009–0.031	0.73–0.87	10–100	0.1–0.2
Mammals	0.68	0.75	100–100,000	0.03–0.10

3. c_{ass}: Energy assimilated versus energy used for growth, E_g – the rest is respirated: This is typically 2% for vertebrate terrestrial herbivores and 10% for vertebrate carnivores.

Fluxes expressed as units of biomass may be converted to fluxes of energy by setting the energy content at 18–21 kJ per gram dry weight. To calculate the energy used for growth:

$$E_g = c\,E_{eaten} = c_g \times c_{ass} \times c_{con} E_{eaten} \quad (12\text{-}7)$$

For herbivores (numbers are for small rodents) only 3–10% of the average annual production is available for biological processes.[35] Typical terrestrial primary productivity range from 1000 kJ m^{-2} to 50,000 kJ m^{-2} corresponding to about 50 gdw m^{-2} to 2500 gdw m^{-2}.

Herbivore biomass, B, (kJm^{-2}) can be related to net above ground primary productivity (NPP, kJ m^{-2} yr^{-1}) by Eq. 12-8:[36]

$$\log_{10} B = 1.52\,(\log_{10} NPP) - 4.79,\ r^2 = 0.6,\ p \approx 0.00,\ \text{df} = 49 \quad (12\text{-}8)$$

Two equations, the upper one for grasslands and the lower for forests related consumption, C, and NPP. F stands for foliage productivity.

$$\log_{10} \text{C} = 1.38\,(\log_{10} NPP) - 2.32,\ r^2 = 0.4,\ \text{p} \approx 0.00,\ \text{df} = 67 \quad (12\text{-}9)$$

$$\log_{10} \text{C} = 2.04\,(\log_{10} NPP_F) - 4.80,\ r^2 = 0.594,\ p \approx 0.00\ \text{df} = 73 \quad (12\text{-}10)$$

5.1.1 Case Study 1: Herbivore Biomass

Problem: A temperate grassland of 1 km^2 has a standing stock of 1600 gww m^{-2} and a production of 600 gww m^{-2} (corresponding to 6t ww ha^{-1}). How many cows (170 kg) can this grassland support?

Solution. 600 gww per m^2 gives about 120gdw per m^2 and about 2400kJ. per m^2 Applying Eq. (12-8) for grasslands we get $\log_{10} B = 1.52$ ($\log_{10}$ $2.4.10^3$) – 4.79 = 0.35, i.e., B = 2.2 kJ m^{-2} yr^{-1} Thus, supportable herbivore biomass on this area is about 110 kg dw or about 550 kg ww. Thus 1 km^2 of grassland could support three cows. By knowing the average size of an animal, the yield of its primary food source and the energy content of the food it is possible to calculate the area required to support an animal sustainably.

5.1.2 Case Study 2: Area Requirements for Species

Problem: A reindeer weighs 55 kg. The reindeer is grazing in a mountainous lichen heath that supplies 3–6 ton ha^{-1} of primary production. The reindeer uses 3% of the primary production. One gram wet-weight of growth gives 1.5 kcal. What will the caloric intake be for the reindeer? How large an area does the reindeer require for food support?

Solution. From Equation 12-6, a reindeer requires $E_m = 0.68 \times 55{,}000^{0.75}$ = 2442 kcal day^{-1}, or 0.9 10^6 kcal yr^{-1}, or 0.6×10^6 g wet-weight growth per year. Since reindeer use 3% of the primary production, it needs 0.6 tons/0.03 = 20 ton ww yr^{-1}. This is the primary production on 3–7 ha of grazing area. This result can be compared to results obtained in Case study 1, which used Equation 12-8 and suggest that a 170 kg cow requires 33 ha of grazing land.

5.2 Species Diversity Models

A useful definition of a **species** is a collection of organisms that can exchange genes, although it does not deal adequately with bacteria and some other organism groups.[37] Species richness is usually defined as the number of species present, irrespective of the number of individuals within each species. Species diversity is a function both of the number of species present and their relative abundance. Thus, measures of species diversity require information on the number of species, n, and also on the number of individuals of each species. This is given by p_i, which is the probability that a random animal belongs to species i. For a constant species richness n, the maximum diversity occurs when all species are represented by the same number of individuals, and the minimum diversity occurs when all species, except one, are represented by one individual and the rest of the individuals belong to one species. This makes it possible to define species diversity in the same way as entropy – or disorder – in physics:

$$H = -\sum_{i=1}^{n} p_i \log p_i \qquad (12\text{-}11)$$

If we define species richness S, as the number of species present, it obeys the following accounting identity:

$$S = S_{inn} + S_{cr} - S_{em} - S_{ex} \qquad (12\text{-}12)$$

Here S_{inn} is species immigration, S_{cr} is species creation, S_{em} is species emigration and S_{ex} is species extinction. For very large areas, often called biogeographic provinces, the pool of species outside the area is the same as the species pool within the area, so $S_{inn} = S_{out}$. Often, one may assume that S is in a state of equilibrium that is characteristic for the area, and which can be predicted from properties of the area, like size, complexity, and the total resources used by the community. This may then be defined as the "carrying capacity".

There are two types of such models. One type explains the number of species present as a function of abiotic factors, like energy availability or spatial heterogeneity.[38] The second type simply expresses an empirical relationship between an area and the number of species found on that area. The average range over which a species can be found varies from 790,000 km^2 for Indo-Malayan mammals to 6.6 mill km^2 for East African mammals. The average range is about 2.2 million km^2.[39]

5.2.1 An Energy Based Model for Low Energy Regions

The *energy* hypothesis states that higher energy availability provides a broader resource base, permitting more species to coexist. Energy availability can be measured as potential evapotranspiration (*PET*, mm yr^{-1}), which is the amount of water that would evaporate from a saturated surface, (see Chapter 16 on Hydrology). Kerr and Parker (1997) showed that energy is important for species richness in cold, low energy regions like Alaska and most of Canada, where evapotranspiration is less than about 1000 mm yr^{-1}. Their empirical model for mammal species richness is:

$$SPP = 0.06\, PET + 4.48\,,\ r^2 = 0.762,\ \mathrm{p} < 0.0001 \qquad (12\text{-}13)$$

In the study, *PET* ranged from 200 to 1000 mm yr^{-1} and number of species from 10 to 55. According to this equation, mammal species richness in an area with 500 mm potential evapotranspiration is 25 species.

5.2.2 A Heterogeneity-Based Model for Temperate Regions

The second hypothesis states that habitat *heterogeneity* is important. Kerr and Parker (1997) measured habitat heterogeneity as change in elevation. In warm, high-energy regions like the continental US and southern Canada where PET > 1000 mm yr^{-1}, they found that heterogeneity was most important. Their empirical model for mammal species richness is:

$$SPP = 0.01\,(h_{max} - h_{min}) + 45;\ r^2 = 0.6523 \qquad (12\text{-}14)$$

Where habitat heterogeneity measured as the height difference, $h_{max} - h_{min}$ ranged from 0 m to 4000 m and mammalian richness from 45 to 90. Heterogeneity was measured as the difference between minimum and maximum elevation within the same 250 km × 250 km quadrant. The same general rule does not hold for trees and land-birds, however.[40]

A medium heterogeneous habitat of 2000 m would support 65 species of mammals on an area of about 62,500 km^2.

5.2.3 Nutrient Based Model for Grassland

Long-term deposition of nitrogen reduces species richness in grasslands. A study in Great Britain showed that for every 2.5 kg N ha^{-1} yr^{-1} of chronic nitrogen deposition, one grass species disappears per 4 m^2 of grassland.[41] The ambient annual deposition of nitrogen was 5 to 35 kg N ha^{-1} yr^{-1}. The reason is that species that are adapted to infertile conditions systematically are reduced.

5.2.4 Species Area Models

The great 19th century scientist Alexander von Humboldt gave ecology its oldest law: larger areas harbor more species than smaller ones.[42] The current version of this law says that the number of species found in a sampled patch area A is a constant power of A, thus:

$$S = cA^z \text{ or ,with a rough approximation, } S_n / S_o = (A_o/A_n)^{0.25} \qquad (12\text{-}15)$$

This rule means that if the area of a suitable habitat is reduced by 50%, the number of species will be reduced to about 15%.[43] However, the values of c and z are not quite constant, but show some predictable patterns. Larger scales of space and time generate species-area relationships with larger z-values. Table 12-7 shows some c and z values that have been reported in the literature.[44] The quadrants used in model 12-14 would contain $N = 0.194 \times 6{,}25{,}0000^{0.326} = 32$ species according to Equation 12-15 with coefficients for mammals. That is considerably less than found by Kerr and Parker.

5.3 Species Extinction and Species Creation Models

Species life times range from about five million to ten million years, with mammals as the least durable. Documented extinctions show that the lifespan of birds and mammal species are around 10,000 years. May et al. (1995) calculate that current deforestation rates of 0.8–2% reduces species lifespan to 200–500 years. The rate of species extinction after an area has been fragmented may give species half-lives of about 50 years.

Table 12-7. Species area curves. Coefficients of the specie area curves $S = c A^z$. A in hectares, c as numbers. Data after Krebs (1995 p. 594, 588-89) and references therein, Galápagos island, West Indies. Generic: May et al. (1995 p. 16). Rosenzweig (2001). Size of areas studied range from 10,000 sqmi to 40,000 sqmi and number of species from 100 to 400.

No	Species	Type of area	c	z	Geographic area
1	Plant	Islands	4.83	0.32	Galapagos
2	Birds	Islands	2	0.24	Western Indies
3	Birds	Mountains > 2300 m	0.059	0.165	Western USA
4		Land-bridged islands	–	0.22–0.36	
5	Reptiles	Islands	0.62	0.30	Western Indies
6		Ocean areas		0.1	
7	Mammals	Mountains > 2300 m	0.194	0.326	
8	Mammals	Part of continents		0.1–0.2	Generic
9	generic	Provinces		1.0	Generic

Combining species area relationships with half-life estimates suggests that current extinction rates are on the order of 1% species per year. With a background rate of 0.40 extinctions per million species per year[45] current rates of mammalian extinction are 17–377 times the background extinction rate, or 7 to 150 extinctions per million species per year. Because of the time lag in extinction, about 10–12 % of all plants and birds are now threatened and are likely to become extinct in the wild in the medium- to near-term future. An example of this "time-lag" is birds in the South Atlantic forests where nearly 90% of the rainforest has been cleared, yet no bird species has so far been shown to be extinct. Extinction rates are extensively discussed in the current literature.[46]

Loss of species appears to occur in three phases. In the first phase the *endemic* species are lost. These are species whose entire range of habitat disappears, for example by being turned into cultivated land. The loss is rapid, but only a modest number of species become extinct. In the second phase, the *sink species* are lost, that is the species that do not reproduce themselves sufficiently for maintaining their population. This reduces the number of species to about 50% of what it was. Phase three extinction reduces diversity to a new steady state.

If we continue to exclude Nature from 95% of the earth's surface, the new steady state will have 5% of current diversity.[47] Contrary to intuitive beliefs, it appears as if species survive preferentially in their historic periphery and not in their present center of abundance.[48]

New species are slowly created, and geographic isolation of a population of species is believed to be the dominant creation factor. To be isolated, there should be less than two to three migrants per generation. However, recent research tends to show that new species may be created even with less isolation. What factors that promote speciation are currently not clear. Some species are more prone to extinction than others; and some species may benefit from human land use and other activities. For example, voles benefit from enhanced understory vegetation in managed forests.[49]

An important question in the field of ecology is whether high diversity enhances or decreases productivity. Some results suggest that as a rule of thumb each halving of diversity leads to a 10% to 20% reduction in productivity. However, the results were based on plant diversity data from grasslands, and may not be generally valid.

5.4 Minimum Viable Populations

The persistence of a population can be formulated as a function of the size of the population, N_m, the population's average growth rate, r_m, and its variance r_V caused by inherent variability and environmental fluctuations.[50] Sometimes population variability is substituted for r_v. Population variability represents both environmental and intrinsic variability. Intrinsic variability is often caused by too high population growth rate compared to its "braking power", so that the population overshoots its carrying capacity, *K*, and then sinks below *K* again. The result may be oscillations like those shown for canopy species in Figure 12-5. To keep the population size at *K*, selective taxation by humans is often required.

The **minimum number of individuals** that will give 95% persistence over 100 years is often called the minimum viable population, *MVP*. Roughgarden (1997) argues that extinction of populations is only possible if $r_V > 2r_m$, otherwise, populations tend toward their carrying capacity. However, Belovsky (1990) found that the variances in climate often were larger than $2r_m$, ranging from $1.43r_m$ to $7.32r_m$. Low values were found from observation of rainfalls in temperate forest and tree growth rings in dry temperate forests. High values were found from observations of rainfall on grassland and tree rings in moist temperate forest. Using an empirical formulae that relate growth rate, r_m to a species biomass expressed in grams:

$$r_m = 18.0\ W_g^{-0.36}\ (\text{no yr}^{-1}) \quad (12\text{-}16)$$

Belovsky found that the *MVP* in low variability and high variability populations respectively was (W in grams, his figures):

$$N_m = 491950 W_g^{-0.36} \quad r_V = 7.32,\ r_m\ (\text{high variability}) \quad (12\text{-}17)$$

$$N_m = 30271 W_g^{-0.40} \quad r_V = 1.43,\ r_m\ (\text{low variability}) \quad (12\text{-}18)$$

This means that for large species, around 500 individuals are required to secure a viable population. Since the exponents are almost equal, the ratio between the number of species with high and low variance in growth rate is about 15:1. If the ratio between females N_f and males, N_m, in the population is uneven, the effective population; N_e, can be calculated as:[51]

$$N_e = 4N_mN_f/(N_m + N_f) \quad (12\text{-}19)$$

For management purposes, it would be convenient to know how large areas are required to support MVP. Generally, larger areas are required for carnivores than herbivores because animal foods are less abundant than plant food (1:10). Population densities are often lower in tropical habitats than in temperate habitats. Belovsky (1990) found MVP and areas required to support viable populations as suggested in Table 12-8. About the same numbers are required in temperate and tropical environments. These areas can be compared to the areas found by Harestad and Brunell (1979) for home range as a function of bodyweight in Chapter 19 on the terrestrial environment. The areas Belovsky (1987) finds for herbivores appear to be supported by empirical observations, whereas areas required for carnivore species probably must be larger. The largest mammalian carnivores (10-100 kg) can be expected to persist 100 years in up to 22% of the current parks, but no park is large enough to guarantee persistence for 1000 years. For the largest mammalian herbivores (100-1000 kg) 4–100% of the parks will allow persistence for 100 years and 0%–22% will permit persistence for 1000 years.

In a later study, Belovsky et al. (1999) found that the initial population size for an endangered population was not as important as having as large carrying capacity as possible, because a population far below the carrying capacity would soon catch up in population size.

Table 12-8. Area requirement for herbivore and carnivore species in temperate environments. Number of mammals in low variability environments. The numbers required in high variability environments are about 15 times higher. Space requirements for herbivores in temperate environments. Carnivores require about 15 times as much space. Data: Belovsky (1987).

Animal weight, kg	Representative species	Required number of individuals for 95% persistence for 100 years	Space herbivore, km^2	Individuals per km^2
0.030	Mouse	8,000	1.6	5000
1.000	Rabbit	2,000	4.0	500
30.00	Dog	500	10	50
100.0	Deer	300	14	22
300	Cow	200	18	11
3000	Elephant	80	40	2

Belovsky also showed that non-linear population dynamics was an important cause of extinction. What does this mean for conservation planning? First, managers must preserve as large K as possible either by conserving as large areas as possible, or by management actions. Secondly, since over-crowding enhances population oscillations, over-crowding could be abated by planned mortality and hindrances to immigration.

Removal or addition of species in an ecosystem may change the food web, but there is no consensus on what an optimal food web would look like, or if simple food webs that resemble food chains are better or worse (e.g., more stable) than others. Recent work tends to suggest that weak interactions tend to enhance the stability of nature.[52]

5.4.1 Case Study 3: Species Extinction

Problem: A protected area is 300 km^2. Is this area sufficient to maintain a herd of reindeer (average weight 55 kg) for 100 years with 95% probability?

Solution. According to the formulae for animals in high variability regions the number of animals is $N_m = 491{,}950 \times 55{,}000^{-0.36} = 9669$ animals. Next, you have to know how large a space a reindeer requires for food support. You can either use the results from Case study 2, or you can use Equation 3 in Table 19-5 (note, that equation is per km^2). N (no ha^{-1}) = 447 $\times 55{,}000^{-0.66}$ = 0.33 ha^{-1}, thus 300 km^2 will support 9900 reindeers, which is more than the 9669 animals required. The observed density of reindeer is 0.013–0.3 ha^{-1}. The results in Case study 2 suggested a density of 0.14–0.3 reindeer ha^{-1}.

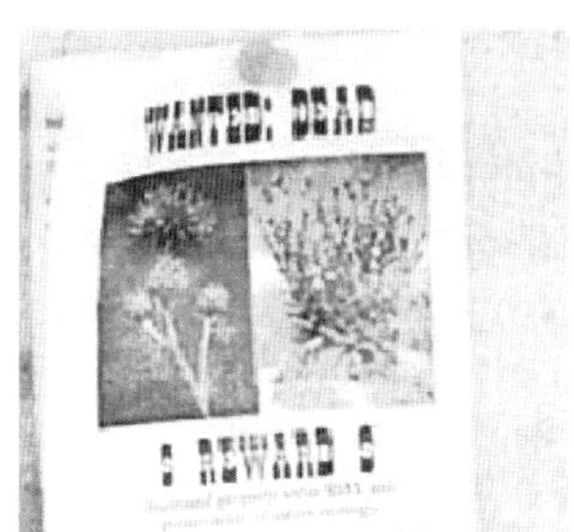

Figure 12-8. Two plant species. Left: The endangered Altai Hawk´s beard. Right: A "Wanted" poster for an invasive weed species photographed in front of the town hall in Netherlands, Colorado, 2005.

6. APPLICATION: THE DEATH OF THE FLOWER ALTAI HAWK'S BEARD

Goodwill astray

Altai Hawk's beard is the name of a flowering yellow Asteraceae (*Crepis multicaulc*) that lives in Northern Russia, the Altai mountains in central Asia, and in scattered distributions in Dsjunania, and some other sites, Figure 12-8. Until 1943, it also existed at one place in Europe, on a sloping hill in Varangerfjorden, Northern Norway. Thus, the plant was abundant in the Altai Mountains (part of this mountain is a World Heritage Site), but rare in Norway.

6.1 Decision Analysis

Below, we describe the basic components of a rational decision analysis process.

Decision problem: How should one protect the Altai Hawk's beard so that it continues to survive in Varanger?

Decision makers: The *decision maker* is the central government, since the Altai Hawk's beard was legally protected in 1919.

Stakeholders: Stakeholders are national agencies responsible for species preservation, environmentalist groups, NGOs, and plant collectors searching for rare species. How to identify stakeholders under different situations is discussed in Chapter 3, section 2.3.

Decision objectives: The *goal* is to protect the plant at least social and economic costs. *Subgoals* are:

- Maximize the number of individuals of the plant.
- Maximize the joy of plant enthusiasts.
- Minimize probability of extinction.
- Minimize costs.

Alternatives: There were two *alternatives*, to leave the habitat as it was, i.e., status quo, or to build a fence around the habitat and set up a sign-post where the plant was designated as protected.

Consequence analysis: A consequence analysis identified the factors that would make the Altai Hawk's beard survive in Varanger; and fencing it was one option. Figure 12-4 shows some issues related to species preservation and Figure 19-4 shows how to formulate optimum and acceptable conditions for plant survival. In Chapter 21, section 7 we show how species can be described according to vulnerability. By fencing in the area and marking it as protected, strollers would be able to notice it easier, and at the same time be warned that they should not pick plants. The cost of putting up and maintaining the fence was estimated. The consequence analysis should result in a consequence and a utility table like Tables 3-5 and 3-6, and primary utility graphs, like those in Figure 3-3.

Preferences: The preferences that were expressed by experts advising the government indicated that the benefits of protecting the plant by a fence would outweigh its cost. In a formal analysis, the preferences should be represented by importance weights for the criteria, as in Table 3-8.

6.2 What Happened?

The plant was discovered in Varanger, protected, and then died. There were two reasons for this. By building a fence and marking it as protected, plant collectors could find it. Secondly, the plant needs light and soil that is stirred. It also needs open space so that the seed will germinate. The fence eliminated grazing, and the plant habitat became overgrown with other vegetation, thus its abiotic environment would not give it sufficient competitive strength to survive. In this case the consequence calculations were erroneous both with respect to plant characteristics, and with the respect to the anticipated behavior of the strollers and collectors. The problem is common in the quest for species preservation: As agricultural practice is changed and large areas in developed countries are left ungrazed, many species will disappear.[53]

---

[1] Haftorn (1971)

[2] Czech and Krausman (1997) p. 1116

[3] Scholin (2000)

[4] Sterner (2003)

[5] O'Neill (1997)

[6] Based on Ando et al. (1998)

[7] James et al. (1999) p. 323, Inamdar et al. (1999): US$ 27 in Tanzania, Cameron $20, Zimbabwe $132

[8] Musters et al. (2000), James et al. (1999), Soulé and Sanjayan (1998)

[9] Burton et al. (1992)

[10] Pimm and Raven (2000)

[11] Hughes et al. (1997)

[12] Van Jaarsveld et al. (1998)

[13] ICUN 2000 Red list available at http:/www.redlist.org/.

[14] zebra mussels: Effler, Boone et al. (1998), Nicholls and Hopkins. (1993); rabbits: Finkel (1999); McEcosystem: Enserink (1999); Injurious spp: Naylor, Williams et al. (2001)

[15] Myers et al. (2000)

[16] Partly after May et al. (1995) Dobson et al. (1997): Table 1, USA ; Rodrigues (2000) p. 241
[17] Dobson et al. (1997)
[18] Nee and May (1997)
[19] Mann (1999)
[20] Pauly and Christensen (1995)
[21] Inverted and redrawn after Krebs (2001)
[22] Tømte et al. (1998), Hsieh et al. (2005)
[23] Marshall and Peters (1989), Sandvik et al. (2004)
[24] Seip et al. (1994)
[25] Krebs (2001), Ricklefs (1990)
[26] Jones et al. (1997)
[27] Johnson et al. (2004) p. 34
[28] Lakes: Scheffer, Hosper et al. (1993); forests Holmgren, Scheffer et al. (2001); saline areas: when trees are removed, the deep roots that formerly supplied fresh water to the soil are cut and the saline water table moves upward, Sarah Lumley, University of Western Australia, personal communication, Bolt et al. (2001)
[29] Holmgren et al. (2001)
[30] See Wachkernagel and Rees (1995), Rennings and Wiggering (1997)
[31] Anderson (1997:147)
[32] Vollenweider (1976)
[33] McNaughton et al. (1989), Moen and Oksanen (1991), Begon et al. (1995) p. 652, 672-678
[34] Data: Seip (1983:401), Østbye and Mysterud (1980) p. 64
[35] Østbye and Mysterud (1980) p. 64
[36] McNaughton et al. (1989)
[37] Rosenzweig (2001)
[38] Kerr and Parker (1997), fish, Oberdorff et al. (1995), zooplankton: Rutherford et al. (1999)
[39] Hughes et al. (1997)
[40] Krebs (1995) p. 527, 535-536
[41] Stevens et al. (2004)
[42] Rosenzweig (1999) p. 276
[43] May et al. (1995) p. 14
[44] Hughes et al. (1997) p. 591
[45] Pimm and Raven (2000), Regan et al. (2001)
[46] Brooks and Balmford (1996), Gibbs (2001), Regan et al. (2001)
[47] Rosenzweig (1999) p. 277
[48] Brooks (2000), Channell and Lomolino (2000)
[49] Stenseth et al. (1998)
[50] Belovsky et al. (1999) p. 1175
[51] Soulé (1982)
[52] McCann et al. (1998)
[53] Lid and Lid (1994), Tutin et al. (1976)

Chapter 13

HARVESTING

Sustainability, footprints, and harvesting economics

We start with the concepts of sustainability and footprint, which means that regions should be self sufficient. We then discuss the economics of sustainable harvesting, and show that many different levels of harvesting effort are sustainable, but some levels are better for the environment than others. The question is then what kind of management policies give the most efficient results.

"Pollution is nothing but the resources we are not harvesting. We allow them to disperse because we've been ignorant of their value".
R. Buckminster Fuller, Inventor/Philosopher

1. INTRODUCTION

Occupation of land for settlement, industry and infrastructure makes land for agricultural purposes increasingly scarce and accelerates soil degradation and other negative aspects of land use. Thus, the current challenge is to develop sustainable land-use practice. This involves improved agricultural practice, and touches on economic, social, political and ethical issues.

The notion of the "sustainable city" suggests a regional strategy where the world as a whole is maintained in a sustainable state by making sure that smaller regions are locally sustainable. McGuire and Childs (1998), for example, claim that paper mills should be small and be located close to the region they serve.[1]

2. BACKGROUND INFORMATION

In this section we discuss themes related to sustainability. A key concept is "footprint" which addresses local sustainability.

Figure 13-1. Regional ecological deficit. The ecological footprint measures how much biological capacity people occupy. To the left, the footprint is larger than a sustainable economy would allow.[2]

2.1 Footprint

The **sustainable footprint** of a population is the aggregate area of land and water on which one can, in principle, produce all the resources the population consumes, plus land required for ecosystem services such as cleaning of water and air, plus land required for shelter in a wide sense – or built land, plus land for absorption of waste. Assessment of the size of the sustainable footprint must be performed continuously, always using prevailing technology. In Figure 13-1 left, the sustainable footprint is larger than the available land, and the situation is therefore not sustainable; to the right, it is.

The footprint concept has become a convenient measure for expressing sustainability requirements, since resource flows and essential life support services that nature provides often can be expressed in terms of such equivalent areas of land. The footprint concept is parallel to, but more general than, the notion of "carrying capacity" used in ecology. Table 13-1 shows estimates of land requirement per inhabitant for a selection of human activities.

Table 13-1. Terrestrial area required to support one inhabitant.[3,4]

Activity	Area required per inhabitant
Hydropower	$0.28\ 10^{-4}$ km^2
Waste	$0.3\ 10^{-8}$ km^2 yr^{-1}
Total requirement for town citizens	0.06–0.11 km^2
Arable land	0.003–0.005 km^2
Forest	0.003 km^2
Pasture	0.018 km^2

The figures above are rule-of-thumb figures for people in developed countries, the actual size of a footprint, F, depends on number of people, P, level of affluence, A, climate, C, and technology, T.[5]

$$F = f(P, A, C, T) \tag{13-1}$$

As the population increases, the impact increases. So does the impact with affluence, which is related to consumption. Technology may work both ways. Better technology allows extraction of deeper ores and longer transport roads. On the other hand, areas required for food production may be reduced, and more waste could be reused or disposed of without affecting the environment. Waste incineration may compensate the need for fire-wood forests. The climate factor is included since harsher climates require larger areas for food production. Estimates of land requirements for the populations of Canada, the Netherlands, and India, are shown in Table 13-2.

In Table 13-2, *CO_2 storage* refers to land required to sequester CO_2 released by fossil energy use. *Built land* includes land used for such things as roads and homes.[6] *Garden* refers to land required for vegetable and fruit production, *Cropland* to land required for crops, *Pasture* to land required for dairy, meat and wool production, and *Forest* to area required to produce wood. We see that the largest footprint contribution comes from forest requirements to compensate for fossil energy use, which is not a sustainable source of energy.

A conservative estimate shows that the average person living in an industrialized country presently needs the equivalent of two to five hectares of productive land to sustain consumption of materials.[7] Table 13-3 shows sustainable footprints for three footprint classes, represented by one country that has a deficit and one country that has a surplus of footprint area in each class.[8]

Table 13-2. Footprint estimates. Land use for the populations of Canada, the Netherlands, India and Sweden. The totals imply that 12% of land areas should be set-aside as wilderness area.[9] Hectare per capita. World supply is the land available for the total world population if shared equally.

Country	CO_2 storage	Built land	Garden	Cropland	Pasture	Energy	Total
Netherlands	2.10	0.04	–	0.45	0.26	0.47	3.32
Canada	2.34	0.20	0.02	0.60	0.46	0.59	4.2
India	0.05	–	–	0.20	–	0.13	0.38
Sweden	2.6	0.70	–	1.20	0.90	1.60	7.20
World supply	0.0	0.20	–	0.70	0.30	1.00	2.10

Table 13-3. Footprint surplus or deficit in selected countries.[10] To get total ecological footprint, multiply the per capita data by the country's population.

Country	Ecological footprint, ha capita^{-1}	Population 1995 10^6	Available bio-capacity, ha capita^{-1}	Footprint deficit (–) or surplus (+)
Less than 3.9 hectare footprint per capita				
China	2.5	1247	0.8	–0.4
Argentina	3.9	35	4.6	+0.7
Between 4.0 and 6.9 hectare footprint per capita				
Japan	4.3	125	0.24	–3.4
Sweden	5.9	9	7.0	+1.1
More than 7.0 hectare footprint per capita				
United States	10.3	268	6.7	–3.6
New Zealand	7.6	4	20.4	+12.8
The world (1)	2.8	5892	2.1	–0.7

(1) There is 1.5 ha per capita productive land on Earth, and only about 0.3 ha per capita in the Central European countries. Of these hectares, less than 0.3 ha is suitable for agricultural production.

Discussion: The footprint concept focuses on resource limitations of land and water areas and household balance, and uses land area as a sort of currency. It would certainly have been a valid concept if it were used for the world as a whole, which for practical purposes is a closed system. However, several assumptions are questionable when it is applied to smaller areas with "open borders". One problem is that different ecosystems and resources require different borders to function meaningfully. Water has catchments areas that may include several regions – the world's longest rivers are more than 1000 km long. Air blows eastward on the Northern Hemisphere carrying pollution to other countries – drift distances of 1000 km is not unusual. Many fish species cover large distances in the oceans. Large mammals like wolves, tigers and reindeer require vast land areas – on the order of 1 km^2 per individual – and define habitats that change with season. Also, many goods and services can be transported over long distances without serious losses, such as electricity, oil, and gas. A tenable position is perhaps that local sustainability is the default requirement, and not to impose it requires argumentation.

2.2 Agriculture and Sustainability

The history of agriculture can be followed along four lines of development:

- *Technology*: new technological tools have been taken into use, like soil preparation equipment, harvesting equipment, irrigation systems, bio-technology, pesticides, and fertilizers.

- *Labor*: Work-hours per harvested unit have constantly decreased, in step with the technology development.
- *Farm level sustainability*: Local farm-level sustainability is becoming weaker as increasingly smaller portions of resources used for agriculture come from the farm itself, and fewer farm products are used on the farm.
- *Soil, land and ecosystem sustainability*: Increasing demand for farm products leads to more intensive utilization of the soil. Pesticides and monocultures affect ecosystems.

We have limited knowledge of the development patterns along these lines, but we currently believe that early agriculture and the most recent agricultural systems in the developed world, say after 1980, have a higher degree of sustainability than in the periods in between.[11] This development has been described by the **Kuznets curve,**[12] which has a ∩-shape representing the environmental load as a function of gross domestic product, GDP, per capita. The picture is not simple, however, since the development varies quite a bit among regions, with highly technical agricultural systems occurring simultaneously with low technology systems.[13]

In developing countries, proteins are often the limiting human nutrient. Thus, the source of protein determines the carrying capacity of the human population in areas that are largely self-sustained. The US recommended daily intake of protein for a 70 kg male is 50 g. Assuming that boneless meat contains 20% protein, this translates into a *minimum recommended daily amount* (RDA) of 0.28 kg meat per day. This is comparable to the actual meat consumption recorded for a range of populations in Africa, Asia and Latin America.[14] Fish consumption, as an additional food source, ranges from 0.02 to 0.13 kg per day per person.

3. HARVESTING

Harvests depend on resource availability, which in turn depends on stock size and yield. We therefore first have a look at prevailing stock sizes of selected non-renewable resources as well as maximum sustainable yields of some renewable resources.

3.1 Stock Size Estimates of Non-Renewable Resources

Estimates of stock life length of fossil fuels vary considerably. Oil stock life length has been estimated from 17 to 80 years. Stock life length of natural gas is estimated at 66 years and for coal at 261 years. Reserve life for coal is estimated to be 3226 years with future technology.

Table 13-4. Estimated reserve life (years) of metals based on current consumption rates.[15]

Metal	Current technology	Future technology
Aluminum	256	805
Copper	41	277
Cobalt	109	429
Molybdenum	67	256
Nickel	66	163
Platinum group	225	413

Nuclear power requires naturally occurring uranium to supply the isotope ^{235}U that releases energy when it disintegrates. A fast-breeder reactor can utilize the more abundant ^{238}U that has a reserve life of probably several hundred years.

3.2 Waste Absorption

Waste deposition may be regarded as the opposite of resource abstraction. Among nature's services is its ability to break down human refuse and store molecular constituents in proportion to the background storage of the molecules. For the mechanism to be sustainable, loadings must not exceed *critical* or *allowable* levels. This is true for most organic products. Some materials may be landfilled or deposited in deep water and may potentially be stocked and saved for unforeseeable time. However, to assess possible time durations for safe deposition, biological and physical conditions, spreading and eruption features, and flows and advection properties at the site, have to be known – and still, unpredictable dynamics may change the expected patterns.

3.3 Resource Life of Renewable Resources

Renewable resources include biomass produced by plants and animals. A sustainable level of extraction of these resources ranges from 5% to 20% of the standing stocks.[16] However, extraction of biomass carries a risk of species extinction. Since the estimated rate of species creation is about five species per year, a sustainable level of biomass extraction would be a rate that worldwide may cause the extinction of 5 species per year. This contrasts with the current and expected rate of reduction, which has been estimated at something between 365 and 65,000 species per year.[17]

4. HARVESTING MODELS FOR RENEWABLE RESOURCES

Good harvesting practice requires knowledge of ecology, economics and sociology. Ecology ensures that one knows at which level stocks are susceptible to extinction, economics helps the harvester obtain an optimal yield, and sociological knowledge may mitigate externalities of the market economy.

Harvesting impacts ecosystems in different ways. Harvesting of old-growth forest removes an almost non-renewable resource, or at least a resource that cannot be renewed on human time scales. Removal of trees or wood also impacts the function of the ecosystem, both during the removal process, and afterwards.

Harvesting theory addresses three major questions:

- What is the maximum sustainable yield (MSY) one can obtain by harvesting from a population?
- Is *MSY* an economically feasible harvesting rate when decision-makers maximize their profit?
- What is the optimal harvesting strategy for a fluctuating population? Is there more than one optimal strategy? Does it require a particular ethical code among harvesters to keep harvesting at the optimal level?

We describe the theory of harvesting of renewable resources by using fish harvesting as an example, but the basic theory applies to all sorts of harvesting of renewable resources, like wood, game, agricultural products and wild berries. We first introduce the concept of sustainable harvest, and then turn to harvesting economics.

4.1 Sustainable Harvesting

In Chapter 12 on ecology, we introduced the **logistic growth model**, which is the most common model of biomass growth. Growth is determined by the differential equation (12-3), which we reproduce here:

$$dN/dt = r_0 N\left(1-\frac{N}{K}\right) \tag{13-2}$$

If we apply the model to a fish stock, r_0 is the intrinsic rate of fish breeding and body mass growth. The model simply says that the growth rate, dN/dt, of fish biomass, N, is almost proportional to N when N is much smaller than the carrying capacity K ($N/K \approx 0$). Thus, growth is small when N is small since there are few fish to breed even though there is abundant food available.

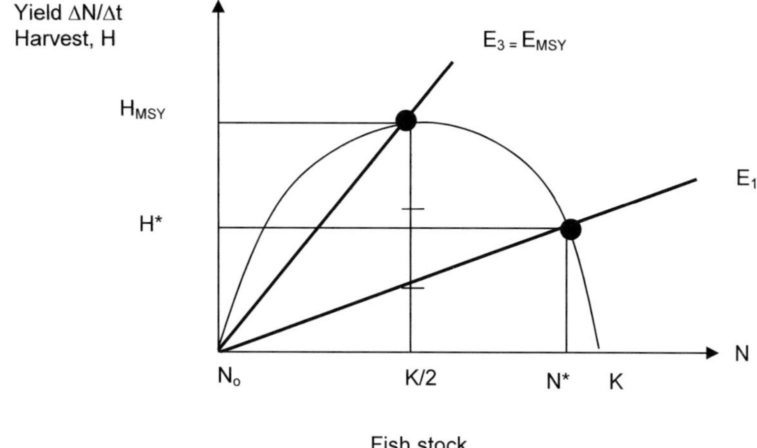

Figure 13-2. Growth rate, or yield, of a fish stock following the logistic growth model. N is stock size and K is carrying capacity. E_1 and E_3 are fishing efforts (e.g., number of fishing vessels). Fishing with effort E_1 at fish stock size N^* gives a sustainable harvest of H^*. H_{MSY} is maximum sustainable harvest and E_{MSY} is the fishing effort that would give this harvest if applied to a stock of size K/2. The dashes at K/2 are explained in the text.

The growth is also small when N is close to K ($1 - N/K \approx 0$), since it is limited by food supply. If we depict the growth rate as a function of N between 0 and K, it has its maximum at $H_{MSY} = r_o K/4$, just in the middle, where $N = K/2$. The growth curve is shown as the bell shaped curve in Figure 13-2.

If fisherfolk continuously reap off the growth, the stock size will remain constant, and this would therefore be a sustainable harvest rate. Thus all points on the curve in Figure 13-2 represent possible sustainable harvesting rates, and the maximum sustainable harvesting occurs when the stock size is kept at half the carrying capacity.

4.2 Harvest Effort

The two straight lines, E_1 and E_3 in Figure 13-2 represent two levels of harvesting effort. Effort E_3 is higher than E_1. You may think of harvest or **fishing effort** as the number of vessels that are fishing. If the harvest, H, is proportional to the size of the stock, N, then $H = E \times N$, which are the straight lines in Figure 13-2. Fishing effort can also be measured as the number of gears used, or other convenient units. Whether a particular fishing effort is sustainable or not, depends on the stock size, however. In Figure 13-2, E_1 is only sustainable when $N = N^*$, and E_3 is only sustainable when $N = K/2$. Let us now explore the relation between fishing effort and sustainable harvest. At carrying capacity, *K*, there is no growth, the fish stock gives no yield, and fishing effort is zero. This is the leftmost point in Figure 13-3. Fishing effort increases to the right, but the higher the effort is, the less is the stock of fish. This is quite obvious, if you think about it: At equilibrium, where we have the same stock size from year to year; we need more fishing boats to keep that stock level low. To keep the sea almost empty of fish, you need quite a number of fishing boats looking for the few fish there are. Therefore, the other *x*-axis represents the fish stock, and points the other way.

If you want to draw a curve with some precision describing sustainable harvest as a function of fishing effort, return first to Figure 13-2 and draw a family of fishing effort lines starting at N_0 and intersecting at equal distances a vertical line going through *K*/2. Since the effort curve E_3 gives about three times the yield of the effort curve E_1 at the same stock size, the effort must be three times as high for E_3 as for E_1. Now, with E_1, E_2, E_3, etc, plotted at equal distances along the effort curve, we can find the corresponding sustainable yields where the effort lines intersect the bell shaped fish yield curve. The resulting graph looks like Figure 13-3.

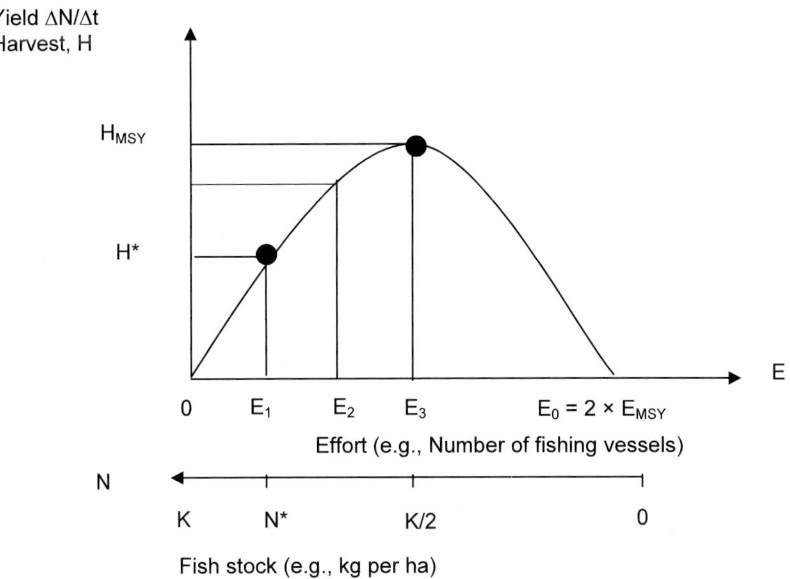

Figure 13-3. Sustainable fish harvest as a function of fishing effort, which can be measured in terms of fishing vessels. The bell shaped curve show the harvest when the efforts, E_1, E_2, E_3, are applied to stocks that exactly gives the sustainable yield. For example, fishing effort E_3 is applied to a fish stock that is half the stocks carrying capacity, *K*/2, and gives then the maximum sustainable yield, H_{MSY}. Note that increasing sustainable harvesting effort, gives decreasing fish stock, as indicated by its opposite direction on the horizontal axis.

4.3 Harvesting Economics

We now turn our attention from physical harvesting to the economics of harvesting.

4.3.1 Assumptions

We assume sustainability, meaning that annual catch of fish always equals annual growth, and the question is what harvesting level is most profitable. The harvest curve in Figure 13-3 can be translated into an income curve by introducing some simple and reasonable assumptions, which still are not really necessary for the qualitative aspects of our conclusions.

First, we assume that the cost of fishing, C, is proportional to fishing effort, E.

$$C = cE \tag{13-3}$$

Similarly, we assume that the price for the fish, p, is constant per unit of fish harvested:

$$I = pH \tag{13-4}$$

The difference $NR = I - C$ is the net revenue, or profit, of the fishing effort.

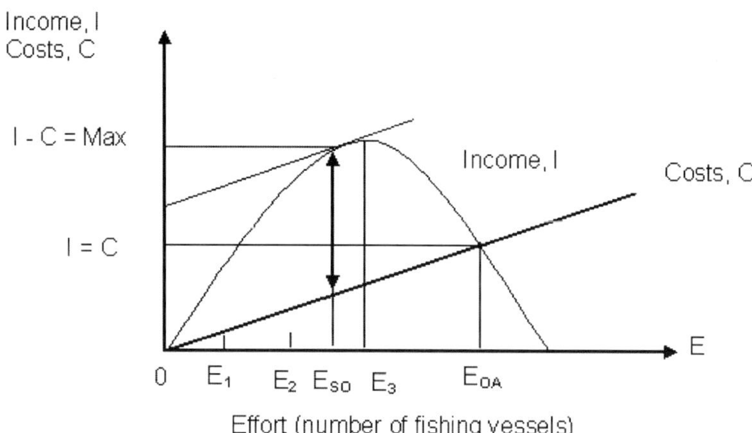

Figure 13-4. Economic yield as a function of fishing effort E. E_{OA} = *Effort at open access*, E_3 = *Effort at maximum sustainable yield*, E_{SO} = *Effort for sole owner at maximum profit.*

4.3.2 Sole Owner

Suppose a sole owner, SO, has all the fishing rights and maximizes his own profit. Which sustainable effort level will he choose? Look at Figure 13-4; starting from zero effort the revenue *I* increases much faster than the costs *C*. He will therefore increase the effort until the profit $I - C$ is as high as possible. This occurs at E_{so} where marginal income is exactly parallel to the cost curve. After that point, the costs will increase faster than the revenues. The sole owner will therefore choose a harvesting rate that is lower than the maximum sustainable rate.

If the harvesting cost, *c*, increases – for instance through taxation – so that the slope of the cost curve increases, the optimum effort becomes smaller and moves towards zero, meaning that the fish stock becomes too expensive to harvest.

If the cost of harvesting is very low, the cost curve approaches the *horizontal axis*, and the economically optimal harvest will approach maximum sustainable yield, Y_{MSY}.

4.3.3 Open Access

What happens if there is open access to the fishing banks? We know from our discussion of game theory in Chapter 6 that this easily leads to non-sustainable over-fishing and depletion of the stock. What actually will happen depends partly on whether the technology is effective enough to make over-fishing and depletion of the fish stock possible. Here, we will assume that the result will be sustainable, and the question is how large the standing stock will be.

In an open access system, anyone may start fishing, and a fisherman will fish as long as the income is greater than the cost. Thus new fishermen will continue to arrive at the fishery as long as there is a positive profit to reap. At E_{OA} in Figure 13-4, fishing cost is exactly equal to the yield, and further increase in fishing effort is therefore not profitable.

We see that the combined fishing effort will be greater than the sole owner fishing effort, and the fish stock will stay at a lower level. But as long as the unit effort costs are positive, the fish stock will be at a fixed, but low level. However, at low levels there is a possibility that the fish stock goes extinct, should unfavorable conditions arise.

4.3.4 Growth Cycles

In reality, growth and relationships between catch and fishing efforts are much more complicated than the simple logistic and linear functions we used. Fish swim often in schools, which tend to make catch a non-linear function of effort with a substantial element of luck. However, our conclusions did not depend on the particular form of the growth curve or catch/effort functions. Most reasonable curves will do. More crucial is the assumption about sustainability; that we assume that we harvest only the growth so that the size of the stock is constant. In practice this will seldom happen. A population does not grow smoothly, but is exposed to fluctuations caused by abiotic and biotic factors. Dynamic interactions between a stock of predators and what they eat often produce cyclic variations. Some species experience high natural mortality rates, *M*, like forage fish that are prey of piscivore fish. This high mortality limits these species' ability to withstand high fishing pressure, *E*, and thus harvest per total fish growth or fish production, *P*: $Y/P \approx E/(E+M)$ tends to be low, about 0.5. For ground fish species, that is piscivore species, *Y*/*P* tends to be about 0.8.[18] It can be shown that with cyclic behavior, harvesting needs to be lower than at maximum sustainable yield.[19]

If a population fluctuates, there is always a risk of extinction. However, many aquatic organisms have mechanisms, like dormant stages, which secure survival even under harsh conditions.[20] The opposite effect, that an organism with a small number of individuals is self-destructive, is called **depensation**. If a species has none of these properties, however, it can be shown that the expected life span of the population is longest if the species is harvested at maximum rate when the stock is above carrying capacity and no harvesting occurs when the stock is below that level.[21]

4.3.5 By-Catch

By-catch of a harvesting process is harvesting of other species than the target species. These species may either be collected, or dumped after they have been killed. For example, seed collectors on the Northeastern coast of India collect about 10,000 seeds of tiger shrimp, but at the same time destroy 1,5 million juveniles of other prawn species, 60,000 fishes, 1,9 mill crabs and other marine resources.[22] In some cases the by-catch may be as detrimental to the marine ecosystem as the catch itself.

4.3.6 Optimal Harvesting of Slowly Growing Stocks

If the growth rate of a stock is so slow that it is less than the interest rate one can obtain in a bank, it will be pay off to harvest it as fast as possible. Whale stocks, for example grow at a rate of 2% to 5% per year; desert bighorn sheep at a rate less than 2%. If a sole owner harvests quickly down to where harvesting costs matches income, OAE, then by selling the yield and putting the money in a bank, she will obtain a higher rent than by continuing to harvest the stock.[23] Pure economics therefore predicts that resources with low growth rates, like large animals and rain forests, will be harvested as fast as possible. Most fish species, however, grow faster than the typical bank interest rate, and are spared. Apart from the ethical side, the prediction above is probably based on too simple assumptions. More complex ecological and economic models that include the matrix of organisms the harvested species belong to may show that the species contribute much more to human life quality than just its commercial harvest value. Thus, sustainable harvesting of renewable resources appears to require ethical behavior to counteract intrinsic deficiencies in open access economies.

4.4 Wildlife Harvest

Hunting by local communities is among the most serious threat to wildlife in many countries. India is a prime example. There are many traditional hunting techniques, like nets, rubble traps and wire snares. Guns, however, are become increasingly popular even among indigenous people. More than 90% of hunters use guns in parts of India.[24]

An equation for estimating the *maximum sustainable harvest* of a local mammalian community in a forest if we know the standing biomass of the species is shown in Equation13-5.[25]

$$P_{max} = 0.6\, B \times N\,(\lambda - 1) \times (1\text{-}M) \qquad (13\text{-}5)$$

B is the average body mass of the individual animal, N is the observed density of each species (individuals per km^2), λ is the finite rate of increase, calculated as $\lambda = e^r$. M is natural mortality, $(1\text{-}M)$ is the proportion of species that can be harvested. It can be set to 0.2 for long-lived species, 0.4 for short-lived species, and 0.6 for very short-lived species, see Table 13-5. Equation 13-5 is similar to Equation 13-3: $B \times N$ corresponds to K and λ -1 to r_o.

Table 13-5. Maximum potential harvest from a mammalian community. Life expectancies are designated by l = long; s = short, and vs = very short. Data: Robinson and Bennett (1999).

Species	Body mass, kg	Density, No. km^{-2}	Intrinsic growth rate	Production, kg km^{-2} yr^{-1}	Potential harvest, kg km^{-2}
Marsupials, vs	1	55	2.92	577.5	346.5
Primates, l	2.4–7.0	25–40	0.07–0.17	7.4–20.5	1.5–4.1
Edentates, s	3.5	21.9	0.69	45.5	27.3
Rodents, s/l	1.5–45	1.6–5.3	1.42	11.6–42.8	3.2–10.0
Perrtsodactyls	160	0.5	0.20	10.6	2.1
Artiodactyls, s/l	25–35	2.6–30	0.9–1.25	22.9–209	9.2–41
Carnivores, s/l	2.2–35	0.04	0.23–0.46	0.1–11.7	0.04–2.3

The potential harvest calculated in this way can be compared to sustainable harvesting both by man and animals. Reported harvests are in the range 0.4 to 3.5 kg ha^{-1} and actual sustainable harvest of game meat from most forests is generally under 2 kg ha^{-1} both for man and for animals.[26]

4.5 Frequent and Short-Term Yield of Seaweed Harvesting

Harvesting of the marine benthic algae (*Ascophyllum nodosum*) occurs at certain intervals. Mechanical harvesting leaves larger portions of the algae intact (say 10%), thus allows shorter periods between harvesting (2-3 years) and gives a larger overall yield than cutting the plant very low. The growth follows the logistic curve. Figure 13-5 shows the yield as a function of how much is left after cropping, and the length of the re-growth period. The maximum yield is obtained with 16% of the plant left for re-growth and 1–2 years between harvesting.[27] If harvesting is expensive, it is advantageous to stretch the re-growth period. Such relationships between growth, harvest intensity and harvest frequency are important for good harvesting strategy.

4.6 Slash and Burn Harvesting

Slash-and-burn agriculture is apparently unsustainable, but has been used in many regions of the world, especially since it is convenient for retrieving new agricultural land from forested areas.

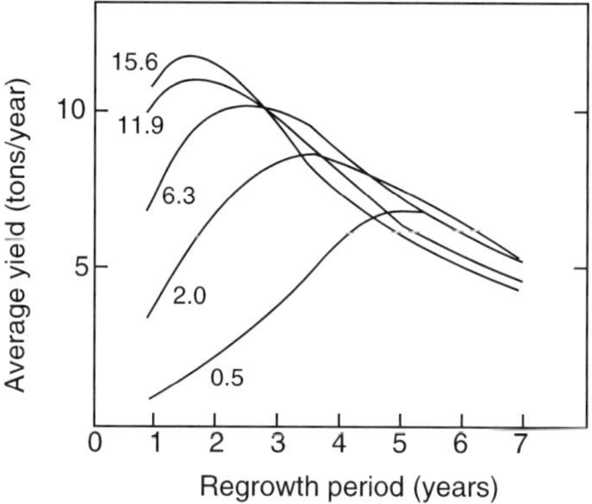

Figure 13-5. Yield as a function of harvesting intensity and harvesting frequency of the marine benthic algae *Ascophyllum nodosum*. The figure at each curve shows harvesting intensity in terms of percentage left after cropping.[28]

Slashed and burnt land grows back within 10 to 20 years, soon enough to repeat the slash and burn process in a sufficiently close area. The method was extensively used in humid Finland, and is presently used in semiarid and sub-humid tropical ecosystems in Latin America, Africa and Asia, and other places.[29] It requires too much land, however, to sustain large human populations. With population growth in former slash-and-burn cultures, the land sharing system collapsed, and the slash-and-burn technique became unviable. This is exemplified by increasing conflicts between cattle herders and agriculturists in Western Africa, and between owners and squatters in the Brazilian Amazon. However, the method illustrates a harvesting regime that is potentially useful for several stocks if properly applied. It may be ecologically sustainable because wildlife in the re-growth areas will experience natural succession regimes in their ecological matrix.[30]

5. HARVESTING POLICIES

There are several ways to restrict harvest to be within sustainable limits. We saw that adding taxes that increase the cost of fishing effort would lower the catches, but be difficult to get acceptance for. Giving subsidies could increase the sustainable catch for a sole owner, but not for open access fishery. Other options are to reduce fishing effort by setting a limit on the number of fishing vessels that may be used, their fishing capacity, the time they are allowed to fish, the landings etc. The use of transferable fish quotas, and futures contracts for a certain percentage of a fisherman's catch at an agreed price at a specified time, are economic instruments that are presently being investigated.[31] Recreational fisheries in east Florida set a bag limit of 1 fish per person and a slot limit on the size of the fish. Another option is to create marine reserves where fishing is not allowed.[32]

To restrict access, the principle "closeness gives rights" are sometimes adopted, giving concession to the feeling many people have that those living close to natural resources are entitled to the ground rent of the nearby waters. However, this would put restrictions on the distribution of individual transferable fishing quotas, which has become a popular instrument for fishing policy.[33]

Similar methods apply to harvesting of wildlife. However, all such restrictions have to be controlled, and few methods have been devised that have been found socially acceptable, economically equitable as well as morally agreeable.

To set fixed quotas is another option: Management calculates the maximum sustainable yield from fish census data, and this *MSY* is adopted as the fixed quota for that fishing season. Fish landings are recorded, and when *MSY* is reached, fishing is closed for the season. To set fixed quotas close to maximum sustainable yield is risky, since the stock may fluctuate to levels lower than that used to calculate *MSY*. If this occurs, the stock will decrease until it is no longer harvestable, or go into extinction.

One method of regulating harvests is to always let a constant fraction of the catch escape back into the water. This method is rather safe since it has a built-in feedback mechanism, which automatically will respond to changing stock densities.

6. CONCLUSION

Sustainable harvesting models of renewable resources are well established, as are mechanisms that can regulate harvesting efforts. Although several models exist, the model based on logistic growth appears to capture the most important features of harvesting processes. However, implementation requires good estimates of stock sizes, which is extremely difficult, causing many resources to be harvested in excess of their sustainable yield.

Major challenges are to find mechanisms that reduce harvesting, maintain, or increase equity among people, and are accepted by the people impacted by reduced harvest. According to the "footprint" concept, the world's resources in 2002 were harvested in excess of supply.

---

[1] MacGuire and Childs (1998)
[2] After Wackernagel et al. (1999)
[3] Data: Norway, Seip et al. (2000), Baltic area: Folke et al. (1997), generic: Pimentel et al. (1995) p. 1117, Ferng (2001), Wackernagel et al. (1999)
[4] Hydro-electric energy gives 1000 GJ ha^{-1}, Wackernagel et al. (1999). Space-use is for dams and corridor space for transmission routes
[5] c.f. Ehrlich and Ehrlich (1990)
[6] Haberl et al. (2001) p. 36
[7] Wackernagel et al. (1995), Wada (1993)
[8] Calculations can be found for Taiwan, Ferng (2001), Italy, Wackernagel et al. (1999), and Austria, Haberl et al. (2001)
[9] After Wackernagel and Rees (1995)
[10] Hall (1994)
[11] Selden et al. (1995)
[12] Selden et al. (1995)
[13] Lumley (2002)

[14] 0.03–0.39 kg capita^{-1} day^{-1}, modal value 0.25 kg day^{-1} capita^{-1}, reported by Robinson and Bennett (1999)

[15] Reserve lives for several metals were estimated by Froch and Gallopolos (1989:96): and Mather and Chapman (1995:145)

[16] Wackernagel et al. (1999)

[17] May et al. (1995)

[18] Mertz and Meyer (1998)

[19] May (1994)

[20] cf. the Golden apple snail., Naylor (1996)

[21] May (1994)

[22] Bhattacharya and Sarkar (2003)

[23] May (1994), Lande (1994)

[24] Madhusudan and Karanth (2002)

[25] Robinson et al. (1999)

[26] Robinson and Bennett (1999)

[27] Seip (1980)

[28] Seip (1980)

[29] Syers et al. (1996), Montagnini and Mendelsohn (1997)

[30] Western Africa: Syers et al. (1996); Amazon: Alston et al. (2000)

[31] Dalton (2005)

[32] Tupper (2002)

[33] Sterner (2003)

Chapter 14

TOXINS

Toxicity of pollutants for humans and ecosystems

We discuss toxicity of pollutants as they relate to humans and to the ecosystem. We discuss concepts like *levels of no concern*, *critical levels*, *safety factors*, *the cancer slope factor*, and *the hazard quotient* for the biotic environment.

●

"Water has a unique curative property which is not found in any other liquid. Adequate intake of water drives out the toxins from our body and make the body healthy." Rig-Veda

1. INTRODUCTION

With the exception of toxins produced for war or terrorism, toxic material is produced for good purpose. However, history has shown that detrimental side effects may overshadow good intentions. A notorious example is DDT (dichloro-diphenyl-trichloroethane), which was used as a pesticide until Rachel Carson in 1962 started focusing on its side effects.[1]

Not all toxic compounds are man-made; they are a natural part of the environment as well. In their struggle for survival, plants and animals develop chemical compounds to make themselves inedible by grazers and parasites. The problem with man-made toxic compounds, however, is that they are often made in large volumes, and may spread over long distances and cause effects in remote areas – like polar and oceanic regions.[2] The most important source of pollutants, in terms of gross damages to environment and human health, is probably coal and biomass burning, road traffic, agricultural pesticides, crop production on contaminated land and food additives.

A major challenge of toxicology is cost-efficient evaluation of the many new compounds that come on the market, for example persistent organic pollutants that may have subtle effects on the environment. It has proved difficult to generalize about the effects of toxic compounds just from their chemical constituents. Another challenge is to inform the public about threats of low, but toxic concentration of pollutants, and that it is necessary to accepts costs to mitigate them.

2. BACKGROUND INFORMATION

Damage to ecosystems and damage to humans are the two main concerns of toxicity evaluation. Concentrations of a chemical are often compared to reference- or background concentrations, which are assumed not to cause damage. Background concentrations are often retrieved from localities far away from human influence, like distant oceans. Since such localities are increasingly difficult to find, localities at the end of a gradient starting from a polluted locality are often used as proxy for unpolluted sites. Similarly, rural areas are used as background reference for urban areas.

2.1 Costs of Toxins

Compounds with potentially toxic effects are developed and used for many purposes, such as pesticides used to enhance crop yield – although there are claims that pesticide use begets more pesticides, and thus in the long run are not cost effective. There are advantages as well as costs connected to the use of toxins.

We do not have a full estimate of the costs to society of unwanted side effects of toxicity, among which are damage to human life quality in terms of shortening of life expectancy, damages to human health, nuisance, unsuitable land, and ecosystem damages. There are 1200 hazardous waste sites on the US National Priority list, or about five sites per million people. A cost of £12.60 per kg of pesticides for damages to health and ecosystems was found in England.[3] Unit costs for pollutants are calculated by averaging negative externality costs. Typical costs range from €5 per kg for volatile organic carbons to €4 million per kg for arsenic and total chromium.[4] However, it appears that even low estimates of costs prior to implementation of abatement measures easily turn out have been twice the actual costs calculated after implementation.[5]

2.2 Attributes of Toxins

Toxins have several attributes related to their use and effects on humans and the environment. Among chemicals that have been discovered with undesirable effects are **Persistent organic pollutants**, or POPs.

Table 14-1. Attributes of toxic, carcinogenic and in general harmful materials. The numbers quoted are suggested as national or international criteria for a toxin to be defined as persistent organic pollutants.[6]

Attribute	Measure
Persistence in the environment	Water half-life (2–6 months); soil half-life (6–12 months); sediment half-life (6–12 months)
Long range transport	Air half life (2 days), vapor pressure > 1000 Pa; remote measurement
Accumulation in biological organism	Bioaccumulation/concentration > 1000–5000, Log $K_{ow} = 5$ K_{ow} is the octanol water partition coefficient, expressing bio-accumulative properties
Toxicity, carcinogenicity, etc.	Acute toxicity; Chronic aquatic toxicity < 0.1 mg L^{-1}, mutagenity concern, based on 24 hr LC_{50} tests, teratogenic (acts on fetus), Endocrine disrupting chemicals.

A list of properties that make them particularly harmful is shown in Table 14-1, and these properties are also relevant for most other toxins.

2.3 Ecosystem Response

Ecosystems are vulnerable when its inhabitant species are susceptible to toxins. Toxins may interact with other stresses in complicated patterns that are difficult to unravel. One important question is how to describe ecosystem responses to toxins. How should we define end-impacts to an ecosystem that is useful from a managerial point of view? Put another way: how should we define the health of an ecosystem?

Total biomass is one possible ecosystem health criterion, but it is probably only good when there is a question of very high toxic loads with correspondingly severe impacts. Protection of total biomass will not protect tigers, or any other top predator. Small toxic effects may cause extinction of predators and grazers, and thereby indirectly cause an increase in the biomass lower down in the food chain. If toxins act on one species that loses competitive advantage, a food web may collapse into a simple food chain, with dramatic effects on the ecosystem.

An alternative health criterion is to describe ecosystem response in terms of concentrations of chemicals in the organisms relative to no-effect concentrations. Common aquatic organisms are used to monitor heavy metal concentrations: blue mussels – *Mytilus edulis*, rocky shore snails like *Littorina litorae*, and the upper ten cm of rocky shore bladder algae – *Fucus vesiculosus* or *Ascophyllum nodosum*. Mercury may be monitored in cod – *Gadus morhua*, or herring – *Clupea harengus*. However, changes in food web structure of the herring gull will modify the ratio between concentrations of toxic material in the environment and concentrations in the bird.[7] Figure 14-1 shows the flow impacts of toxins on an ecosystem.

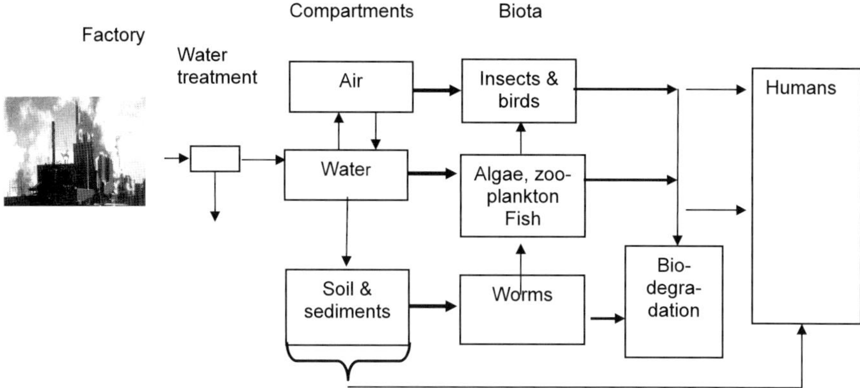

Figure 14-1. The flow from emissions to impacts. Toxic substances are discharged into water after being treated in a wastewater treatment plant. Some of the toxic substances volatilize into air, some adsorb to sediment particles. Air, water and soil are called "compartments". Arrows indicate transfer of toxic compounds. Within each compartment the toxic substances may act on different species. With time many toxic materials will be increasingly fixed in the soil matrix.[8]

Assessment of ecosystem risk: What kind of test species should be used to assess ecosystem risk? Some authors argue that the most vulnerable species should be chosen as sentinel species since protection of those species will result in protection of all species in the ecosystems. Others argue for selection of dominant species. It is apparently difficult to give general rules, and in practice it seems to become a matter of convenience.

2.4 Human Response

Toxic material is introduced to the body through breathing, digestion and dermal contacts. Human consumption patterns are particularly important for people's exposure to contaminants. One sometimes describes the food basket of a population to illustrate typical diets.[9]

Morbidity and mortality: Epidemiological studies provide data that indicate the extent of mortality and morbidity as a consequence of contact with toxic materials. Air pollutants are classified as carcinogens or as chemicals with other toxic effects. The concentration level of different compounds in a specific area varies extensively over time. It is not clear whether short-term peak concentrations or the accumulated dose over a longer time period such as months is most important. We also distinguish between acute and delayed effects. Carcinogenic effects are usually delayed. During the long latency time for cancer development – up to 30 years for lung cancer – the exposure typically varies considerably because exposed persons change occupation, smoking habits, etc. This makes it necessary to define *cancer probability* for life time periods of 70 years.

Nuisance: Bad smells are difficult to measure objectively. One usually identifies exposure zones where the intensity is above an acceptable threshold. To predict the impact of alternative projects, one then has to estimate the number of people within the exposure zone. This is similar to calculation of noise impacts; see Chapter 10 on environment and life quality

2.5 Toxicity Measures

Toxicity tests are usually applied to a standard test organism. Four parameters are important:

- The concentration of the chemical compound (mg m^{-3}).
- The duration of the experiment (hours).
- The proportion of the population that respond.
- The seriousness of the response (morbidity, mortality).

A standard measure of the toxicity of a chemical compound is called *Twenty-four hours 50% lethal concentration*, or **24h-LC_{50}**. It is the concentration that causes 50% mortality in the test-organisms after 24 hours. A test lasting more than 72 hours is often called chronic exposure.

A *lethal dose* (LD) indicates how deadly a given toxin is. It is usually recorded as dose (in grams) per kilogram of subject body weight that makes a given percentage of subjects die. The most common lethality indicator is the **LD_{50}**, a dose at which 50% of subjects will die.

No Observed Effect Level, **NOEL**, is the highest concentration without observable effect on the organism. *Lowest Observable Effect* Concentration, **LOEC**, is the lowest concentration with observable effect. One usually observes the effect on growth rate, but sometimes also the effects in terms of malformation and reproduction potential.

3. HEALTH STANDARDS

Quality standards differ depending on whether we talk of ecosystems or human health, but human standards are sometimes applied to ecosystems.

3.1 Ecosystem Standards

For ecosystems, it is often possible to define acceptable load levels where there is a balance between load and removal – either through degradation or some other type of self-cleaning – which keeps the ambient concentration below NOEL levels. The standards may be defined for particular parts of the vegetation. Crop protection requires an SO_2 concentration that is less than 30 μgSO_2m^{-3}, forests and natural vegetation less than 20 μgSO_2m^{-3}, and lichens less than 10 μgSO_2m^{-3}. This compares with the human health guideline for annual mean SO_2 concentration, which is 50 μgSO_2m^{-3}.[10]

3.2 Human Health Standards

Health authorities set standards for maximum human exposure to concentrations or doses of toxins and other types of pollutants. The standards are based on presumed effects on humans with regard to mortality, morbidity and nuisances. Such relationships are notoriously difficult to quantify, and for practical purposes one often assumes a linear relationship between dose and effect.

Exposure to concentrations of air pollutants can be measured in microgram per cubic centimeter ($\mu g\ cm^{-3}$). Exposure to pollutants received orally is expressed as Chronic Daily Intake (CDI) and measured as milligram per kg bodyweight per day, ($mg\ kg^{-1}\ day^{-1}$). It may also be measured relative to a reference dose that is acceptable.

Increased mortality as response to exposure is often expressed as per cent increase from a base mortality. Morbidity is indicated by the probability of acquiring a certain illness during a certain period. Increased mortality, and probability of morbidity for various degrees of illnesses caused by particulate matter, PM_{10}, is shown in Table 14-2.

Table 14-2. Mortality and morbidity as a function of concentration of particulate matter, PM_{10}.The mortality figures are percent increase in mortality per $\mu g\ m^{-3}$ PM_{10}. The morbidity figures are increased annual probability of acquiring the illness per $\mu g\ m^{-3}$ PM_{10}. Low, central and high estimates. Results summarized by Wang and Smith (1998) p. 15.

Health impact	Estimated increase in health impact per $\mu g\ m^{-3}$ PM_{10}		
	Low	Central	High
Mortality (percent)	0.04	0.1	0.3
Morbidity (probability)			
Respiratory hospital admission	0.7×10^{-5}	1.2×10^{-5}	1.6×10^{-5}
Restricted activity days > 16 yr old	0.04	0.06	0.09
Acute bronchitis < 16 yr old	0.8×10^{-3}	1.6×10^{-3}	2.4×10^{-3}
Respiratory symptoms	0.09	0.18	0.27
Chronic bronchitis > 16 yr old	3×10^{-5}	6×10^{-5}	9×10^{-5}

Concentrations that cause a life-time risk below 10^{-6} or 10^{-7} are often called "Virtually safe doses", Tolerable risk levels", or "Levels of no concern". Other human health standards are given in Chapter 21 on air environment. Updated guideline concentrations can be found on WHO's internet pages.

3.2.1 Relationship Between Concentration and Exposure Time

Normally, people are exposed to pollution concentrations for only limited durations. WHO guidelines allow exposure to 100,000 μg m^{-3} carbon monoxide, CO, for 15 minutes; to 60,000 μg m^{-3} for 30 minutes; to 30,000 μg m^{-3} for one hour, and to 10,000 μg m^{-3} for eight hours. An exposure time of eight hours is often used as a standard.

3.2.2 Safety Factors

When a threshold is established based on experiments or epidemiological data, a *safety factor* is usually applied when a standard is defined. As the name says, a safety factor will reduce what is considered a safe concentration of a pollutant. A factor of 10 is often used to translate from average to sensitive humans, from animals to humans, and from critical effects to individual organ defects, teratogenity. Factors of less than 10 are often used to translate from Lowest Acceptable Exposure Level, LAEL, to "No observable effect level", NOEL; from short term to long term exposures and from a minimum to a complete database.[11]

There are several reasons for applying a safety factor. One is to transfer results from a non-normal laboratory situation to normal situations. In the US EPA IRIS database for chromium (VI) (9-3 1998), a factor of 10 was introduced to account for the expected interspecies variability, in lieu of specific data; a factor of 10 was also introduced for expected inter-human variability (old people, children.), a factor of 3 was included since the study on which the data were based had duration shorter than human lifetime (70 years). For this particular chemical, an additional modifying factor of 3 was introduced to the previous factors to account for concerns raised by a particular toxicity study. Altogether an uncertainty factor of $10 \times 10 \times 3 \times 3 =$ 900 was applied to the LC data obtained for chromium (VI).

Mortality and morbidity models often overestimate to be on the safe side. This is acceptable in some cases, but the practice poses difficulties for cost/ benefit calculations if some costs and benefits are expected values and some are worst-case values.

4. DOSE-RESPONSE MODELS

Dose-response models relate end-impacts to doses or concentrations of toxic compounds, including particulate matter. We first describe a laboratory experiment to estimate lethal concentrations or doses. Thereafter we discuss dose-response models of human health. Toxic effects on plants, animals or the ecosystem are discussed more extensively within the relevant chapters where we can address subtle effects, like thinning of birds' eggshells.

4.1 Laboratory Experiment to Measure LC_{50}

We want to estimate the 50% lethal concentration of a certain toxic compound. We have put 100 specimens of Daphnids – a kind of one-millimeter long zooplankton – in each of six test tubes and added different concentrations of the toxic compound. The concentrations and the resulting mortality rates after 24 hours are shown in Table 14-3. How would you go about estimating LC_{50} for the Daphnids?

Table 14-3. Result of laboratory experiment to estimate LC_{50} of a toxic compound.

Test tube	1	2	3	4	5	6
Concentration (mg L^{-1}) for 24 hrs	1	2	4	8	16	32
Number of dead Daphnids	1	17	45	67	89	87

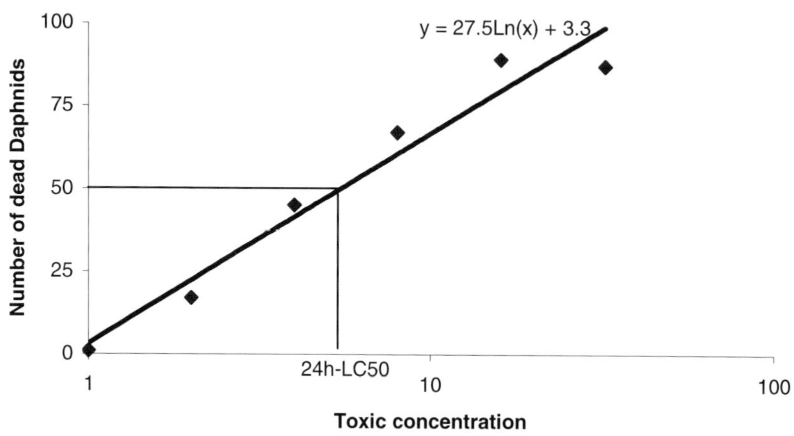

Figure 14-2. Estimation of 24h-LC_{50} according to the data in Table 14-3.

Solution: Draw a graph showing the number of dead individuals on the y-axis as a function of the logarithm of the concentration (x-axis) and fit a straight line through the points. This can be conveniently done in a spreadsheet. Determine the concentration corresponding to y = 50% (50 specimen dead) as shown in Figure 14-2, $x = \text{Exp}((50 - 3.29)/27.54) = 5.5$ mg L^{-1}.

4.2 Dose-Response Models for Human Health

There are two methods for calculating health hazards from toxic chemicals. For carcinogenic substances and for particulate matter with diameter less than 10μ, PM_{10}, there is assumed to be no safe level. For such substances, excess cancer risk over a lifetime is assumed to be a linear function of the dose as shown by the upper line in Figure 14-3. The slope is called the toxicity slope factor, TSF. For non-carcinogenic substances, TSF can be used with a threshold, meaning that the straight line through the origin would be shifted to the right. However, the Hazard quotient method is more commonly used and will be explained below.

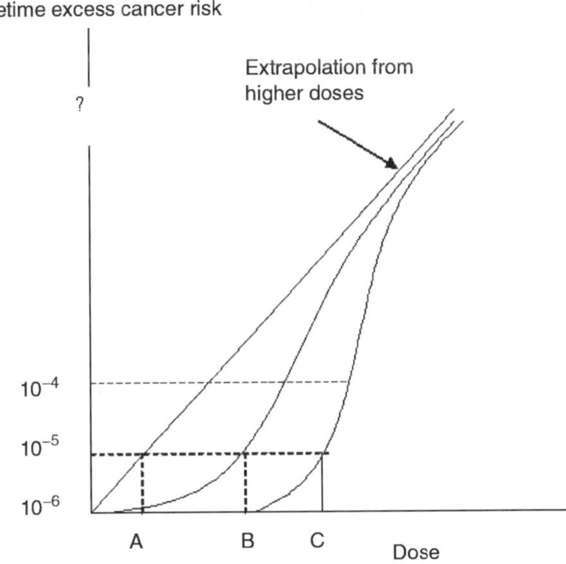

Figure 14-3. Hypothetical dose-response curves for chemically induced toxicity at low doses. A: Linear; B: Sublinear; C: Sublinear and the curve intersects the dose axis at a positive value, which means that there is a threshold before adverse effects start.[12]

Table 14-4. Cancer slope factors, *CSF*, and typical Chronic Daily Intake; *CDI*. Data Kowalczyk (1966:22), Goldhaber and Chessin (1997), Smith et al. (1992).

Compound, letter shows source	Slope factor (*CSF*) mg kg^{-1} day^{-1}	*CDI* mg kg day^{-1}
Arsenic	15	–
Benzene	2.9×10^{-2}	2.5×10^{-4}, well water
Chlordane	1.3	6.0×10^{-4}–6.0×10^{-5}
Arocolor 1260 [(1)]	1.9	

[(1)] Arocolor 1260 is a PCB formula. PCB is polychlorinated biphenyls. Other PCB formulas have proven to be either less carcinogenic or not carcinogenic.[13]

Toxicity slope factors: Toxicity slope factors, or cancer slope factors (CSF), measure how much cancer risk increases with the dose. CSF is defined by the US Environmental Protection Agency as the "plausible upper-bound estimate of the probability of a response per unit intake of a chemical over a lifetime (usually 70 years)". Some typical slope factors are given in Table 14-4. The slope factors reported often give the upper 95th percentile of the confidence limit of the slope, and not the expected slope.

CSF estimates are often based on experiments on laboratory animals. To extrapolate from dose-responses in animals to man is made complicated by difference in weight, life span, genetic homogeneity, metabolic pathways and metabolic rates, bioavailability and toxic mechanisms.

Exercise: Calculate the number of cancer deaths averted (*CDA*) from consumption of fish in the Great Lakes, USA, when the toxic load in a lake is reduced by 50%.[14] There are 1.10^6 consumers of fish, eating 20g fish per day. The concentration of PCB in fish is 0.66 mg per kg wet weight. Reducible loadings from point sources are 1% of the total loading to the Great Lakes (total 580 kg DDT $year^{-1}$ versus 4.6 kg DDT $year^{-1}$ from point sources); expected decrease in point loading is 50%. The cancer slope factor is 1.9 mg kg^{-1} day^{-1}. The time frame is 70 years.

Solution: We use the formulas:

$$\textit{cancer deaths averted (CDA)} = \textit{number of consumers} \times \textit{\% source contribution to total loading}/100 \times \textit{expected decrease in loading} \times \textit{risk}$$

$$\textit{risk} = \textit{chronic daily intake (CDI)} \times \textit{cancer slope (CSF)}$$

$$\textit{CDI} = \textit{fish consumption rate} \times \textit{fish concentration of chemicals}$$

Calculation: Chronic daily intake, *CDI*, is $0.020 \text{ kg day}^{-1} \times 0.66 \text{ mgww kg}^{-1} = 0.0132 \text{ mg kg ww}^{-1} \text{ day}^{-1}$. The risk is $CDI \times CSF = 1.3\ 10^{-2} \text{ mg kg}^{-1} \text{ day}^{-1} \times 1.9 (\text{mg kg}^{-1} \text{ day}^{-1})^{-1} = 2.47 \times 10^{-2}$. $\text{CDA} = 1.10^6 \times 0.01 \times 0.5 \times 2.47\ 10^{-2} = 124.$

Note that the risk is the probability of developing cancer during a lifetime (70 years) caused by a chronic daily dose. Over a period of 70 years, for a population of 1 million people, 124 deaths are averted because of a reduction in point source PCB of 50%.

4.2.1 Hazard Quotient Method

The hazard quotient method gives the Hazard Index, $HI = CDI/RfD$. This is the ratio of the exposure level, *CDI*, to the chronic reference dose, *RfD*, which is the maximum dose that can be consumed daily without adverse effects. A procedure for the development of *HI* is shown in Table 14-5: When determining the daily intake of phenol from drinking water and fish, various sub-populations must be considered (the old, children, men, and women) as well as (seasonally) varying phenol content of food. On the basis of a "Food basket" the daily intake for various sub-populations can be estimated. The intake values for phenol given in Table 14-5 gave a hazard quotient of 0.3. Since this is less than 1.0, it shows that the total intake of phenol is safe in this case. However, daily intake from other toxic substances has to be added to the food intake of phenol from food.

Table 14-5. Hazard index calculated for phenol through drinking water and fish consumption. Phenol has an octanol water partition coefficient, K_{ow} = 30 and is intermediate between non-polar and polar compounds.

Route	Observed chronic daily intake, *CDI* ($mg\ kg^{-1}\ day^{-1}$)	Chronic reference dose, *RfD* ($mg\ kg^{-1}\ day^{-1}$)	*HI= CDI/RfD*
Drinking water	0.1	0.6	0.18
Fish	0.08	0.6	0.12
Hazard Index			0.31

5. MITIGATION

We will discuss mitigation measures for discharges of pollutants in subsequent chapters. However, the term *co-benefits* have become popular, and refers to extra benefits associated with the removal of a pollutant. For example, efforts to reduce discharges of CO_2 – which affects the global environment, but is not really a pollutant – often remove other substances as well. The less fossil fuel that is burnt, the less sulphur dioxide is discharged to the environment because all fossil fuels contain some sulphur.[15]

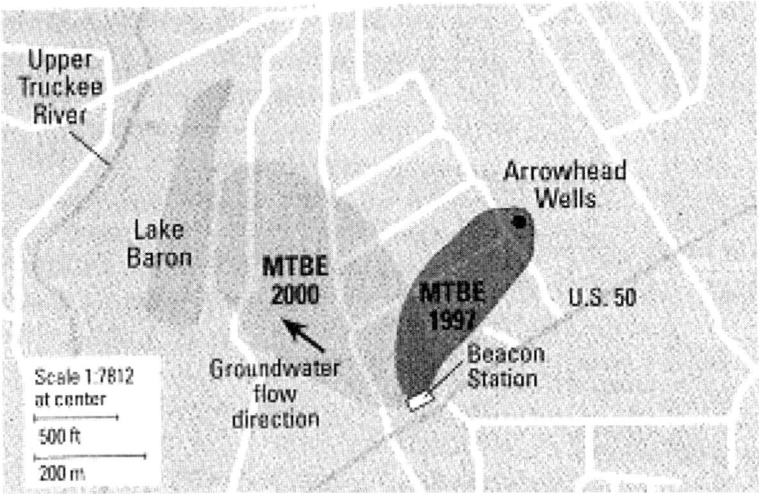

Figure 14-4. Toxic plume changes position. Pumping pulled the 1Methyl-tert-butyl -MTBE plume (dark shade) to the Arrowhead Wells; after the Arrow Head Wells ceased pumping in 1992, the MTBE plume changed direction (Light shade). After Dernbach (2000).

6. APPLICATION: CONTAMINATED GROUND – THE LEGACY OF AN INNOCENT TIME?

In the film *Erin Brocowich* (1999), the actor Julia Roberts plays the role as a lawyer for people living in an area where water is contaminated with chromium in the form of Cr VI, which is the toxic form of chromium. To emphasize her point about greed versus pollution, she makes references to the Love Canal case from the 1980s. The Love Canal was intended to connect the upper and lower ends of the Niagara River, but was abandoned as a canal and used as a dumpsite for Hooker Chemicals and Plastics Corporation from 1942 to 1952.

6.1 Decision Analysis

Below, we propose how a rational decision process might have proceeded.

Decision problem: Should the land abandoned by the Chemical Company be used as development land for private and public housing?

Decision-maker: The *decision maker* is the company that bought the land from Hooker Chemicals and Plastic Corporation and wants to develop it.

Stakeholders: The stakeholders are Hooker Chemicals and Plastic Corporation, the municipality that wanted free land for housing, the people that wanted to buy the houses, and federal and state regulatory agencies in charge of environment and health. See Chapter 3, section 2.3.

Decision objectives: The main objectives are to maximize safety and profit. Sub-goals are:

- Maximize the number of houses.
- Minimize health effects.
- Maximize profit to Hooker Chemicals and the development company.

Alternatives: The most important alternatives are:

- No survey, develop the land now.
- No survey, do not develop the land.
- Survey the land first for possible future toxic effects, and then decide whether to develop the land.

Consequence analysis: To survey the ground, samples would have to be taken from soil and water (Chapters 15–17, 18) and the measured concentrations related to data for allowable human exposures, (Chapter 14, section 3). This process would cost money that could not be recovered if the ground proved to be too costly to clean, illustrating the ethical dilemma posed by choice of the depth of Environmental Impact studies (Chapter 4, section 2 on EIA steps). References to models for leakage of toxic material from contaminated ground can be found in Linkov et al. (2002). Uptake and assessment of toxic material in humans is dealt with in this chapter (two methods; Sect. 4). To minimize health effects toxic material could either be removed, sealed, or a combination of both measures could be undertaken. However, plumes of toxic water may move naturally, or because of pumping operations, so that they reach drinking water wells, (Dernbach 2000) and this brings in the topic of decisions under uncertainty (Chapter 2, section 2). From the analysis should result a consequence and a utility table like Tables 3-5 and 3-6, and primary utility graphs, like those in Figure 3-3.

Preferences: Preferences might partly be expressed by the Environmental Protection Agency. Health effects could be defined as a step function, either the ground would comply with the health regulations, or not (this Chapter; section 3 and Chapter 4; section 2). In the latter case, categorical exclusion might stall the land from being developed. In the former case, the housing development corporation's preferences apply and could in principle be expressed in monetary units.

6.2 What Happened?

Several homes and a school were built on the Love Canal land in 1977. After a series of complaints, reports on increased numbers of birth defects and miscarriages, studies were undertaken of the health of the residents. In 1980, President Carter declared Love Canal a disaster area. The movie picture Erin Brocowich described a similar situation, occurring decades later. Here are three possible interpretations of why such "slow leaks" were allowed to occur for such a long time.

1. The incident fell between "crisis" and "the customary order of things" so no one really cared, in particular not the press.
2. It was the result of planned evil – or perhaps.
3. It was a legacy of an innocent time.

---

[1] Rachel Carson (1962)
[2] Rodan et al. (1999)
[3] Mourato et al. (2000)
[4] Tellus (1991)
[5] Watkiss et al. (2004)
[6] Rodan et al. (1999:3483) and others
[7] Cod and herring: Knutzen (1999) p. 569-581; food web structure: Herbert et al. (2000)
[8] After Chow et al. (1988)
[9] Definition of food baskets also called market basket or total diet study, can be found on Food and Drug Administration. (http://www.cfsan.fda.gov/~comm/tds-toc.html)
[10] Nigel Bell, Imperial College, personal communication
[11] Ahlborg (1996)
[12] The term "level of no concern" is discussed in the EPA IRIS data base
[13] Smith (1997)
[14] The case is based on a study by Smith (1997), but we select other numbers
[15] Aunan et al. (2004)

Chapter 15

SOIL AND SEDIMENT

Physics and chemistry of soils and sediments – soil loss and formation

We give an outline of soil and sediment properties and their relevance for agriculture and husbandry, and as part of river and lake environments. We focus in particular on soil composition and adress the question of why different soils have different properties. We also discuss formation and erosion of soils.

●

"Chuma chilli mthaka" Wealth is found in the soil. Malawi proverb

1. INTRODUCTION

Soil is the mother of all vegetation. But she is at times a poor mother. If you pass a fallen tree, you may notice that the soil is stratified in layers in the hole where the roots used to be, the top layer often being darker than beneath. Soil properties determine to a large extent growth forms on arable land, and thus the yield of food and fibers in the agricultural industry. Soil characteristics together with external inputs like energy, precipitation, fertilizers, pesticides, and farming practice determine agriculture productivity, soil pollution, and erosion.

Soil erodes constantly even without man. Hurricanes or typhoons cause landslides and volcanic eruptions cover tracts of land, but the extent of such accidental damages is probably small compared to the chronic erosion. Soil erosion increases when the soil is used for agriculture, through plowing, sowing or planting and harvesting processes. Removal of plant material also changes chemical and physical properties, leading to loss of nutrients and minerals that are required for plant production.

Approximately 30% of the surface of built land is sealed, reducing infiltration of water and increasing overland flow, which in turn may pollute rivers and lakes or cause floods.[1]

2. BACKGROUND INFORMATION

The earth's land surface of 14,900 Mha (1.49×10^8 km^2) contains about 1500 gigatons of carbon down to 5 m below the surface. Soil storage probably amounts to about 5×10^{12} tons. Of the surface area, about 1.1×10^8 km^2 (78%) is unproductive and 0.33×10^8 km^2 is productive land. This gives an average of about 0.03 km^2 surface area per capita, and 0.0062 km^2 productive land per capita with a world population in 2005 of 6.4×10^9.[2] Soil is unproductive because it is ice covered (10%), too cold (15%), too dry (17%), too steep (18%), too shallow (9%), too wet (4%), or too poor (5%).[3]

The sustainable rate of soil loss through erosion is about 9.3×10^9 tons per year, but the predicted erosion rate in 2040 is between 45 and 60×10^9 tons per year. The desired reduction is therefore about 85% on a global scale.[4]

The rate of soil loss is normally expressed in units of mass or volume per area and time unit. Annual soil erosion rates from vegetated surfaces may be of the order 0.005 ton per hectare in low relief areas, and 0.45 ton per hectare in steep parts. Loss from agricultural land is in the range of 50 to 500 tons per hectare. The major cause of soil erosion in such areas is removal of vegetation, for example during crop harvesting or logging. Soil erosion rates differ among countries. Typical erosion rates are shown in Table 15-1. The highest erosion rates are in semi-arid and semi-humid regions of the world, like China and India. The term **non-point** *source* is often used about pollution from agricultural areas. Since agriculture is normally an open activity that cannot be confined to a small area, pollution and supply of eroded soil to waters is difficult to control.

2.1 Soil Classification

Soil is the top 1 to 2 m of the surface that contains some amount of organic matter. Normal soil contains about 20–30% water, a similar amount of air, about 5% organic matter, and 45% minerals.

Table 15-1. Annual rates of erosion in selected countries. Erosion units in ton per hectare. Data: compilation by Morgan (1995).

Country	Natural	Cultivated	Bare soil
China	0.1–2	150–200	280–360
USA	0.03–3	5–170	4–9
Australia	0.0–64	0.1–150	44–87
Ivory Coast	0.03–0.2	0.1–90	10–750
Nigeria	0.5–1	0.1–35	3–150
India	0.5–5	0.3–40	10–185
Ethiopia	1–5	8–42	5–70
Belgium	0.1–0.5	3–30	7–82
UK	0.1–0.5	0.1–20	10–200

Soil types are extremely diverse, and there are several classification systems. The FAO-UNESCO system uses 28 basic groups. The US Soil Taxonomy issued by the Natural resources conservation service classifies soils in 12 orders; spodsols, alfisols, ultisols, oxisols, mollisols, aridosols, entisols, inceptisols, andisols, vertisols, histosols and gelisols. When the terms are used, they are also often followed by a brief description, because at this level they cover a rather wide range of properties and require qualification to be meaningful in a specific context. There are so many terms for soil that it may be comforting to know that soil names have been used with "profound divergences" up to the present.[5] Furthermore, adjectives used to moderate the major soil groups have connotations that are more mnemonic than scientific, such as RENDZIC from Polish colloquial rzediz, connotative of noise made by a plough over shallow stony soil.

Soils may also be classified according to how they were formed. This gives six major groups: Moraine soils, sediment rock, weathering soil, avalanche soil, shifting sand and boggy soil. This classification not only reflects the origination of the soil, but gives information about its properties as well.[6]

2.2 The Value of Soils

The value of soils is difficult to assess. For one thing, the prices of different soils depend on climate. Hedonic valuation is one possible method, which can be applied to regions with different soils. Since adequate soil is a prerequisite for cropland, a starting point for estimating the value of soil is to evaluate the services it offers. Table 15-2 shows a global estimate of the value of three kinds of services provided by cropland.

Table 15-2. Summary of average global value of cropland and its services. Data Constanza (1997).

Type	World area	Pollination	Biological control	Food production
Unit	10^6 ha	US\$ ha^{-1} yr^{-1}	US\$ ha^{-1} yr^{-1}	US\$ ha^{-1} yr^{-1}
Cropland	1400	5446	2117	665

3. SOIL CHARACTERISTICS

Six major aspects of soil are important for its properties as an agricultural resource, for confinement of landfills, and as a fundament for built land:

1. *Texture*: characteristic sizes of particles in the soil. A soil texture chart is shown in Figure 15-1.

2. *Structure*: the packing of particles in the soil, for example as large scale sheets or as spheres.
3. *Chemistry*: chemical species in the soil and how soil is able to attract or retain chemical species like potassium, phosphate, and nitrogen.
4. *Organic matter content*: the amount of organic matter in the soil.
5. *Drainage properties*: the velocity and direction of water movements through the soil.
6. *Porosity*: the amount of air in the soil.

3.1 Texture

Soil texture refers to the proportion of minerals of different size that constitute the soil. The texture classification scheme in Figure 15-1 shows how the size fractions termed clay, silt, and sand combine to form different soil types.

Clay is particle size less than 2 μm. It is a fine-textured soil that usually forms very hard lumps or clods when dry and is quite plastic and sticky when wet. The amount of clay less than 0.002 mm in a clay sample can be tested by rolling the soil out in threads. For example, with fine clay content in the range 5–15%, the treads will be about 3 mm thick. Clay has a shiny surface when smoothed out wet.

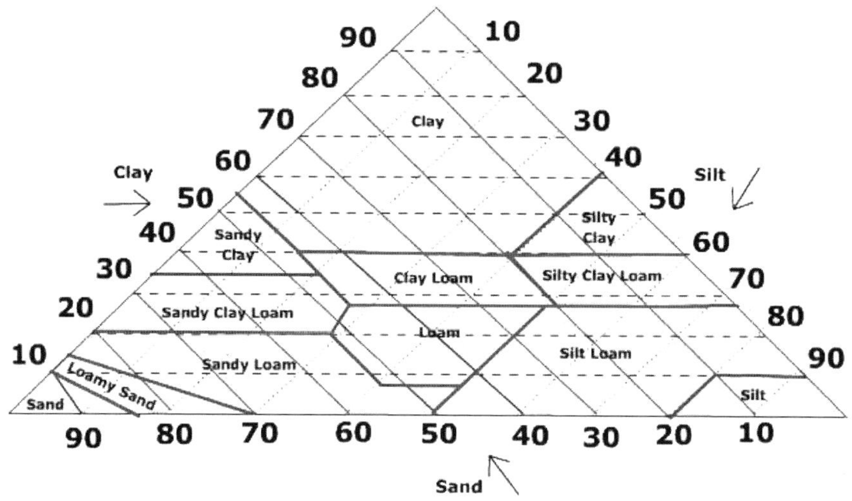

Figure 15-1. The soil texture triangle, showing the percentages of clay (particles less than 0.002 mm) silt (0.002 to 0.06 mm) and sand (0.06 to 2 mm) in the basic texture classes. The chart is read as follows: From triangle sides read along lines slanting to the left. For example, a soil with 60% sand, 30% silt, and 10% clay would be a sandy loam; notice that the percentages sum to 100%.[7]

Silt is 2 to 60 μm in size. Silt loam is a soil having a moderate amount of fine grades of sand and only a small amount of clay. Dry silt appears cloddy, but the lumps can easily be broken, and it feels soft and floury in powder form. If you press silt with a finger, it shows the fingerprints.

Sand is 60 to 2000 μm in size. It is loose and single grained. The individual grains can easily be seen and felt. Squeezed in the hand when dry it falls apart. Sand feels gritty, and single grains adhere to the fingers.

The word *colloidal* refers to the size of particulate matter. The colloidal size is intermediate between matter that is visible under an optical microscope and invisible molecules (less than 0.0002 mm).

3.1.1 Other Sediment Classification Schemes

Sediments of the water environment are often classified categorically without direct reference to scientific soil nomenclature. To transfer from categorical terms to size classes Malavoi and Souchon[8] applied the following conversion factors: mud: <0.002 mm; fine sand: 60 mm–0.5 mm, coarse sand: 0.5 mm–2 mm; fine gravel 2–8 mm; coarse gravel: 8–16 mm, fine pebble: 16–22 mm, coarse pebble: 32–64 mm, fine cobble: 64–128 mm, coarse cobble: 128–256 mm, boulders 256–1024 mm, and bedrock >1024 mm. In lowland streams benthic macro invertebrates colonize the upper 25 cm of the sediments, which contain about 245 g Phosphorus per m^2 and 23 kg Carbon per m^2.[9]

3.1.2 Surface Area of Soil Particles

The surface area of the particles is related to their size. The surface is important because it is where most chemical and physical reactions occur, making the extent of these reactions proportional to the surface area. A cube with one cm sides has a specific surface of 6 cm^2 per cubic centimeter; a cube with one mm sides has a specific surface area of 60 cm^2 per cubic centimeter. The ratio between the specific surface area and the side of a cube (r) or the radius of a sphere (r) is $S = k/r$, where k is 6 and 3 respectively. However, the specific surface of clay (and humus) is relatively much larger than of sand and silt, because the particles are shaped like plate or lath. Thus, for most sand and silt soils, the clay content determines the specific area. If we assume a specific area of 100 $m^2\,g^{-1}$ and a bulk density of 1.25 kg L^{-1}, then one hectare of a silt loam soil to 20 cm depth has a total particle area of 25 million hectares.[10]

Texture classes: For practical purposes it is common to define three texture classes that consist of mixes of the three basic texture classes: clay, silt and sand:

- *Coarse texture*: sands, loamy sands and sandy loams with less than 15% clay and more than 70% sand.
- *Medium textured* e.g., sandy loams, loams with less than 70% sand.
- *Fine textured*, e.g., clay, silt with more than 35% clay.

3.2 Structure

Soil structure refers to how the particles are packed. Particles may be packed in laminar sheets, with small and large particles interchanging, or in any other structure. See Figure 15-2. Directional properties are particularly important since they may enhance water flow in one direction, and limit it in another.

Soil texture influences soil structure. Sandy soils are generally porous, but they may appear massive if they contain a certain amount of silt and clay. With still more silt and clay they become aggregated. Silt is tight, has small pores and drains poorly. Table 15-3 summarizes soil characteristics across size fractions (soil textures).

Table 15-3. Soil characteristics across soil textures.[11]

Soil type	Sand	Silt	Clay
Size	60–2000 μm	2–59 μm	<2μm
Dominant mineral	quarts	quarts and feldspars	mica, vermiculite, montmorillonite, kaolinite, and amorphous colloids
Shape of particles	spherical or cubical	spherical or cubical	plate or lath shaped [(1)]
Specific surface area, $m^2 g^{-1}$	0.001–0.1	0.1–1.0	10–400
Structure	homogeneous	structured	homogeneous
Organic matter content, %	0–40	0–40	0–40
Hydraulic conductivity and internal drainage	high and excessive	medium good	low and poor
Water retention	low	medium	high
Runoff potential	low	medium	high
Chemical properties	2–3 meq/100g		≈100 meq/100g
Cation exchange capacity [(2)]	low, zero		high
Anion exchange capacity [(3)]	often low		often high

[(1)] The smallest clay particles are amorphous.

[(2)] It retains Na, K, Mg, and Ca.

[(3)] It retains phosphate, sulfate, and nitrate.

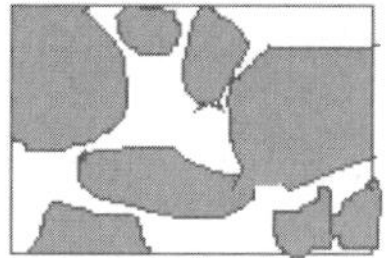

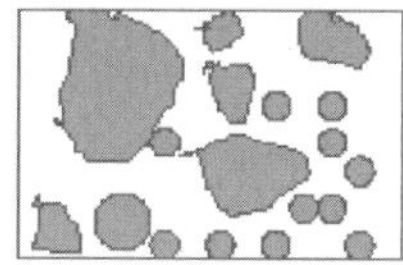

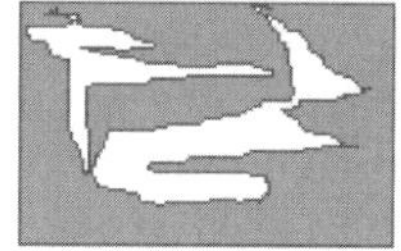

Well sorted sendiments with high porosity

Mixed sediments with high porosity

Rock porus by solution (e.g., limstone)

Figure 15-2. Soil (rock) structure.

3.3 Chemistry

Silica and silicates make up the largest part of the mineral content of the soil. Generally, the largest particles are highest in silicon Si; the finer ones contain more potassium K, calcium Ca, and phosphorus P. Soils have two important chemical properties. The *cation exchange capacity* of the soil describes its ability to retain positively charged ions, like Na, K, Mg, and Ca. The other property is the ability to retain anions, like phosphate, molybdate, sulfate and nitrate. These properties are important for agricultural use of the soil, and for the chemical properties of the water flowing through the soil. That the chemical characteristics of soil are important for the vegetation growing on it, is shown in Figure 15-3, which shows that grass grows luxuriantly on a lawn where fertilizer containing phosphorus and nitrogen has been applied in high concentrations.

The lawn has been fertile-ized in a zig-zag pattern, and growth reflect high fertilizer concentration

Figure 15-3. Growth as a function of soil properties.

3.4 Organic Matter and Organisms

Organic matter content is one of the major factors for plant growth. It is related to the vertical structure of the soil profiles.

3.4.1 Soil Horizons

A vertical transect through soil will often show distinct layers or **horizons**. Most soil scientists use an alphabetic nomenclature for the horizons, but definitions vary widely.[12] A system with seven horizons is common. The thickness of the horizons may vary from a few centimeters to more than a meter. See Figure 15-4.

- *Horizon A*: Mineral horizon formed at or adjacent to the surface. An accumulation organic matter that is turned to humus through biological activity. It is normally darker than the underlying horizon.
- *Horizon B*: Mineral horizon in which rock structures is faintly evident. B-horizons may vary greatly and can usually only be identified if you know the overlying and the underlying horizons.
- *Horizon C*: Horizon of unconsolidated material from which the upper layers are presumed to have formed. Designates parent material, but normally weathered or pre-weathered. Roots may be unable to penetrate, but the soil can still be dug with a spade.
- *Horizon H*: Organic horizon formed from organic material deposited on the surface, saturated with water. 12–18% organic matter.
- *Horizon O*: Organic horizon, as H, but unsaturated and with more than 20% organic matter.
- *Horizon E*: Horizon underlies an A, O or H horizon, lower content of organic matter, lighter color.
- *Horizon R*: Layer of continuous indurate rock. Digging with a spade is impracticable.

3.4.2 Organic Matter Content

Organic matter is an integral part of soils and it affects its physical and chemical properties. All organic substances in the soil, living or dead, are part of the organic matter. A typical elemental composition of a mineral soil is shown in Table 15-4. Organic matter is a source for food and energy for microorganisms, it is a nutrient source for plants as it decomposes into inorganic forms, it stabilizes soil aggregates, it improves the water holding capacity and water conducting capacity of soils, and it has an impact on runoff and erosion.

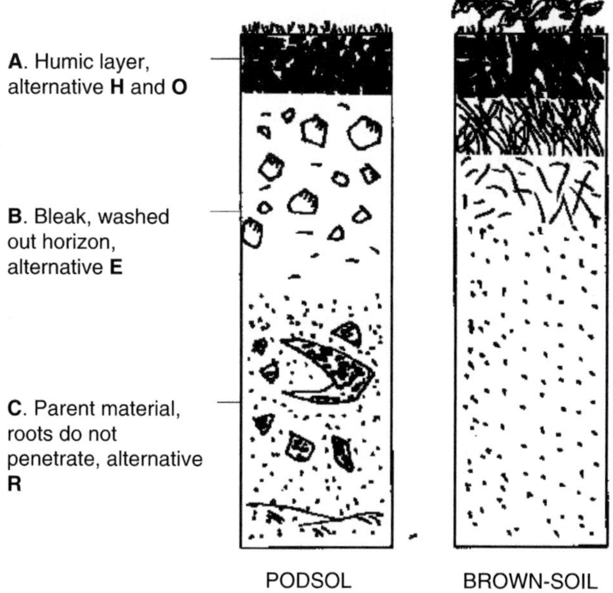

Figure 15-4. Soil horizons.

Organic matter content of 0–1% is very poor humic and can be found in semi-dry areas of Africa. Between 1 and 2% organic matter content is characterized as low humic. Above 2% organic carbon, equivalent to about 3.5% organic content and up to about 12% organic matter content, stabilizes the soil. Above this concentration, no general rule applies.[13] 2% to 4% organic matter content give the medium humic soil that can be found in Mediterranean Europe. 4%–8% give high humic, and 8%–20% very high humic soil.

Table 15-4. Chemical constituents of mineral soil. Loss by leaching in normal vegetation, Europe.[14]

Element	Percent by weight	Loss by leaching, kg ha^{-1} yr^{-1}
Carbon	5–60	50–250
Oxygen	32–38	–
Hydrogen	3–4	–
Nitrogen	4–5	5–25
Phosphorus	0.4–0.6	0.5–2.5
Sulfur	0.4–0.6	–
Potassium	0.2–3	7
Magnesium	0.001–0.005	42
Calcium	0.004–0.002	50

Organic soil with organic matter content greater than 20% can be found as the black soil areas of Northern Russia. Strongly humic >14% soils are formed as biological material degrades. Degradation is enhanced by moisture and aeration, and by a high portion of roots.

Color is a useful characteristic of soils, although there is no one-to-one correspondence with the functional properties of the soil. In temperate regions, brownish-black and dark brown colors, especially in the upper layers (A-horizon) indicate high levels of organic matter. In the lower layer (B-horizon) pale brown to reddish and yellowish colors indicate well-drained soils, whereas dark brown and blackish colors indicate poor drainage. Grayish colors indicate permanently saturated soils in which iron is in the ferrous form.

3.4.3 Soil Organisms

An array of organisms inhabits soils and contributes to ecosystem services. They shred, mix and digest coarse organic matter into smaller particles, transport particles to other areas, break down chemical compounds and decay particles of organic matter that is resistant to decomposition.

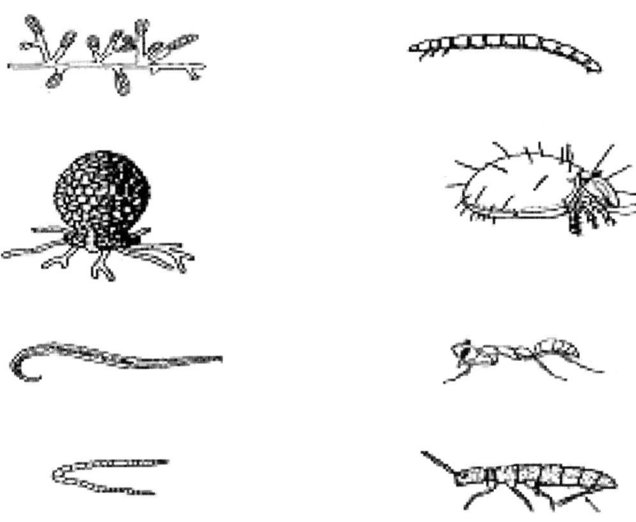

Figure 15-5. Soil organisms. Left column from top to bottom: Fungi, protozoa (principal microbial grazer); nematodes (1-1.5 mm) (principal microbial grazer); Enchytraeids, - look like earthworms, but are smaller; Right column: Insects (a beetle larvae) mites, ants (termites are major decomposers in tropical terrestrial systems); collembola or springtails. Bacteria are not included, but are very important.[15]

Many soil organisms feed on soil micro fauna like fungi and bacteria. This has two important effects: The animals release nutrients, and secondly a suitable grazing pressure is exerted on the micro fauna and stimulates it to increase its decomposing work. The end products are CO_2 – or if no oxygen is present, CH_4 – nitrogenous compounds, and other nutrients.[16] Soil organisms appear to be abundant and are therefore no limiting factor with regard to agricultural yield, in contrast with the structural and chemical properties of the soil. A selection of soil organisms is shown in Figure 15-5.

3.5 Drainage Properties

Water flow properties are characterized by hydraulic conductivity, permeability or infiltration rates. Water flow is normally measured in meter per second or millimeter per hour. Typical values are shown in Table 15-5. Water conduction occurs at two distinct levels. Interstructural voids are likely to be larger than textural voids and thus more important for saturated through-flow. A wide range of infiltration rates may be found in apparently uniform soil. In clay under old oak woodland, a range from 0.04×10^{-3} ms^{-1} to 1.0×10^{-3} ms^{-1} was reported in one study.[17]

Table 15-5. Soil characteristics. Specific yield is the percentage of water that will be moved by gravitational forces.[18]

Material	Porosity (%)	Specific yield, %	Hydraulic conductivity, m sec^{-1}
Crushed stone	>30	>20	10^{-4}–10
Coarse gravel	24–36	22	10^{-4}–10^{-2}
Coarse sand	31–46	25	10^{-7}–10^{-3}
Clay	36–60	3	10^{-11}–10^{-9}
Shale	0–10	3	10^{-13}–10^{-9}

3.6 Porosity

Soil properties can be characterized by four volumes and one weight: volume of solids, V_s, volume of voids, V_v, total volume of un-dried soil, $V_t = V_v + V_s$, and dry volume of solids without air or water, V_d. The weight is the dry mass of soil volume, M_d. **Porosity** is defined as the volume of voids, V_v, compared to the total volume, V_t as shown in Table 15-6.

Table 15-6. Soil attributes.

Parameter	Legend	Definition	Range
Porosity	n	V_v/V_t	Sand: 35%–45%. Clay: 50%–55%
Void ratio	e	V_v/V_s	–
Bulk density	ρb	M_d/V_t	0.7 kg L^{-1} for peats, 1.1 kg L^{-1} for clays, 1.7 kg L^{-1} for sands and loams
Particle density	ρ_p	M_d/V_d	2.65 kg L^{-1} for most soils

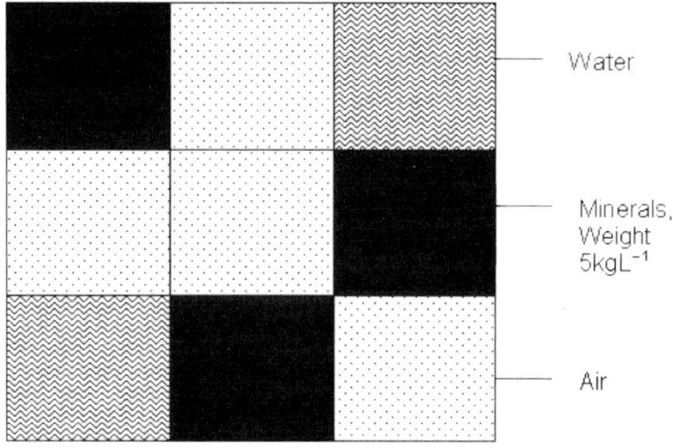

Figure 15-6. Schematic representation of a soil sample.

Specific yield is the amount of aquifer, which is free to drain by gravity. It is less than porosity, because some water is attached to the particles by surface tension. Some characteristic porosity values are given in Table 15-6.

3.6.1 *Exercise*: Calculation of Soil Properties

Required: Calculate porosity, void ratio, bulk density and particle density for the soil in Figure 15-6. What type of soil is this sample closest to?

Solution: *Porosity* = V_v / V_t = 6/9 = 0.66. *Void ratio* = V_v / V_s = 6/3 = 2. *Bulk density* = M_d / V_t = 15/9 = 1.7. *Particle density* = M_d / V_d = 15/6 = 2.5. This soil is closest to clay.

3.7 Soil Density

The density of dominant soil minerals such as quarts and several of the feldspars is about 2.65 kg per liter. Some soil constituents like apatite and hematite has higher densities (3.2 kg L^{-1} and 5.3 kg L^{-1} respectively). The density of soil organic matter (humus) ranges from 1.3 to 1.5 kg L^{-1}. Typically, one ton of fertile agricultural topsoil contains 1–6 kg total nitrogen, 1–3 kg total phosphorus, and 2–30 kg potassium. A severely eroded soil may have 0.1–0.5 kg of nitrogen.[19] A hectare of typical sandy forest topsoil down to 0.5 m will contain about 5000 m^3 of soil, 50–180 tons of humus, about 50 kg nitrogen and about 25 kg of phosphorus. A hectare of grassland silt down to 0.7 m will contain much more humus, about 1300 tons.

3.8 Geographic Characteristics

Soil properties vary on a regional scale. Fennoscandia, the northern parts of East Europe, the northern part of Asia, Canada and Alaska have all been covered by ice caps, with characteristic soil qualities of the early ice covered land (podsols).

The vast Savannas north and east of the Black Sea and further east into Asia have thick (1 m–1.5 m) layers of soil with a high content of mold. This black soil is also found in North American prairies, and some places in South-America, India, Africa and Australia. Lateritte is the most typical soil in tropical regions with high precipitation. The color is red, and when it dries, it easily forms brick-like lumps. Chemical processes and leaching often makes these soils poor in organic matter. The driest areas of the world, which are found in North Africa and further eastward into Asia, have shallow soil with very little content of organic matter. Possibly the most widespread feature of tropical soils is the extent of sudden precipitation and heavy rainstorms that lead to extensive leaching, so that most soluble plant nutrients are carried down below the root zone. Consequently, many soils in warm and humid areas have a low level of fertility (phosphate and nitrogen are commonly deficient like the soil of the Brazilian Amazon). The downward movement of soil nutrients often causes formation of hard-pans which impede drainage and root penetration.

4. SOIL QUALITY STANDARDS

Among design criteria for quality soil are:

- Resistance to erosion.
- Crop yield (agriculture).
- Low leaching (waste depots).

Optimal soil properties depend upon climate. For example, in semiarid climates small and medium pores are required to store water in the soil, whereas in humid regions storage is not as important as large pores for aeration. The function of pores is summarized in Table 15-7.

Table 15-7. Desirable soil characteristics.[20]

Pore size	Diameter, mm	Function of pores	Best pore distribution in semiarid climate
Large	>0.06	Aeration and infiltration	Medium amount
Medium	0.06–0.01	Conduction of water	Medium amount
Small	0.01–0.0002	Storage of water	Large amount

Few standards are defined for soil quality, but some qualitative criteria have been applied:

- *Resistance to erosion*: Resistance to erosion increases dramatically with vegetation. However, the yield of the plants need not be high to give good erosion resistance.
- *Agricultural yield*: Soil quality can be expressed in terms of porosity, aggregation, cohesiveness, and permeability for water or air. Since the chemical and biological processes in soil are related to pores, porosity is an important parameter. Soils with more than 12% expanding clay may have a structure favorable for plant growth. It is important that the soil is exposed to freezing and thawing and wetting and drying. Soils with more than 35% clay are usually massive and not favorable for plants. An exception is tropical soils that contain non-expanding clays.
- *Soil depth*: Even though a series of factors determine the suitability of a soil for different growth forms and different species, soil depth alone appears to be an important factor and it often determines the main group of species that will grow in a habitat. For example, in Norway, pine dominates if soil depth is less than 20 cm, whereas spruce dominates at soil depths of more than 20 cm.[21]
- *Landfill liners and built land*: Clay liners for landfills require negligible permeability at values lower than 10^{-9} m s^{-1}. Soil types suitable for built land preferentially consist of coarse sediments.

A summary of soil quality indicators from New Zealand is shown in Table 15-8. The original tables were for different soil types and different soil uses. For example pasture, cropping and horticulture require higher total nitrogen (100 μg g^{-1}) than forestry (40 μg g^{-1}). The figures in the table must therefore be considered as general guidelines.

Table 15-8. Provisional targets for soil quality indicators in New Zealand. The five upper indicators are provisionally established. The lower five are on an experimental basis.[22]

Indicator	Unit	Too low	Optimal or target	Redundant or excessive
Total carbon	%	Depleted = 3	Normal = 5	Ample = 12
Total nitrogen [(1)]	$\mu g.g^{-1}$	Low = 20–50	Ample = 120–200	Excessive > 200
Soil pH	-	Slightly acid = 4	Optimal = 6	Very alkaline ≥ 6.6
Bulk density	ton m^{-3}	Loose < 0.7	Adequate = 1.0	Compact > 1.2
Macroporosity	%	Low ≤ 8	Adequate = 20	High > 30
Rooting depths	cm	< 50	Optimum >70	-
Earthworms	No m^{-2}	< 500		Detrimental >1200
C:N ratio	-		Optimal < 15	> 20
Aggregate stability	mm mean weight	Poorer < 2	> 2mm	More is better
Topsoil depth	cm	Prod. drop off < 10	> 10	More is better

[(1)] Anaerobically mineralized

5. DOSE-RESPONSE RELATIONSHIPS

Dose-response models for soils are described under separate processes of the hydrological cycle. As far as we know, no simple mathematical models relate parent rock material, precipitation, and other physical and chemical factors to soil type. In this section, we briefly discuss generic values of soil loss and soil formation. Particular measures for reducing soil loss during agricultural production are discussed in Chapter 20 on agriculture.

5.1 Soil Loss

Topsoil thickness depends on soil formation and erosion. Erosion rates depend on land use, and are summarized in Table 15-9. We see that erosion differs by three orders of magnitude between a forested area and an area where the surface is disturbed by construction work, and the uncertainties are considerable.

The Universal Soil Loss Equation, USLE, considers total precipitation, and its distribution with season, type of soil, vegetation and morphology of the field. A description of the model is available on the Internet.[23] The single most important factor is probably vegetation, since a dense vegetation and a well developed root system may reduce erosion to one thousandth of the erosion of bare soil. Perennial crops give less soil erosion than annual crops. For example pure stands of rubber and oil-palm trees, or tea, may well be considered as soil-conserving types of land use.[24]

Simulation models for soil erosion are abundant. The European Soil Erosion Model and the CREAMS model from USDA are two examples, but will not be treated here.[25]

Table 15-9. Erosion rates from land uses. 1 km^2 = 100 hectare.[26]

Land use	Erosion rate, ton km^{-2} yr^{-1}	Ranges, several authors
Forest	8.5	
Grassland	85	10–2100
Abandoned surface mines	850	
Cropland	1700	1800–5100
Harvested forest [(1)]	4250	1400–4100
Active surface mines	17,000	40–10,200
Constructions	17,000	

[(1)] The erosion rate of a burnt forest returns to pre-fire conditions with a half-life of 20 years for total phosphorus and 11 years for dissolved phosphorus.[27]

Figure 15-7. The Årungen watershed, Norway. The area is divided into subareas with characteristic slopes, vegetations, and soil properties. Left: Cultivated (dark) and uncultivated (white) land. Right: Erosion rates. Dark > 10 ton ha^{-1} yr^{-1}, medium grey > 5 ton ha^{-1} yr^{-1}, light grey > 1 ton ha^{-1} yr^{-1}.

5.2 Soil formation

Ideally, the rate of soil loss should not exceed the rate of formation from parent rock and decomposing vegetation. The rate of formation depends on many local factors, such as fertility and drainage characteristics. Soil formation rates are reported from 1 to 1½ ton ha^{-1} yr^{-1} or even larger.[28] One ton ha^{-1} yr^{-1} is 0.13 mm yr^{-1}, or less than 3 cm in 200 years. Land may ***tolerate*** a soil loss equal to the formation rate, and soil-loss tolerances for U.S soils are reported to be in the range from 2.2 to 11 ton ha^{-1} yr^{-1}. Agriculture in the US mostly takes place on soils with soil-loss tolerances of 11 ton ha^{-1} yr^{-1} or more.

6. CONCLUSION

Soil is the most important natural resource for mankind and sustainable usage is possible. Still, much of current use is unsustainable, both in terms of its usage as a growth medium, and because of soil loss. We have discussed soil texture, soil structure, and soil organic material content, and schemes used to classify soil and its change with depth. Precipitation causes overland runoff and erosion, which can be reduced by, for example, sustaining vegetation, terracing or reducing plowing. Measures like crop rotation, fertilization, and irrigation are used to maintain plant growth quality of the soil. Such mitigation techniques are discussed in Chapter 20 on agriculture.

---

[1] Statzner and Sperling (1993)
[2] Search internet on world population
[3] Mather and Chapman (1995:48)
[4] Weterings and Opschoor (1992) p. 25–28
[5] FAO (1994) p. 12
[6] Låg (1976)
[7] Waite (2000) p. 186 www.uwsp.edu/geo/faculty/ritter/ glossary/s
[8] Malavoi and Souchon (1989)
[9] Statzner and Sperling (1993)
[10] Kohnke (1970)
[11] Data: Kohnke (1970), Morgan (1995)
[12] FAO (1994)
[13] Morgan (1995) p. 31
[14] Data Kohnke (1970:69) and others
[15] After Brussard et al. (1997)
[16] Freckman et al. (1997)
[17] Knapp (1978) p. 46
[18] Data Kohnke (1970:69) and others
[19] Pimentel et al. (1995) p. 1118
[20] Data Kohnke (1970:139)
[21] Låg (1976)
[22] After Sparling et al. (2003)
[23] Search: Universal Soil Loss Equation
[24] Ruthenberg (1980) p. 260
[25] ESEM: Morgan (1995); CREAMS Knisel (1980)
[26] Data: O´Riordan (1995) p. 323, Mann and Tolbert (2000), McEachern et al. (2000:80)
[27] McEachern et al. (2000) p. 80
[28] Low number: Pimentel (1996); high numbers: Troeh (1999)

Chapter 16

HYDROLOGY

Rain, runoff, evaporation and storage

We describe the hydrological cycle and storage compartments for water, and the movement of water between the four processes: precipitation, infiltration, evaporation, and runoff. We also discuss run-off, river-flow, drought, desertification, and erosion in a hydrologocal perspective.

"But methought it lessened my esteem of a king, that he should not be able to command the rain." Samuel Pepys (1633- 1703) diary on July 9, 1662

1. INTRODUCTION

Water precipitates down on earth, is absorbed by soils and plants, flows in rivers, is stored in lakes and seas, evaporates, builds clouds, and precipitates again. This is called the hydrological cycle. It provides many services; provision of water for domestic use and agriculture are among the most central. Some services are not under human control, like precipitation, but others can be modified or changed.

Man influences the hydrological cycle in several ways, sometimes causing conflict between human needs and environmental goals. Examples are:

- Dam building and dam demolition.
- Change of river course.
- Change of surface texture.

Man is an impressive dam builder, with more than 36,000 dams that are more than 115 meters high in operation worldwide.[1] One of the most spectacular is the construction of the Three Gorges Dam on the Yangtze

River. It will provide China with a substantial fraction of the electricity it needs, and hopefully put an end to downstream floods that have claimed more than 1 million lives in the past 100 years. It provides storage of water for agriculture, industry and municipalities. But it will also inundate 632 km^2 land and require 1.2 million people to relocate. A particularly unwanted side effect is large areas that will become periodically inundated. These areas between low and high water levels are mainly without natural growth and perceived as unaesthetic. Furthermore, silting and sedimentation are major problems when water velocities are reduced. This, together with other aspects, has made the decision to build the dam controversial.

Through dam building, the global mean age of river water has tripled to well over one month. Such aging leads to significant changes in net water balance, flow regimes, re-oxygenation of surface waters and sediment transport. The amount of stored water on the continents is shown in Table 16-1. South America has the largest storage area per inhabitant (3.5 m^2 $capita^{-1}$), North America the largest storage area per land area (49 m^2 km^{-2}).

Changes in river courses occur naturally, and by man made constructions. Large dams increase evaporation before the water reaches the sea, but the impact on the global hydrological cycle is probably not very large. Changes in soil surface structure may profoundly alter the water holding capacity of the soil, and if care is not taken, turn vegetated land into barren soil.

In this chapter we describe the four basic hydrological processes: precipitation, runoff, infiltration and evaporation, and the most important factors that impact them. Thereafter we give, for each process, one or two simple equations that allow us to make predictions about the size and the timing of the processes. More complex models have been developed, but our equations serve well as first estimates.[2]

Table 16-1. Water reservoirs of the continents. Data: Van Jaarsveld et al. (1998), Capacities calculated.

Continent	Number of reservoirs, 1000	Sum of maximum capacities km^2	Capacity per capita m^2 $capita^{-1}$	Capacity per land area m^2 km^{-2}
Africa	43	915	2.1	30.5
Asia	201	1480	0.6	33
Australia/Oceania	19	54	3.2	6.8
Europe	88	430	0.6	41
North America	175	1184	3.3	49
South America	96	806	3.5	45
Globe	622	4869	–	

2. BACKGROUND INFORMATION

Water moves among three interconnected spheres; the hydrosphere that includes oceans, lakes, rivers and shallow groundwater; the atmosphere, which includes the air column up to approximately 250 km, and the lithosphere, which includes the soils.

2.1 The Hydrological Cycle

The mother of all water on land is precipitation. Four other processes describe the fate of this water: evaporation, runoff, infiltration into soil, and ground water flow. A water molecule, starting in the ocean, may initiate a journey through the atmosphere by evaporating from the ocean surface into the air, then falling down on the ground as precipitation, running off as overland flow until it reaches a river, and then following the stream into the ocean again. It is not very meaningful to calculate the cycle time of an average molecule in the hydrological cycle – the cycle patterns vary too much, but we have estimates of the time duration of many sub processes.

A schematic picture of the hydrological cycle is shown in Figure 16-1. The global land area precipitation amounts to 111 Mton = 111×10^{12} tons per year. Evaporation from sea, ice, and continents to the atmosphere consumes one quarter to one half of the solar energy falling on Earth.

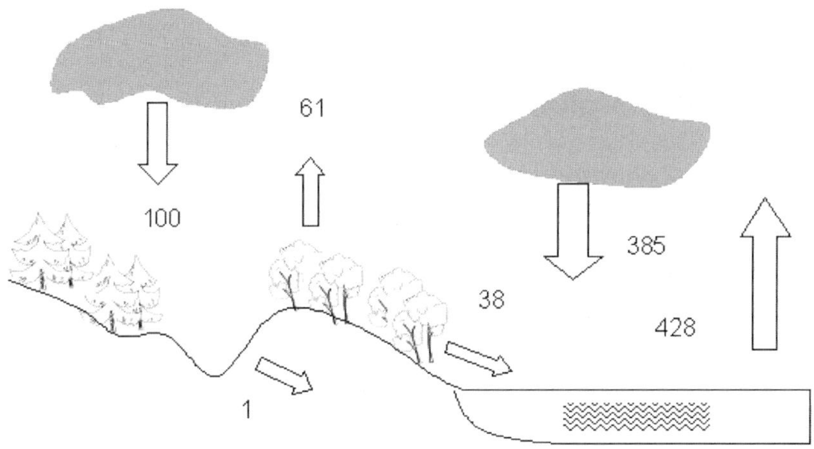

Figure 16-1. The hydrological cycle. Precipitation over land is per definition 100%. About 38% ends up as overland flow, 61% evaporates and 1% becomes ground water flow. After Chow et al. (1988) and others.

The total global precipitation amounts to about 500×10^{12} tons of water each year, or on the average 1 ton per m^2 per year. The residence time for water in the atmosphere is about 9 days.

Water is abundant, but still in many places a scarce resource. The polar ice stores about 68% of the freshwater on the earth, whereas the next largest storage is ground water with 30%. Lakes and rivers only store 0.26%. The storage of freshwater in the Polar Regions is predicted to contribute substantially to the expected 30 cm to 100 cm increase in sea level following global warming. The basic mass balance equation for the hydrological cycle on a given area is:

$$P = R + E \pm \Delta S \pm \Delta G \tag{16-1}$$

- P = precipitation (mm hr^{-1}).
- R = stream runoff (mm hr^{-1}).
- E = evaporation (mm hr^{-1}).
- ΔS = change in surface land storages (soil, lakes and rivers) (mm hr^{-1}).
- ΔG = change in groundwater (mm hr^{-1}).

A hydrological year runs from October 1 to September 30 at northern latitudes. Over a hydrological year we may approximate: $\Delta S = \Delta G = 0$. In that case:

$$\textbf{\textit{Precipitation}} = \textbf{\textit{Runoff}} + \textbf{\textit{Evaporation}} \tag{16-2}$$

2.2 Hydrological Services and Negative Externalities

Table 16-2 shows estimates of the value of some services supplied by hydrological processes. For example, natural water acts as sewage treatment plants by decomposing organic matter. However, hydrological processes also cause major catastrophes like hurricanes, flooding, and landslides. Some disasters have even caused more than 500,000 deaths: The 1900 drought in India, the drought in the Soviet Union in 1921, and floods in China in 1928, 1931, and 1939.[3] The probability of dying in a flood or in a tornado is roughly 1 in 35,000 for an average American.[4]

Table 16-2. Summary of average global value of the hydrological cycle. Water regulation is described in section 4.4. Data Constanza et al. (1997).

Type	World area 10^6 ha	Water regulation US\$$ha^{-1}$ yr^{-1}	Water supply US\$ ha^{-1} yr^{-1}	Waste treatment US\$ ha^{-1} yr^{-1}
Lakes, rivers	200	5,446	2,117	665
Wetlands	330	15	3,800	4,177

Damage costs of flooding depend upon local conditions. When the Mississippi river flooded in 1993 because of a doubling of normal precipitation, 50,000 km^2 land was covered by floodwater, 50,000 houses were damaged and the cost was estimated at US\$ 10×10^9. The damage per house was US\$ 200,000, but damages also included lack of domestic water supply, agricultural loss, barge traffic closure and road closure.[5]

3. STREAM HYDROLOGY

When we examine a catchment area, we see a web of interconnected streams and rivers. Horton initiated the quantitative study of stream networks in 1945 by developing quantitative models that connected hydrological properties of streams. These models could be used to check if a description of a watershed is reasonable.[6] Horton developed a system for ordering stream networks and derived laws relating the number and length of streams of different orders. **Stream order** is a hierarchical classification system of stream channels in which headwater streams with no tributaries (the smallest reaches) are defined as first order streams. These smallest recognizable channels normally flow only during wet years. When two first order streams join, the resulting reach is a called a second order reach; when two second order reaches join, the result is a third order reach, and so on.

Higher order arises only when two reaches of the same order join, a smaller order stream flowing into a larger order reach does not change the order of the largest. See Figures 16-2 and 16-3.[7] Note that this corresponds to addition of logarithms. Similar rules apply to addition of noise measured in decibels from several sources.

The **bifurcation ratio**, $R_B = N_i/N_{i+1}$ is the ratio of the number N_i of channels of order i to the number of channels of order i+1. This ratio is relatively constant from one order to another and often takes a value between 3 and 5. The average length of streams of successive orders are related by the **length ratio** $R_L = L_{i+1}/L_i$. The areas drained by streams of different orders can be described by a **stream area ratio** $R_A = A_{i+1}/A_i$. Typical values for bifurcation ratios, length ratios, and area ratios are discussed in the exercise below. A fourth parameter useful for stream analysis is the drainage density and the **average length of overland flow**, L_0. **Drainage density** means drainage efficiency and is the sum of ratios of total lengths of stream channels and catchment areas, as in Figure 16-2.[8]

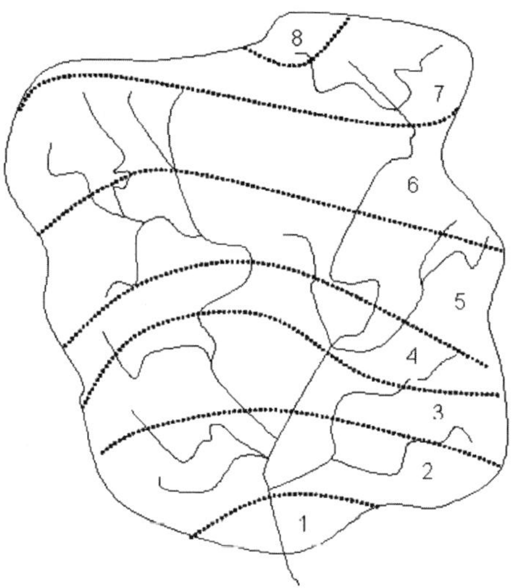

Figure 16-2. Hydrograph showing a catchment area with isochrones, which are lines of equal travel time (dashed lines).

$$D = \Sigma\Sigma\, L_{ij} / A_i \qquad (16\text{-}3)$$

The first sum is over the orders i, and the second sum is over all streams of the same order, $j = 1$ to N_i. The lower the drainage density, the higher the average length of overland flow:

$$L_0 = 1/2D \qquad (16\text{-}4)$$

Horton's stream laws describe the most likely combinations of channels in a network if a random selection is made from all possible combinations.[9]

3.1 Exercise: Stream Characteristics in a Catchment Area

Stream characteristics are important information for wildlife and fishery managers.

Required: Prepare the characteristics of the watershed in Figure 16-2 by calculating bifurcation ratio, length ratio, and area ratio. K = 7.5 hr is a storage constant, equal to the time it takes for the center of a mass of a flood wage to pass from the upstream end of the reach to the downstream end. T_L = 8 hr is the catchment lag – the maximum travel time through the catchment.

Figure 16-3. Streams. The left stream is of order 4 (width 7–8 m) and the right of order 5 (width 30–50 m). Note the revetments to the left of the flow direction in the smaller stream. (Left: Jura mountains; Right: Haut Provence, France).

The catchment is 250 km^2. Typical values for real stream systems are R_B = 2–5, R_L = 2–3, and R_A = 3–5. Is the catchment area of Figure 16-2 typical?

Solution: By examining the catchment area, you will find 16 streams of order 1, five streams of order 2, two streams of order 3 and one stream of order 4, which is the largest order in this catchment area. The bifurcation ratios range between 2 and 3.2, which is on the low side. It is hard to calculate the length ratio from the figure, but since the catchment area is about 20×12.5 km = 250 km^2, L_2/L_1 is probably around $5 \times 4/16 \times 2 = 0.6$, which is far from the typical ratio of 2-3. The drainage ratio cannot be inferred from the figure. Compared to typical catchments, the one in Figure 16-2 appears rather artificial.

4. DOSE-RESPONSE RELATIONSHIPS

Hydrological dose-response models are of three main types.

- *Time series models* relate future events to past events. Non-linear simplex type models are efficient for predicting future events from past observations.[10]
- *Regression models* relate future events to morphological, physical and chemical properties of the catchment area. These relationships are purely data driven; there is no guarantee that the parameters of the model can be interpreted mechanistically or in terms of cause effect relationships.
- *Cause-effect models* are based on general principles from physics or chemistry.

The models we present here are time series and regression models. Unfortunately, it was not the custom at the time many of the regression models were published to also report statistical significance.

4.1 Precipitation

Precipitation, or rainfall, is probably the single most important factor for lowland climates. From a managerial point of view, it is important to know its patterns and effects. How is rain distributed with time? What is the time between a heavy rainstorm and peak runoff?

Important precipitation variables are intensity (mm hr^{-1}), duration (days), return period (years), and aerial extension. By **return period**, we mean the average number of years it takes before a rainfall with the same duration and intensity returns. Since there is much uncertainty and variability connected to precipitation, we must describe our knowledge in statistical terms.

We may do that with time series and probability distributions. If we observe monthly precipitation at a location over a long time period, and make a histogram of our observations, we will see that it tends to be normally distributed, with lower variance in temperate than in tropical regions.

A **humid month** is defined as a month where rainfall exceeds evapotranspiration. Semi-arid climates have from 2 to 4.5 humid months per year. A semi-humid climate has 4.5 to 7 humid months. Subclasses of semi-humid climates are defined with reference to rainfall distribution over time, like bimodal rainfall patterns. A humid climate has seven or more humid months – usually more than 1400 mm rainfall annually. Very humid climates have more than nine wet months.[11]

Precipitation is in the form of rain, slush, hail or snow and occurs when air rises, expands and cools sufficiently for water vapor in the air to condense. Condensation increases if the atmosphere is rich in nuclei such as dust larger than 0.1 mm. Characteristic attributes of precipitation are listed in Table 16-3.

Table 16-3. Precipitation characteristics. Data Kiely (1997) p. 154, Jones (1997) p. 41 and others.

Name	Intensity mm h^{-1}	Drop size mm	Terminal velocity, mm s^{-1}	Areal extension km^2	Duration
Cloud droplets	–	0.002–0.02	0.13–13		
Drizzle	<1	0.1–0.2	279–720	25,000–250,000	A few days
Light rain	< 2.5	0.5–5.0 [(1)]	720–9000	2300–4600	<12 hr
Moderate rain	2.5–7.5	>0.5		100–400 mesoscale	< 180 min.
Heavy rain	> 7.5	>0.5		<10, cellular	min.
Storm	30–40	>0.5			
Haile				<10, cellular	
Snow				25,000–250,000	A few days

(1) Maximum size

4.1.1 Precipitation Data

Precipitation and other hydrological processes are convenient to record as time series, for example as a series of rain intensities (mm hr^{-1}). Such time series may be more than 100 years long.[12] Records of Nile floodings actually stretch back several thousand years. Area distribution, frequencies, and intensities are used to calculate the amount and distribution of precipitation in a certain catchment area over a certain period. Precipitation is measured with rain gauges at suitable sites and distances of approximately 1000 m. A gauge is a funnel with an opening, which is protected against wind. Snow retains water for a prolonged time compared to the residence time allowed by infiltration and run-off under non-freezing conditions.

Snow density varies from 0.005 in new fallen snow to 0.6 in old, highly compressed snow. Since the density varies with depth, snow packs must be sampled extensively before water equivalents can be obtained.

Fog can contribute substantially to the water supply of forest areas – of the order of 50% to 100%. The contribution of dew to soil humidity may be estimated by collecting it in shallow depressions in the earth lined with, for example, ceramic tiles.

Table 16-4 shows typical rain intensity data for continents. South America has the largest precipitation, Africa the smallest – just about one third of South America.

4.1.2 Intensity-Duration-Frequency Curves

Rainfall patterns differ with regions, with high rainfall intensities of short duration in semi-arid regions, and low intensity rains of long duration in temperate regions.

Table 16-4. Typical rain regimes. Water processes averaged over continents. Data from Baumgartner and Reichel (1975), Hornberger et al. (1998) and other sources.

Location	Land area 10^6 km^2	Annual precipitation mm yr^{-1}	Evaporation mm yr^{-1}	Runoff mm yr^{-1}	Temperature °C
Europe	10.0	650	371	278	10
Asia	44.1	870	525	345	13
Africa	29.8	767	642	124	21
Australia	7.6	447	420	27	20
N. America					
-USA	24.1	880	549	331	15
-Canada		710	443	266	0
S. America	17.9	1640	992	648	25
Antarctica	14.1	169	28	141	–33
Total	148.9				

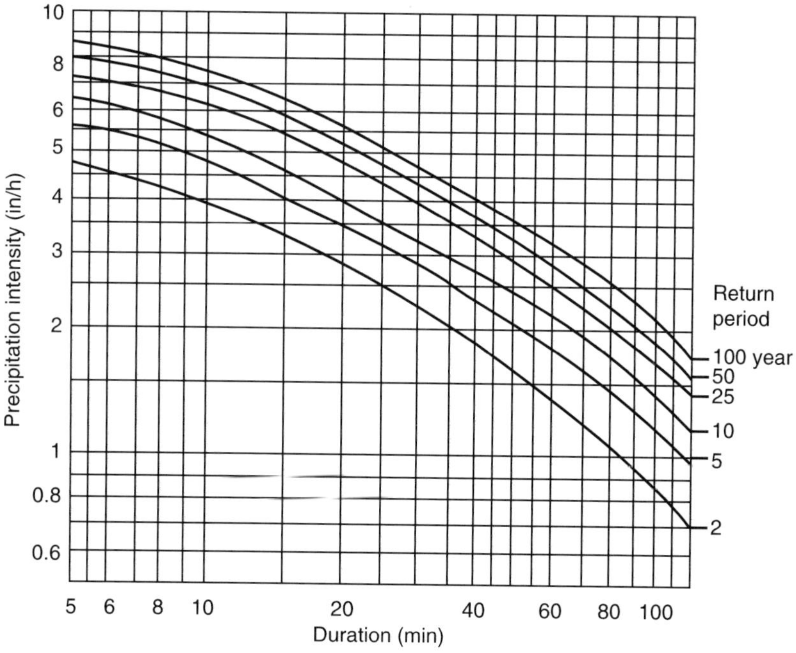

Figure 16-4. Intensity-duration-frequency curve for Chicago, USA.[13]

To calculate the right dimensions of constructions like river revetments, sewer networks, bridges, and dams, it is important to know how often combinations of rain intensities and durations will occur. **Holland's equation** is helpful here. It is a statistical relation between rainfall intensity P (mm hr^{-1}), duration D (minutes) and frequency N (average number in 10 years). A graphical illustration of Holland's equation is shown in figure 16-4. The parameters must be set locally. The equation for the British Isles (1967) is:

$$P = 25.4(D/60N)^{0.318} \tag{16-5}$$

Exercise: Suppose you are in the British Isles, and want to know the maximum rain intensity that would last for 30 minutes with a return period of 10 years. What is it?

Solution: Holland's equation (16-5) gives:

$$P = 25.4\,(30 \times 60^{-1})^{0.318} = 20.38 \text{ mm per 30 minutes or 41 mm hr}^{-1}$$

This is above the intensity of a rainstorm as it is defined in Table 16-4.

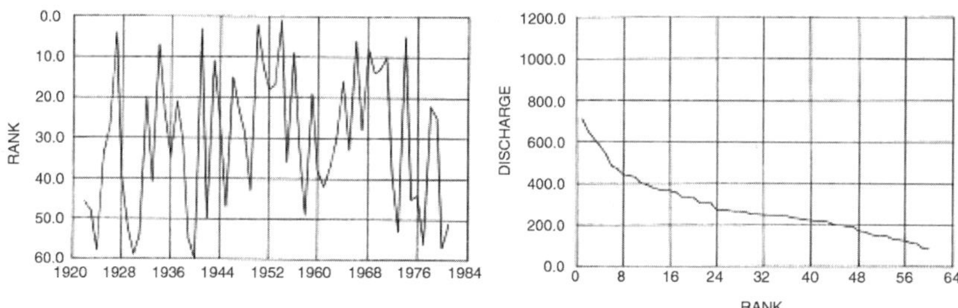

Figure 16-5. English River, Canada, discharges. Left: The annual maximum discharge as a function of time for the English River 1922-1981. Right: the discharge-rank plot for this data.

4.1.3 Estimation of the Return Period

If we have a long time series of observations, we can estimate empirically the return period for different combinations of intensity and duration. The **return period**, T_r, is the average time that elapses between two events that equals, or exceeds a particular level, Q.

The method is the same regardless of whether we deal with rainfalls or water flows in rivers, and we shall illustrate it with an example from the English River in Canada. Observed water flows in the river near Sioux Lookout are shown in Figure 16-5.

Although we have data for the whole period 1922–1981, we shall for the purpose of this illustration only use the data for the 15 years 1922–1936 that are shown in Table 16-5.

To estimate the return periods, we first rank the data, assigning rank $n = 1$ to the largest flow, $n = 2$ to the next largest, and so on. See Table 16-6. We then estimate the return periods with the following simple formula:

Return period: $T_r(x) = (N+1)/n$ (16-6)

N is the number of recorded years ($N = 15$), and x is the size of the flow that was assigned rank n in the recorded data.

Table 16-5. The first 15 observations of flow data for the English River near Sioux Lookout from a 1922-1981 data series.[14]

Year	*1922*	*1923*	*1924*	*1925*	*1926*
Flow $m^3 s^{-1}$	195	172	111	243	164
Year	*1927*	*1928*	*1929*	*1930*	*1931*
Flow $m^3 s^{-1}$	583	224	149	88	131
Year	*1932*	*1933*	*1934*	*1935*	*1936*
Flow $m^3 s^{-1}$	331	220	470	272	242

Table 16-6. Ranked peak flows for the English River.

Rank	Year	$P(x) > Q$	Return period yr	Flow Q
1	1927	0.0625	16/1 = 16.00	583
2	1934	0.1250	16/2 = 8.00	470
3	1932	0.1876	16/3 = 5.33	331
4	1935	0.2500	16/4 = 4.00	272
5	1926	0.3130	16/5 = 3.20	264
6	1925	0.3750	16/6 = 2.66	243
7	1936	0.4370	16/7 = 2.29	242
8	1928	0.5000	16/8 = 2.00	224
9	1933	0.5650	16/9 = 1.77	220
10	1922	0.6250	16/10 = 1.60	195
11	1923	0.6900	16/11 = 1.45	172
12	1929	0.7500	16/12 = 1.30	149
13	1931	0.8130	16/13 = 1.23	131
14	1924	0.8740	16/14 = 1.14	111
15	1930	0.9380	16/15 = 1.07	88

The formula is based on an observation that the general shape of discharge-rank plots like those in Figure 16-6 has a shape similar to $1/x$. For flood series, discharge rank plots often appear as shallow curves suggesting that there is a heavy tail of large floods. The resulting estimates return periods are shown in Table 16-6. We see from the table that the return period for the largest flow in the 15-year data set is 16 years, with flow of 583 mm. The table also shows estimates of the probability that a flow x in one time period (one year) will exceed a high flow Q with return period T_r. The probabilities are calculated according to formula (16-7).

$$P(x \geq Q) = n/(N+1) = 1/T_r \tag{16-7}$$

The probability $P(x \geq Q)_n$ that x will exceed Q in the span of n years is found by first calculating the probability that it will *not* exceed Q in n years:

$$P(x < Q)_n = [P(x < Q)]^n = [1 - P(x \geq Q)]^n \tag{16-8}$$

The probability $P(x \geq Q)_n$ becomes:

$$P(x \geq Q)_n = 1 - [1 - \mathrm{P}(x \geq Q)]^n = 1 - (1 - 1/T_r)^n \tag{16-9}$$

Exercise: Sewage overflow

Problem: Assume that a sewage system is designed to overflow on the average every 10-years. What is the length of the time-period that gives a chance of 50% for overflow?

Solution. The requirement is that $P(x \geq Q)_{n=10} = 0.50$. The return time is $T_r = 10$. The time-period for an anticipated overflow is n, which can be found from equation 16-9:

$$P(x \geq Q)_n = 0.50 = 1 - (1 - 1/T_r)^n$$

This gives $n = \log(0.5)/\log(9/10) = 6.5$; there is an even chance that there will be a sewage overflow within a time-period of 6.5 years.

4.2 Evaporation

Evaporation returns water from liquid or solid phase to vapor in the atmosphere. Water is also transported from biological material like plants and animals to the atmosphere through evapotranspiration, which means first transpiration, the evaporation. **Interception** is the portion of precipitation that is retained by vegetation. We distinguish among:

- Evaporation from water surfaces, E_0, (mm hr^{-1}).
- Potential evapotranspiration, *PET*, (mm hr^{-1}).
- Actual evapotranspiration, evaporation when water is in short supply, *AE*, (mm hr^{-1}).

Evaporation differences among the continents were illustrated in Table 16-4. There is least evaporation in South America (60% of rainfall) and highest in Australia (90%). Evaporation depends upon:

- The depth of the water table.
- Capillary effects raising the water above the water table, which is related to soil pore characteristics.
- The local heat budget.
- Vegetation.
- The pattern of precipitation.

Evaporation from a lake or an open water surface is in its simplest form a function of the difference in saturation and actual vapor pressure, wind speed, and available energy as heat:

$$E_0 = 0.11 u_{2m} (e_a - e_s) \qquad (16\text{-}10)$$

Here, e_a is actual- and e_s saturation water pressure in the air (mbar). (mmHg converts to mbar by dividing by 0.760). The wind velocity, u_{2m}, is measured 2m above the water surface, 2m s^{-1} being often used as a standard.

Since wind requires a certain open distance to obtain speed, the wind velocity in (16-10) is often substituted with the open water area, A (m^2). Thus, the area of the lake is often used as an independent parameter for the calculation of evaporation.

$$E_0 = 0.291A^{-0.05}(e_a - e_s) \quad (16\text{-}11)$$

Potential evapotranspiration, *PET*, is not observed, but calculated:

$$PET = gE_0 \quad (16\text{-}12)$$

The parameter g depends upon vegetation and hours of daylight and therefore tends to be lower at higher latitudes. *PET* is often used as a measure of the amount of energy hitting a certain location. Studies show that at *PET* values less than 1000 mm yr^{-1} – corresponding to average PET values north of the USA, energy is the main determinant for species richness.[15] Evaporation uses about 50% of the incoming solar energy.

Vegetation increases evaporation as shown in Table 16-7. Evaporation is typically 20% to 40% of precipitation from grassland and 40% to 70% from forests. Typical evaporation ranges in European countries measured in millimeter per year are: for spruce forests 580 to 800, grass 420 to 520, and open water 600 to 610 mm yr^{-1}.

4.2.1 *Exercise*: Evaporation from Lake Mjøsa

Problem: Lake Mjøsa, Norway, has an area of 300 km^2. Saturation vapor pressure is 15 mmHg and the actual vapor pressure is 9 mmHg. Wind speed is measured up to 2m s^{-1}. Calculate the evaporation.

Solution: Using Eq. 16-11 we find:

1. $0.291 \times (300 \times 10^6)^{-0.05} = 0.11$.
2. $e_s - e_a = (15\text{–}9)/0.760 = 7.9$ mbar.
3. $E_0 = 0.11 \times 2 \times 7.9 = 1.74$ mm day^{-1}.

1.74 mm day^{-1} gives about 634 mm yr^{-1}, which compares well with observed evaporation rates from open waters in northern Europe, which are around 600 mm yr^{-1}.

Table 16-7. Evapotranspiration rates from different vegetation types.[16]

	Larch, birch, Douglas fir	Spruce	Maquis shrub	Beech	Pine	Pasture	Bare earth
Ratio to bare soil	9	8.5	8	7.5	5	3	1

4.3 Infiltration

Infiltration, or hydraulic conductivity, is movement of water from the surface into the soil. The rate at which infiltration drainage occurs is largely a function of pore diameter, connectivity and tortuosity. These factors determine the longer-term saturated infiltration capacity of the soil. Infiltration is measured in mm hr^{-1}, which is the same unit as for precipitation. If precipitation is less than infiltration, no water will be stored on the soil. The actual infiltration rate depends upon several factors. Ice or frost makes the soil surface impermeable, giving overland flow for small precipitation rates. High rain intensity gives larger infiltration rate than low intensity. In addition, infiltration decrease exponentially with slope, being typically 1/3 at 30% slope compared to level surface.

Typical infiltration rates for unsaturated soil were given in Table 15-5 in the previous chapter. Infiltration capacity may decline during rainfall due to expansion of clay particles as they absorb water or clogging of pores by slaking of particles from aggregates. Thus, although field measurements may in a certain situation give nominal infiltration capacities of more than 200 mm hr^{-1}, runoff may occur if it rains only 20 mm hr^{-1}.[17] As rain continues to fall, the soil will be soaked with water and the infiltration rate decreases to some minimum value with time. Equation 16-13 shows an expression describing this relationship:[18]

$$f = f_c + (f_o - f_c)e^{-kt} \qquad (16\text{-}13)$$

Here, f is infiltration rate, f_0 is initial infiltration rate, f_c is final infiltration rate at soil saturation, and k is an empirical constant. Figure 16-6 shows how the infiltration rate decreases with the duration of a rainfall. It intersects a constant rainfall curve and if the rainfall continues until time T, the soil infiltration rate is reduced so much that the rainfall intensity exceeds infiltration, and overland flow will occur. Such runoffs may easily cause soil erosion.

4.3.1 *Exercise*: Overland Flow

Assume that it is raining intensely – at a rate of 30 mm hr^{-1} – on dry sandy loam. The maximum infiltration rate for the soil is $f_0 = 80$ mm hr^{-1} and the minimum is $f_c = 10$ mm hr^{-1}.

The coefficient describing the soakedness of the soil is $k = 0.40$ hr^{-1} (See Equation 16-13).

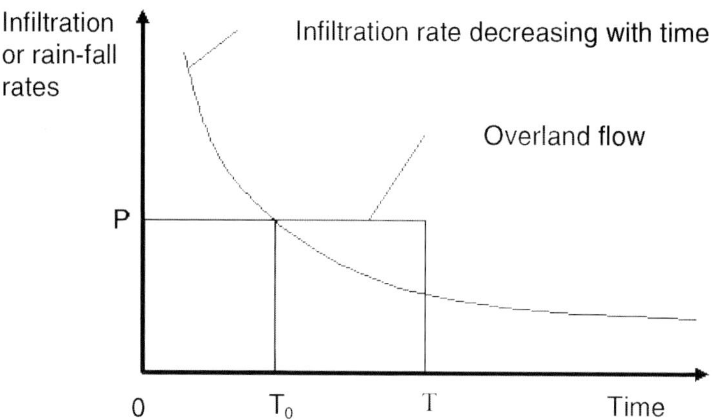

Figure 16-6. Overland flow. If precipitation intensity, P is greater than the infiltration rate, and precipitation duration is at least T_o, water will flow on top of the soil.

Problem: How long time will it take before overland flow begins? If the rain lasts for six hours, what is the amount of overland flow?

Solution: Since $f = 10 + (80 - 10)e^{-0.4t}$ overland flow occurs at $e^{-0.4t} = (30 - 10)/70 = 0.29$, and $t = 3.13$ hours.

The total amount of infiltration is:

$$f_{tot} = f_{30}(T - T_0) - (f_0 - f_c)(e^{-kT_0} - e^{-kT})/k = 30 \times 2.9 - 70 \times (-0.195)/0.4 = 52$$

The total amount of rain is 30 mm $\times$ hr^{-1} $\times$ 6 hr = 180 mm.

4.4 Storage

Water collects in soil, depressions on the surface, ponds, dams, reservoirs, lakes and rivers. The meandering patterns of many rivers help store water and reduce the probabilities of floods, and models for routing water through channels have been developed.[19] Models relating runoff to land characteristics are described in Chapter 20 on agriculture.

4.5 Runoff

Runoff is the same as overland flow – flow of water through the uppermost layer of the soil, or the aeration zone. Runoff can in its simplest form be thought of as a sheet of water – like a paper sheet – being transported down a slope. *Overland* flow is sometimes called **sheet flow** or **Hortonian flow**. On smooth surfaces, sheet-flow velocities may be as high as 0.1 m s^{-1}, since there is little frictional drag. Sheet flow has only small erosion effects. It is rarely stable, because raindrop splashes form rills together with local irregularities. Rills may later develop into larger gullies. Runoff is measured in terms of mm s^{-1} or m^3 s^{-1}.

Overland flow may also occur as so-called soil profile through-flow. Movements occur through voids, including pipes created by roots and burrowing animals. The characteristic velocity is slow, often of the order 0.5 m hr^{-1}.[20] Rainfall will create runoff if it cannot infiltrate the soil, either because the infiltration capacity of the soil is lower than rainfall intensity, or because the soil is saturated with water or is frozen. The first form is most common in semiarid areas with high rainfall intensities and a sparse vegetation cover. Saturated overland flow is more typical of moist temperate environments where rainfall intensity is low and rainfalls have longer duration. Overland flow provides a major input of water, sediments and solutes to streams and lakes.

4.5.1 Rare Runoff Events

There are at least four runoff properties of interest:

- Annual average runoff.
- Extremely low and extremely high runoff.
- Lag time between peaks of intense rain.
- Peak runoff.

Extremely low runoff is of interest if the runoff feeds a river that is used for irrigation or as a recipient for waste. Extremely high runoff can cause flooding. Table 16-8 shows typical return periods used for designing infrastructure installations.

Table 16-8. Typical design return periods.

Impact	Design return periods, years
Sewer networks	2–7
Sewer network in urban area with basements	50–100
River structures	30–50
Bridge	100–500
Dams	1000–100,000

Typical floods with 10 years return period are 1.5 times mean annual flood, Q_{MAF}. Floods with return periods of 25 years may be 1.5 to 2.3 times Q_{MAF}, and typical floods with 100 years return time 2 to 3.5 times Q_{MAF}. Floods with 500 years return time are 2.5 to 5 times Q_{MAF}.[21] Events with a 50-year return period or greater are likely to cause system disturbance or a permanent change in channel morphology.[22]

Figure 16-7 shows for example that, in a region in the UK where the annual rainfall is 1000 mm yr^{-1}, an urban sewer network with design overflow time of 50–100 years would have to cope with a maximum daily rainfall of 75–80 mm. When we plot empirical runoff volumes versus time, we usually observe that rare events occur in excess of what would be expected from the normal distribution.[23] In the same manner, contaminated catchments may retain pollutants longer than would be expected from a normal distribution of dilution processes.[24] The normal distribution is therefore not a good model for such processes, and the gamma distributions are often used instead. Gamma distributions arise through combinations of two or more random series of events spread over a continuous period, such as random rainfall events and random prior soil moisture conditions.[25]

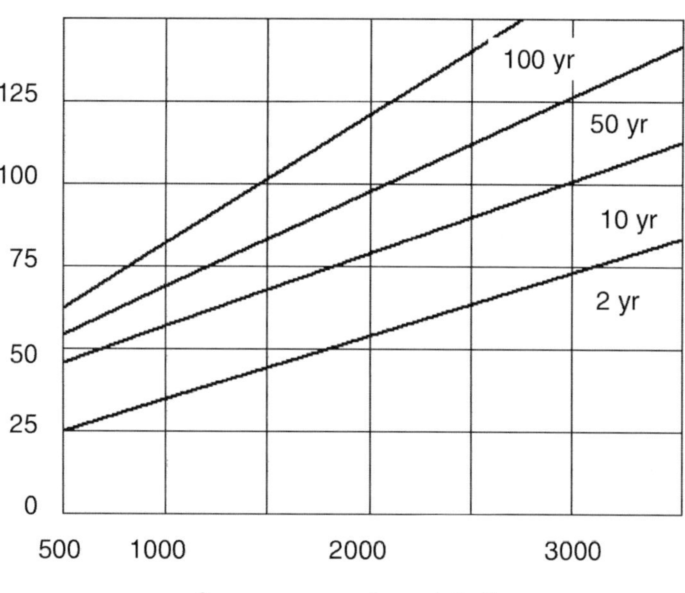

Figure 16-7. Relationship between one-day maximum rainfall with a given return period and average annual rainfall. Data for UK. After Institute of Hydrology cited by Wilson (1983) p. 21.

4.5.2 Runoff Models

We present two runoff models. The first is a simple model relating runoff to precipitation intensity and the area of the catchment:

$$Q_p = 0.278\ CIA \tag{16-14}$$

Q_p is peak river flow ($m^3\ s^{-1}$). C is the runoff coefficient. I is rain intensity, mm hr^{-1}, and A is the catchment area, km^2. C is between 0.7 and 0.9 for streets, and 0.95 for heavy industrial areas. For arable land, see Table 16-9.

A more complete runoff model is shown in Equation 16-15. The parameters represent factors that contribute to runoff; see Table 16-10.[26]

$$Q = C \times A^{0.94} \times F^{0.27} \times S^{0.16} \times SO^{1.23} \times R^{1.03} \times (1 - L)^{-0.85} \tag{16-15}$$

Table 16-9. Runoff coefficient from arable land. Data summarized by Thompson (1999) and others.

Land use	Sandy	Loamy	Clay
Arable, flat, (<2%)	0.05	0.10	0.15
Arable, steep (>7%)	0.15–0.20	0.40	0.50
Pasture	0.15	0.35[(1)]	0.45
Woodland	0.10	0.30	0.40

(1) 0.23–0.35 for steep (>7%) grassed lawns on clay

Table 16-10. Parameters for Equation 16-15. Stream slopes are often characterized by their mid slopes, that is, by removing the top 15% and the last 10% of the length of streams. The remaining parts tend to have relatively constant slopes.

Legend	Definitions	Typical values
Q	Mean annual flood	
A	Catchment area, km^2	1–1000 km^2
F	Stream frequency or number of stream junctions per km^2	1.0 for an area with many small streams
SO	Soil index range 0–1.0 where 1.0 gives highest runoff	$SO = 0.4$ for soil with medium infiltration rate
S	Stream slope between 10% and 85% location, m/km	$S = 1.4$ m.km^{-1} in a flat landscape; 5m.km^{-1} in a rolling landscape
SAAR	Standard average annual rainfall	10 mm – 3000 mm
R	Net 1 day rainfall with a 5-year return period. Can be calculated from average rainfall	$R = 2.48 \times$ SQRT (SAAR) – 40 = 42.4 mm when average annual rainfall is 1100 mm
L	Lake area, fraction of catchment occupied by lakes	0.0–1.0
C	Regional multiplier	0.018–0.02, Ireland and Scotland

4.5.3 The Unit Hydrograph

The unit hydrograph is a plot of stream discharge versus time. It characterizes the hydrological properties of the catchment. It is defined as the catchment outflow resulting from a direct runoff generated uniformly and with a constant rainfall over a unit time, such as an hour. The unit hydrograph assumes that the flow rate, Q, is proportional to the rainfall and that this does not vary with time. The principle of superposition is thus valid, and runoffs from two separate rainfalls can be added. The time duration of urban hydrographs tend to be shorter than rural ones, Figure 16-8.

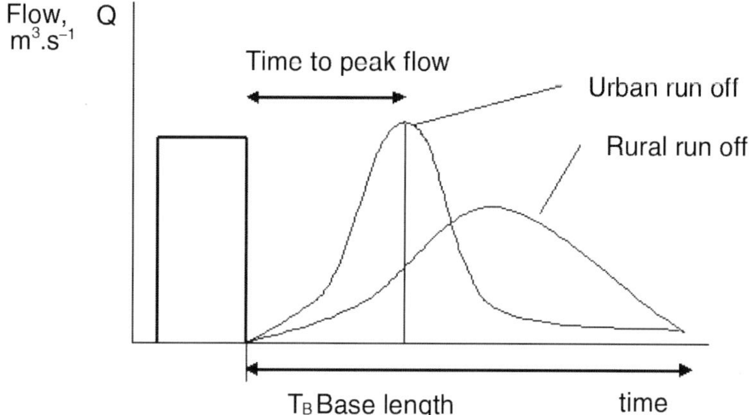

Figure 16-8. Unit hydrograph for urban and rural catchment.

Irrigation practice makes water susceptible to increased evaporation. Creating storage on land or increasing the meandering pattern of rivers can increase infiltration.

5. MITIGATION

The hydrological cycle can be changed in several ways on a small scale. Planting forests may increase evapotranspiration from soils. It also increases the infiltration capacity of the soil.

6. CONCLUSION

The hydrological cycle consists of four processes: precipitation, evaporation, runoff, and storage. We have presented simple models for each of these processes. To design river bridges or sewage networks it is important to be able to predict extreme events. Models to estimate extreme events based on previously recorded information are presented. Changes that modify the way water flows may have important effects on many of the services water flow offers. Building dams, for example, may give control over extreme flows, and make water useful for hydropower, but it also changes the way particles suspended in the water are transported and it changes evaporation, two processes that are important for natural and man-made ecosystems both in the water and in the land close to the water.

---

[1] Van Jaarsveld et al. (1998)
[2] Seip and Botterweg (1988)
[3] Burton et al. (1993) p. 11
[4] Morgan (2000)
[5] Search Mississippi river flood
[6] Horton (1945)
[7] Chow et al. (1988) p. 169, Cushing et al. (1995) p. 739
[8] Chow et al. (1988)
[9] Schreve (1966)
[10] Sugihara and May (1990)
[11] Ruthenberg (1980)
[12] Pilon et al. (1985)
[13] After Chow et al. (1988) p. 454
[14] From Pilon et al. (1985) p. 72-73
[15] Kerr and Parker (1997)
[16] Data after Jones (1997) p. 60
[17] Morgan (1995):31
[18] Horton (1933)
[19] Jones (1997)
[20] Briggs and Courtney (1994) p. 268
[21] Data for Great Britain; Jones (1997) p. 180
[22] Padmore et al. (1998)
[23] Jones (1997) p. 108
[24] Stark and Stieglitz (2000) p. 493
[25] Jones (1997) p. 109
[26] Details can be found in NERC (1975) or Kiely (1997)

Part 4

THE NATURAL ENVIRONMENT

Overview

Chapters

17. LAKE: Surface water environments and water quality.
18. RIVER: Water environment – rivers and wetlands.
19. TERRESTRIAL ENVIRONMENT: Habitat suitability in natural terrestrial environments.
20. AGRICULTURE AND LAND USE: Agriculture and land use with emphasis on developing countries.
21. AIR: Emission, dispersion and effects of airborne pollution.
22. CLIMATE: Anthropogenic emissions, climate and ecosystems.

Although the Natural Environment can be considered as a single entity, it is useful to describe its physical realms separately. This does not mean that we should not treat the Natural Environment as a continuous entity, but that proper decision making requires us to look across the boundaries of those individual realms.

The chapters are organized as follows:

Introduction: What the chapter deals with and what it leaves out.

Background information: An outline of our current basic knowledge and terminology, and a short overview of values and negative externalities.

Quality standards: A description of standards set for managerial purposes to safeguard humans as well as the environment from deteriorating into an unhealthy state.

Dose-response: This section uses *dose* and *response* as metaphors for human effects on the environment.

Mitigation: This section outlines mitigation measures for particular environments as examples for inspiration. Hopefully, the book will induce creativity so that new ways of mitigating human impacts can be found.

Application: This section provides a framework for a possible case study.

Chapter 17

LAKE

Surface water environments and water quality

We describe surface water environments and discuss methods for monitoring, classification, and prediction water quality in lakes. In the following chapter, we address problems related to rivers, streams and wetlands. Man affects surface water by withdrawing it (e.g., for urban use and agricultural irrigation) and by discharging it polluted back into surface water. Surface waters are also polluted by non-point sources like agricultural runoff. Eutrophication and acidification are the most endangering forms of pollutants. We give an outline of water quality criteria and present six models for calculating the effects of nutrient loads on the lake water ecosystem.

●

"Moody and withdrawn, the lake unites a haunting loveliness to a raw desolateness." Dale Morgan

1. INTRODUCTION

Freshwater environments consist of lakes, rivers, and wetlands. In this chapter we discuss lakes in particular, and leave rivers and wetlands for the next chapter. A **lake** is a water-body where the water is retained for more than a week. Lakes range in area from small ponds to several thousand square kilometers, and store about 0.26% of the world's fresh waters with depths ranging from less than a meter to almost a kilometer. The world's largest lake is the Caspian Sea, 438×10^3 km^2, and the next eight in rank range from 80×10^3 to 20×10^3 km^2.

1.1 Natural and Man-Made Impacts

Flooding of areas near the shore is a common kind of natural impact associated with lakes. Flooding inundates low-lying land and increases soil erosion. However, flooding may also be advantageous when near-shore soils become nutrient enriched.

Human impacts are related to use of water or the catchment area. Some uses require water in pure condition – "blue-water", whereas others are much less sensitive to water quality, and require only "green water". Since we are so dependent on freshwater supplies, competition for the same water resources often causes conflicts.

Abstraction of water includes withdrawal for industrial cooling processes, irrigation, and hydropower production with damming. This may conflict with in-stream uses, like use of water as a recipient for waste, non-point erosion and commercial recreation. Leisure fishing and swimming are other in-stream uses, as well as using lakes for visual recreation, and as political boundaries.

Human impacts include physical modifications, like dredging or land filling projects, river canalization and flood control measures. Among the 12 use categories mentioned above, there are 72 possible conflicts, but probably only 22 contribute to actual conflicts.

1.2 Water Substitutes

Water is a commodity that – with few exceptions – is impossible to replace. One exception is water used for hydropower that may be substituted by fossil fuel or nuclear energy.

Water can be used with different efficiencies. Multiple uses in wetlands allow waters to be used both as a purification agent, as a medium to sustain ecological systems, and as a recirculation pump that generates precipitation.

1.3 The Value of Water Ecosystems

The average reported annual marine and freshwater edible food-catches in the period 1988–1991 was 94.3 million tons, excluding 27 million tons of discarded by-catch. This is 8% of the annual global total of 1516 million ton aquatic primary production.[1] Inland fishery contributes about 12% of all fish consumed by humans. Irrigated agriculture supplies about 40% of the world's food crops, and hydropower provides nearly 20% of the world's electricity production.[2] In Colorado, USA, the market price for a permanent supply of 1000 m^3 raw water is about US$9000; or US$450 yr^{-1} if a 5% interest rate is used. Still rental rates can be as low as US$10-36 yr^{-1}.[3]

Table 17-1. Water pollution damage costs. The annual cost of water pollutant damage in Germany (west, before 1986; G), The Netherlands (N), and USA. na = not available. Data: Schultz (1986) (G), and Opschoor (1986) (N) Cited by Pearce and Turner (1990) p. 123. USA: Cropper and Oates (1992).

Damage	10^6 US\$ yr^{-1}	US\$ $capita^{-1}$ yr^{-1}	Country
Freshwater fishing	100	1.7	G
Groundwater damage	2900	48	G
Recreation	na	≈ 30	USA
Water pollution, total	3000	50	G
Water pollution, total	100–300	6.5–19.4	N

Water is becoming a scarce resource in many parts of the world. Some 1.7×10^9 people reside in highly stressed river basins where water availability falls below 1000 m^3 per capita and year.

In addition to water shortage, large bodies of freshwater are polluted. Estimated values of the damage caused by water pollution in some selected countries are shown in Table 17-1. The marginal values per household and year for preservation of striped shiner and squawfish – two threatened or endangered species – are US\$6 and \$8.5; for steelhead salmon it is \$31, and for Atlantic salmon \$8.[4]

2. BACKGROUND INFORMATION

Open water ecosystems are of particular importance in lakes, but macrophyte vegetation plays a crucial role in shallow lakes – lakes that are less than 10 m deep. The major *taxa* in freshwater ecosystems include phytoplankton algae, which are green free-floating plants containing chlorophyll that build organic biomass from light and nutrients. In turn, zooplanktons graze upon phytoplankton and smaller species of its kind. Whereas herbivorous grazers in terrestrial habitats may graze up to 30% of green plants, grazers in lakes may graze more than 50% of the plants.[5] Fish graze on zooplankton and benthic organisms living in the sediments. This food chain gives rise to successional sequences in lake ecosystems.

Species succession in deep oligotrophic to mildly eutrophic northern latitude lakes follows a relatively predictable sequence. First diatoms start to grow, as shown in Figure 17-1. These are unicellular algae requiring dissolved silica as building material for their surrounding silicon shells. After some time, the silica resources in water are exhausted, the diatoms die and sink to the bottom and are replaced by yellow algae (*Chrysophyceae*) and green algae, (*Chlorophyceae*). Next, zooplankton starts grazing on all algae present, leading to a period of low algal population density – this is the **clear water period**, when lake transparency is not representative of the general water quality of the lake.

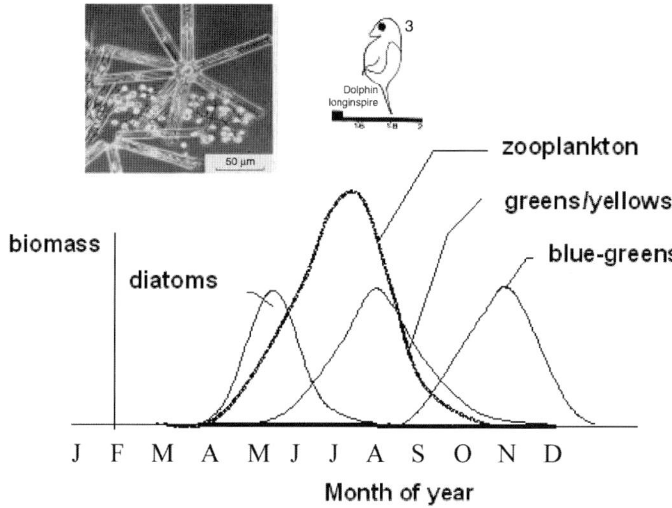

Figure 17-1. Succession in eutrophic lakes. In mildly eutrophic lakes in temperate regions algal succession proceeds as shown. The period when zooplankton graze down most of the phytoplankton is called the "clear water phase".

The final stage belongs to the blue-green algae (*Cyanobacteria*) that grow after the decline of the zooplankton population. Blue-green algae have the ability to thrive at the low light intensity in deep water, are buoyant, and are not easily eaten by zooplankton. They can thus out-compete other species of phytoplankton that require more light and are less grazing-resistant. Eutrophic and hypereutrophic lakes often follow another successional sequence than mildly eutrophic lakes.

Fish species can be classified as: *white* or *sports* fish (principally *Centrarchidae*), *forage* fish (*Cyprinidae*, *Cyprinodontidae*; *Atherinidae*, *Poecilidae*), and *rough* or *trash* fish, like gizzard shad (*Dorosoma cepedianum*). White fish generally tolerate less pollution in terms of nutrient loads than trash fish. They are also supposed to taste better than trash fish.

There is no agreement within the aquatic sciences on whether phyto-plankton abundance is determined primarily by the amount of nutrients loaded on the lake (bottom-up control) or primarily by grazing (top-down control). The preliminary conclusion seems to be that bottom-up control is the only important controlling factor in the long term – more than two years. This implies that the only way to reduce unwanted algal growth is to reduce nutrient inputs to the lake, either by reducing external phosphorus loads, reducing internal loads by chemical means, by biological means, or by dilution.[6]

Figure 17-2. Lake water quality. Upper left) Clear water, saturated oxygen, little growth on stones, little pelagic plankton, salmon present; upper right) slightly reduced transparency, decreasing oxygen with depth, slippery stones, algal growth, some phytoplankton, salmon present; lower left) low transparency, loss of oxygen at deep water, strong growth on stones, much plankton, lowered suitability for salmon; lower right) low transparency, no oxygen in deeper layers, bacteria and fungi abundant, large algal growth, blue-green algae dominates, no salmon.

2.1 Lake Description

In this section, we describe lakes and their ecosystem. There is an important difference between deep and shallow lakes.

2.1.1 Morphometry

Lake morphometry describes the physical form of the lake, which is an important determinant of hydrological and biological regimes. Shallow lakes are often nutrient-rich because of past inputs of culturally derived nutrients. However, many lakes, particularly in developed countries have now become cleaner because of environmental control, in the sense that they are less loaded with macronutrients and toxic compounds. Lake water quality is illustrated in Figure 17-2. The figure was used to make consequences of nutrient reductions vivid for decision makers.

The trophic level of a lake is related to lake morphology and lake catchment. In an investigation of 18 lakes, total phosphorus TP (mg m^{-3}), and lake mean depth z (m), was found to be significantly related:[7]

$$\text{Log}\,(TP) = -\,1.13\log(z) + 3.29,\ r^2 = 0.66;\ p = 00001,\ n = 18 \qquad (17\text{-}1)$$

In the same lakes, total phosphorus TP (mg m^{-3}) was related to lake catchment area C (km^2) and lake volume V ($10^6 \times$ m^3):

$$\text{Log}\,(TP) = 0.38\log(C/V) + 1.57;\ r^2 = 0.59 \qquad (17\text{-}2)$$

Average lake depth z, and lake area A (km^2) were also related. Equation 17-3 was estimated for Norway and Equation 17-4 for Minnesota.

$$\text{Log}\,(z) = 0.46\log(A) + 0.728;\ r^2 = 0.54;\ p = 0.0005,\ \text{n} = 18 \qquad (17\text{-}3)$$

$$\text{Log}\,(z) = 0.21\log(A) + 0.89;\ r^2 = 0.61;\ n = 20 \qquad (17\text{-}4)$$

Table 17-2. Characteristics of lakes of three different trophic states, with mass ratios for phosphorus and nitrogen. After Welch (1980:132) Kiely (1997:240), Seip (1994), OECD (1982), Marshall and Peters (1989), McCauley et al. (1989), Praire et al. (1989), Bays and Chrisman (1983).

Attribute	Oligotrophic	Eutrophic	Hypereutrophic
Lake size	>30 ha	≈ 30 ha	<30 ha
Lake mean depth	>10 m	<10 m	<2 m
Littoral area	stones	macrophytes	–
Sediments	stones, silt	muddy, nutrient rich	muddy, nutrient rich
Secchi disk depth	<5–3 m	1–2 m	<1 m
Water color	green or blue	yellow or green	green or reddish
Chl–a	2–4 mg m^{-3}	6–10 mg m^{-3}	20 – >200 mg m^{-3}
Ratio Chl–a$_{max}$: Chl–a$_{av}$ $^{(1)}$	2	3–4	–
Total phosphorus	10 15 mg m^{-3}	20–30 mg m^{-3}	>200 mg m^{-3}
TP/ Chl–a slope	Small	1:1	small
Critical TP loading, 10 m deep lake	<0.2 g m^{-2} yr^{-1}	≈ 0.2 g m^{-2} yr^{-1}	>>0.2 g m^{-2} yr^{-1}
– 100 m deep lake	≈ 0.8 g m^{-2} yr^{-1}	≈ 0.8 g m^{-2} yr^{-1}	
TN:TP ratio (by mass)	40	20	5
Correlation log TP and log TN	0.32	0.28	0.18
Total fish abundance	5 kg ha^{-1}	60 kg ha^{-1}	>250 kg ha^{-1}
Limiting growth factor	P	P	N, light

$^{(1)}$ Nutrient enrichment increases the amplitude of seasonal growth in increasingly nutrient rich lakes.[8]

2.1.2 Productivity

Lakes can be characterized by their **productivity** in terms of algal growth during the productive season. There is a strong correlation between concentration of phosphorus and algal productivity in most lakes. **Oligotrophic** lakes are nutrient-poor, and **eutrophic** lakes are nutrient-rich. Since deep lakes in general are also large and oligotrophic, it is possible to classify lakes in three groups, as shown in Table 17-2. The lake characteristics associated with oligotrophy and eutrophy are based on data from lakes in northern temperate regions; and may differ somewhat from other regions of the world. The last category of lakes shown in Table 17-2 is called hyper-eutrophic, and is more loosely defined than the first two.

2.1.3 Deep Lakes

At northern latitudes, water usually has a uniform top to bottom temperature of 6–8°C during the cold period of the year, and a stratified period extending from about March to October. **Stratification** occurs when the water is heated by the sun and warm air, and an upper warm layer of water forms atop a colder layer beneath as shown in Figure 17-3. The upper warmer layer is called the **epilimnion**, and the colder layer below the **hypolimnion**.

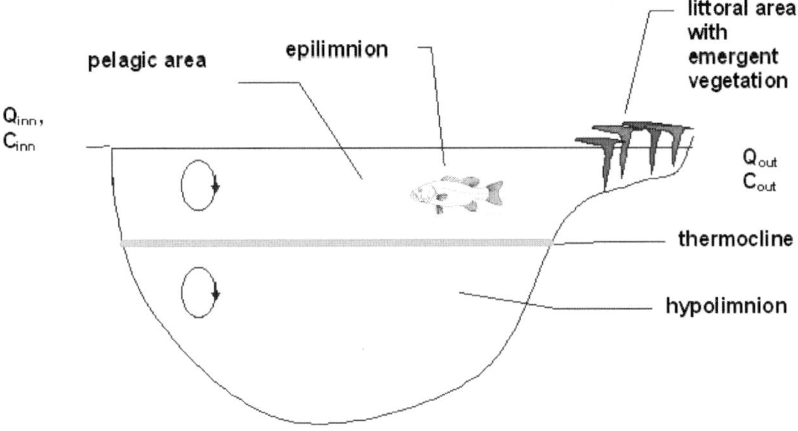

Figure 17-3. Lake stratification during summer.

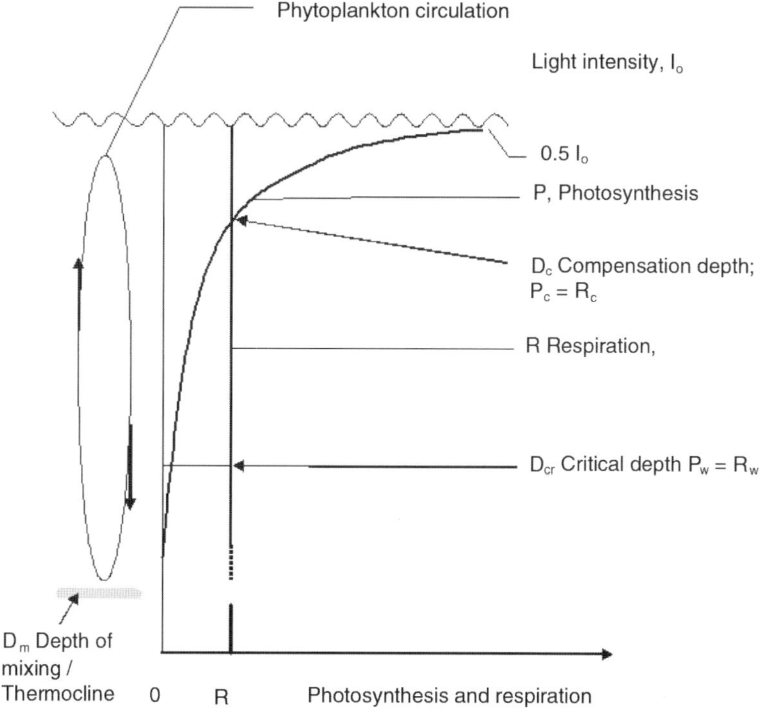

Figure 17-4. The relationship between the compensation depth, D_c, the critical depth, D_{cr}, and the depth of mixing, D_m. Secchi disk depth is about 2 times critical depth. In the situation here, the phytoplankton stays too long beyond the critical depth and no growth occurs.

The transitional layer in between is called the metalimnion, and contains the thermocline. As heating occurs, a thermal gradient with depth is set up. If this gradient is around 0.8–1.0°C per meter or higher, little mixing occurs across the temperature gradient or **thermocline**.[9] This corresponds to a change in the density of water of about 0.01g m^{-3}. Typical exchange values for epilimnion water are a fraction, 0.10–0.15 per month, or eddy diffusion coefficients of 0.04–0.06 $cm^2\ s^{-1}$ assuming a 10 m mixing length.[10]

The average thermocline depth, D_t (m), during the stratified period, is related to the **fetch**, which is the maximum effective length of the lake, L_{MEL} (m), defined by the straight line connecting the two most distant points of the shoreline over which wind and waves may act without interruptions from land or islands. A log-linear estimate of the relation based on observations of 167 lakes is shown in Equation 17-5.[11]

$$\log D_t = 0.34 \log L_{MEL} - 0.25,\ r^2 = 0.9,\ n = 167 \qquad (17\text{-}5)$$

During the stratified period, the thermocline deepens because of the turbulent forces of wind. 90% of the variation in annual average thermocline depth, D_t, in Lake Mjøsa – a 100 km long lake in Norway – could be explained by maximum monthly average air temperature, T, and the maximum monthly average wind velocity, W, during the months May – September as shown in Equation 17-6.[12]

Maximum epilimnion depth for this large lake is 50–70 m.

$$D = -3.1T + 4.6W + 35.8; r^2 = 0.9; p = 0.004, n = 8 \quad (17\text{-}6)$$

Residence time of the water in a lake is defined as lake volume divided by annual outflow. It may extend from only a few days to decades. 70 days to 5480 days have been observed in nine Swiss lakes ranging from 0.06 to 11.8 km^3 in total volume.[13] However, in deep lakes epilimnion water residence times may be 20% to 10% of the overall residence time. Since phytoplankton require both light and nutrients to grow, the first vernal phytoplankton bloom in deep lakes occurs when the first stratification begins, that is, when phytoplankton begin to circulate only within the epilimnion, where photosynthesis exceeds respiration. See Figure 17-4.

If lake stratification is strong and prevails during the summer, water and free-floating organisms are effectively isolated from the lower layer. Phytoplankton growth and sinking dead organic material may then exhaust the nutrient pool in the upper layer. At the critical depth, D_{cr}, daily average water column photosynthesis, P_w, is equal to daily average respiration, R_w, and there is no net growth. Blue-green algae or blue-green bacteria (*Cyanobacteria*) are first on the list of species one wants to avoid. They are responsible for most surface water nuisances.

They are often filamentous, forming large dense algal mats on the water surface, and may clog water intake pipes. As for drinking water, they may clog filter beds; giving the water bad odor and taste, and some are toxic to animals and man.[14]

2.1.4 Shallow Lakes

Shallow lakes have by definition depths less than 10 m. They usually do not stratify, because wind and wave action prevent it. Thus, water in shallow lakes is often exposed to nutrient-rich sediments, which may release nitrogen and phosphorus into the water column. Shallow lakes tend to be located in agricultural and urban areas with high runoff and nutrient load. Large parts may be covered with macrophyte vegetation, Figure 17-5. Phytoplankton growth starts when the ice cover disappears.

Figure 17-5. Shallow lake with extensive wetland areas that provide habitat for birds.[15]

Marshall and Peters (1989) developed a model for phytoplankton blooming day, J (Julian day), as a function of summer mean chlorophyll, *chl-a* (mg m^{-3}), and annual mean temperature, T (°C):

$$J = 159 - 23\log(chl\text{-}a) - 33\log T,\ r^2 = 0.54,\ p < 0.005,\ n = 57 \qquad (17\text{-}7)$$

In shallow lakes, there appears to be two alternative stable states, at least over ecological time scales. Lakes having similar nutrient loads can exhibit either:[16]

- A phytoplankton-dominated, turbid-water state.
- A macrophyte-dominated, clear-water state.

In the first state, dissolved nutrients support extensive phytoplankton growth, causing the water to become turbid or muddy. In the other state, uptake of phosphorus by aquatic macrophytes plays an essential role in keeping a nutrient balance within the lake that sustains clear open waters. For example, when East Bay of Lake Taihu, China, was in the clear water state, about 600,000 tons of aquatic plants were harvested every year; that amounted to 264 ton of phosphorus or 57% of the external phosphorus loading.[17] The plant material was dried and used for soil improvement. However, preferences for the two states depend upon lake use.

2.2 Withdrawal of Water

Water shortages are common in most of the world. An estimate of available water per capita can be obtained by dividing annual net precipitation – annual precipitation less annual evaporation – by the number of people. Estimates for Holland and Norway are shown in Table 17-3.

Table 17-3. Water availability. Evaporation is estimated from precipitation and temperature.

Country	Population 10^6 inh.	Area $10^3 km^2$	Precipitation mm yr^{-1}	Evaporation mm yr^{-1}	Net water available m^3 $capita^{-1}$ yr^{-1}
Norway	4.3	324	960	677	23 000
The Netherlands	15	37	770	453	780

Of all people in the developing world, 28% (range 18%–46 %) have no access to safe water, whereas all people in the industrialized world do.[18] In industry, water use per produced unit is often used as an environmental assessment criterion. In paper production, for example, water use per 1000 kg of paper product should be less than 87 m^3.[19]

Withdrawal of water for industrial purposes such as cooling and manufacturing, or for domestic use, reduces river flow. This may cause lower flow than required for effective self-cleaning, oxygen-depleted zones, riverbed destruction, etc. Often, water withdrawal permits are granted on the condition that a limiting low-flow of water is guaranteed. Water withdrawal on a continent basis is shown in Table 17-4. A person requires between one and two cubic meters of drinking water per year, a small amount compared to what is available per person even in many arid countries. Urban life requires about 250 m^3 per capita and year. Water withdrawal ranges from 245 m^3 per capita and year in Africa to 1600 m^3 per capita and year in North America.

Table 17-4. Water withdrawal in cubic kilometers = 10^9 m^3.[20]

Region	Population 10^6	Area of irrigation, 10^6 ha	Irrigation	Domestic use [(1)]	Industry	Total per capita, $m^3 \times capita^{-1} \times yr^{-1}$
Europe	496	17	110	48	193	707
Asia	2,932	140	1,300	88	118	
Africa	589	11	128	10	6.5	245
North America	411	29	330	66	254	1581
South America	279	8.5	70	24	30	444
Australia & Oceania	26	2.0	16	4.1	1.4	827
USSR	282	20	260	23	117	1418
World total	5,015	227.5	2,206	263.1	759.9	644

[(1)] About 40% for consumptive use

Water withdrawal partly reflects the regional share of global runoff. For example, total river runoff is 14,550 km^3 per year in Asia with a population of 3630 mill people (60.5% of global population) whereas runoff in Africa is 4320 km^3 per year, with 750 mill people (12.5%).[21] Abstraction of water is essential to support forestation, agriculture, and animal husbandry. Water requirement for irrigation is about 12,000 m^3 of water per hectare and year.[22]

In the dry and market oriented countries of Latin America and Mexico, beef production requires large water resources. 30 m^3 of water is required to produce 1 kg beef.[23] In a natural habitat, water requirement for mammals is 70 mL $\times$ (kg wet weight)$^{0.75}$ per day. Thus, a roe deer of 25 kg requires $70 \times 25^{0.75}$ mL = 70×12 mL = 860 mL per day; in other words, a minimum of one liter per day.[24]

To clean water for domestic use, water purification requires a water surface of 0.06 m^2 per m^3 of clean water for infiltration and protection of the infiltration area. In Sweden, the domestic water consumption is 122 m^3 per capita and year, meaning that 7 m^2 of water surface per capita is required for clean water provision.[25]

2.3 Discharges into Water

Chemical substances that are discharged into water can be classified as nutrients and toxic chemicals. For the last class, toxicity, persistence and bioaccumulation properties are important attributes. In the USA, wastewater generation is about 246 L per capita and day.[26] Chemical oxygen demand COD, biological oxygen demand, BOD, total phosphorus TP, and total nitrogen TN, are all terms that characterize discharge of organic matter into water. A paper mill discharges per 1000 kg of paper products about 0.03 kg TP, 0.6 kg TN, 5 kg total suspended solids, TSS, or 30 kg COD. Other sources of nutrients are atmospheric deposits and runoff from land.

3. WATER QUALITY STANDARDS AND ASSESSMENT

Ecosystem "health" is the end-impact criterion for habitat quality assessment. Health is difficult to define, but measurable toxic effects at susceptible stages of the most vulnerable species can be used as indicators.

3.1 Water Quality Standards

Since water has many uses, quality classes are formulated in terms of nutrient concentrations. Quality class I may be suitable for, say, swimming, quality class II designates lower limits for fish farming, etc. Note that there are separate criteria for the surface layer in summer – the productive period, and winter with little or no algal growth. Criteria for oxygen concentration are given both in terms of average levels and low limits, because low levels are particularly detrimental to fish. However, there tends to be a constant relationship between mean and low concentrations.[27]

Water quality standards are related to four major sources of water degradation, discharges of nutrients and toxic substances, water withdrawal and changes made along the shoreline.

3.1.1 Critical Load

Water quality refers to physical, chemical and biological characteristics. Phosphorus is considered the most limiting nutrient in lakes in Europe and North America. It is usually in short supply relative to nitrogen, and will increase growth of phytoplankton if supplied to lake water. Freshwater eutrophication control efforts worldwide therefore focus on regulation of phosphorus loading.[28] Discharging nutrients may cause over-enrichment of the waters, or eutrophication. This may lead to an abundance of blue-green algae (*cyanobacteria*). But water bodies have to some extent self-cleaning ability, and can tolerate a certain rate of nutrient inputs without changing their ecological character. This rate is called the critical load and is approximately 1 gTP per m^2 and year. The degree of eutrophication can be measured as the percent of the area of the lake where total phosphorus concentration exceeds 15 mg m^{-3}, which is the upper limit concentration for oligotrophic lake water. In estuarine and coastal marine waters, both nitrogen and phosphorus are generally in short supply, and several authors have suggested that regulating both nitrogen and phosphorus input to coastal zone waters may best control eutrophication.[29] Some freshwater and brackish water algae can fix nitrogen, and changes in the ratio of total nitrogen to phosphorus in the water column may cause some algal species to become toxic.[30]

3.1.2 Toxic Pollution

Effects of toxic chemicals are often indirectly described as the area of the lake where concentrations of the chemical are above the safe level. Toxicity and safe levels are related to such notions as No Observable Effect Level, NOEL, and Levels of No Concern, LNC. These terms were discussed in Chapter 14 on toxicity.

3.1.3 Water Use

Abstraction of water for use in agriculture and industry can endanger both surface-water and groundwater sources. To prevent that, governmental guidelines may include upper limits of water use per unit produced. Water use can be measured as the fraction of available fresh water extracted per year. On a global level, 54% of available fresh water is currently abstracted for human use.[31]

3.1.4 Habitat Suitability

Assessment of habitat suitability includes the open water body, the shoreline, and the mouths of tributaries entering the lake. Shorefront work may destroy important habitats, like macrophyte beds, that are important both for fish and birds. Intensive use of boats may cause re-suspension of sediments that are rich in nutrients or even toxic compounds.

3.2 Water Quality Assessment Methods

There are four methods for assessing water quality: **in-situ** measurements including caging techniques, mesocosms, bioassay techniques and chemical analyses. *In-situ* measurement is the ultimate assessment method, since we want to know how all exogenous driving forces – including pollution – affect the ecosystem we wish to manage. We are also interested in how water affects humans when it is used for drinking, swimming, or irrigation purposes. On the other hand, *in-situ* observations are difficult to reproduce, time-consuming, and expensive. They are also susceptible to natural climatic variations, uncontrolled physical and chemical factors, and to interactions between the test species and other species present in the system. Chemical analysis is often used as a substitute for the other techniques in routine monitoring of water quality.

3.2.1 In-situ Observations

Passive in-situ observation means monitoring of selected ecosystem attributes. Sampling is done periodically or with a certain effort, and results presented as quantity per unit area, per unit volume, or per unit of sampling effort. Before starting a monitoring program, it is useful to perform screening tests to determine:

- *How large sample you should take to obtain sufficient accuracy for the parameter to be estimated*: It depends on the standard deviation σ of the population; therefore you first need an estimate of σ, which you can obtain through a pilot sample. Let us assume that the pilot average is $\bar{x}$ and the standard deviation s. You want a proportional precision p, which is the ratio between the mean and the error margin. The *error margin* is half the width of the confidence interval. The sample size, n, can now be calculated with formula 17-8, where $t_{\alpha/2}$ is related to the Student distribution.

$$n = \left(\frac{t_{\alpha/2}s}{p\bar{x}}\right)^2 \qquad (17\text{-}8)$$

- Example: You have a pilot sample with average 100 and $s = 40$. If you want a 95% confidence level, then $t_{\alpha/2} = 2.0$. Finally, if you require p = 5%, then $n = (2 \times 40/(0.05 \times 100))^2 = 256$. You may include your pilot sample among the estimated 256 observations, but you should be aware that there is no guarantee that the final precision will be of the required size. This time we got $n = 256$; another pilot sample might give $n = 900$. The variation is considerable. In practice, samples are often taken that are smaller than scientifically required because of monitoring costs, or parameters with less validity, but cheaper to sample are used.
- *How often samples must be taken to detect change over time*: For example, to detect a two-fold real change in average summer chlorophyll between two years in a lake, more than four samples must be taken during a summer. It will suffice to detect a three-fold change to establish whether there has been a transition from oligotrophy to mesotrophy.[32]
- *How much time it takes to achieve a steady state*: The question of how much time it takes to achieve a steady state for the mass balance, given inflows and outflows, will be discussed below. However, the time for achieving a steady state with respect to species assemblages is often not known. At the time of observation a species may not be present because a successional sequence finalizing development of the species has not been completed.[33]

If you want to know if the amount of algal biomass is below a certain critical value, you may estimate it by one of three different algal sampling programs with progressively decreasing costs:

1. Identifying and enumerating the number of algal cells in a water sample. The cost is €250 per sample.
2. Measuring the amount of the photosynthetic pigment chlorophyll-a in the water sample. The cost is approximately €35 per water sample.

3. Measuring the **Secchi disc depth**, SD, which is the depth where a black and white, circular plate about 20 cm in diameter disappears from view. The cost is €2.5 per sample. The clear Crater Lake in the US has a SD of 40 m; in contrast, highly eutrophic carp ponds can have a SD of 20 cm.[34]

The biomass of phytoplankton can be calculated with any of the three methods, but with different degrees of accuracy. You will find mathematical models in the section on dose-response relationships.

Another monitoring technique is **catch per effort**. As an example, one could place fishnets at selected locations for say, 1 hour, and record the number of fish caught per net per unit time as *catch per unit effort*. One can estimate the changes in number of fish in the lake by assuming that *catch per unit effort* is proportional to the number of fish in the lake.

Active in-situ observations, also called *active field tests, or caging observations* may be carried out as follows: A specified number of a species of interest, such as fish or mussels, is caged and positioned at a susceptible location in the water. Thereafter, the response of the species is monitored for a specified length of time. Observations could be number of deaths, intensity of sublethal effects, or carcinogenic effects. The number and size of lesions or tumors are often recorded as indicators of carcinogenic effects. Examples of sublethal effects include the proportion of females to male individuals in hermaphroditic organisms, or the number of eggs per female.[35] Behavioral traits, such as coughing in fish, have also been used as response variables.

3.2.2 Mesocosms and Bioassays

Mesocosms are large enclosures placed in-situ so that part of natural physical and chemical processes can act on the system, An extended version of mesocosm were the famous experiments by David Schindler in Canada where basins of whole lakes were separated by soft walls. **Bioassays** are smaller enclosures sited in laboratory environments. They are often used for toxicity testing, establishing the relationship between potentially toxic chemicals, biota, and human health. Toxicity tests aim at finding concentrations of a chemical compound that just starts to affect a certain organism. The 24h-LC_{50} test (Lethal Concentration over 24 hours) is one such test. Human epidemiological studies relate potential sources of toxic material to human morbidity and mortality. Table 17-5 shows as an example the relationship between fecal coliforms that results from municipal and industrial discharges and human use of water.

Table 17-5. Biological water quality characteristics. Data: Legget and Bockstael (2000:137).

Compound	Characteristics
Fecal coliforms	$<4\ (100\ \text{mL})^{-1}$ is acceptable. $> 200 \times (100\ \text{mL})^{-1}$ requires beach closing; $>10^3\ (100\ \text{mL})^{-1}$ should be avoided.

3.2.3 Water Physics and Chemistry

Water physics and chemistry parameters describe the thermal regime in lakes, concentration of oxygen and other dissolved substances, alkalinity, and the pH level. Water quality parameters also include total phosphorus and total nitrogen. Phosphorus measurements are sometimes restricted to concentration of dissolved orthophosphate, PO_4-P, or Molybdenum Reactive Phosphorus, MRP. Similarly, nitrogen measurements are sometimes restricted to concentrations of dissolved nitrate and ammonia.

However, none of these dissolved inorganic forms alone have been shown to be a better predictor of algal biomass than total phosphorus. Table 17-6 gives an overview of the most important nutrients that affect lake ecosystems.

4. DOSE - RESPONSE RELATIONSHIPS

There has been much discussion about which factors cause eutrophication of lakes,[36] and current belief is that input of phosphorus and nitrogen to a great extent govern eutrophication of surface waters.

4.1 Lake water quality models

We present ten dose-response relational models for lake water quality. The models relate water quality parameters to the most important factors.

Table 17-6. Nutrient water quality characteristics. BOD_5 is Biological oxygen demand, see Chapter 18 on rivers.

Characteristics	Common sources	Characteristics
Total organic material measured as BOD_5	Pulp and paper mills	0 is optimal, > 15 mg L^{-1} is unwanted.
Nitrogen, TN	Outflows, agriculture, sewage outlet	30 is optimal, Background is 1–10,000 mg m^{-3}
Nitrates	Outflows, agriculture, sewage outlet	5 is optimal, >30 mg nitrate L^{-1} is unwanted. For drinking water the maximum concentration is 50 mg nitrate × L^{-1} or 11.3 mg nitrate L^{-1}.
Phosphorus, TP	Outflows, agriculture, sewage outlet	2–5 mg m^{-3} is optimal, Background is 1–1000 mg m^{-3}

4.1.1 Lake Area and Depth → Critical Concentration

Vollenweider (1976)[37] identified *critical TP loading* for lakes, L_c, in terms of permissible supplies of TP, measured in mg per square meter lake area per year.

$$L_c\,(\text{mg m}^{-2}\text{yr}^{-1}) = 100\text{–}200(Z\rho)^{0.5} \approx 50 \times Z^{0.6} \qquad (17\text{-}9)$$

Here, Z (m) is average lake depth and $\rho = 1/\tau$ is flushing rate, the reciprocal of the water residence time, τ, in years. The coefficients 100 and 200 are empirical fits for a line separating oligotrophy from eutrophy.[38] The critical load gives a concentration of total phosphorus in lake water less than about 10 mgTP m^{-3}. Andersen (1997) has shown that this load corresponds approximately to the load level where zooplankton can no longer control phytoplankton biomass – technically at the level where the stable phyto-plankton/zooplankton state bifurcates. Critical nutrient loads for total nitrogen are about 5–10 times higher. Although criticized, these values are widely used as a standard for lake water quality.

Since the ratio of TN to TP is approximately 5 in water flowing off agricultural land[39] one may set the limit of TN concentration in waters flowing into marine waters to 50 mg m^{-3}.

4.1.2 River Nutrient Concentration → Lake Nutrient Concentration

OECD (1982) reports models that relate the concentration of TP in in-flowing water to the concentration of TP in the lake. The inflow concentration is flushing corrected by including a function of water residence time τ_w.

$$TP_{lake} = 1.55\left[\frac{TP_{in}}{1+\sqrt{\tau_w}}\right]^{0.82} \qquad (17\text{-}10)$$

$$TN_{lake} = 5.34\left[\frac{TN_{in}}{1+\sqrt{\tau_w}}\right]^{0.78} \qquad (17\text{-}11)$$

4.1.3 Lake Nutrient Concentrations → Lake Chl-a Concentrations

Lake nutrient concentrations limit the growth of phytoplankton, or algae measured as *chlorophyll-a*. In lakes where phosphorus concentration is less than 200 mg m^{-3}; phosphorus is most often the limiting factor, whereas nitrogen is most limiting in nutrient-rich lakes.[40] Lakes with TP concentrations less than 20 mg TP m^{-3} have often molar N:P ratios greater than 36 in the water.[41] A molar ratio of 36 equals an equivalent weight based ratio of 16.

The first models relating the annual mean concentration of TP in lake water to the growth season average algal abundance were developed by Sakamoto (1966), and Vollenweider (OECD 1982). Equation 17-12 states that the logarithm of algal biomass in a lake is related linearly to the phosphorus concentration of the lake.

$$\log \mathit{Chl\text{-}a} = -0.60 + 0.96 \log \mathit{TP} \tag{17-12}$$

This simple *TP-Chl-a* model appears to capture a high portion of the relationship between *TP* and *chl-a*, but the relation was later shown to have a more sigmoid form.[42] This model has been used as the decisive argument for diverting sewage from lakes and implementing control measures on agricultural land that is close to lakes.

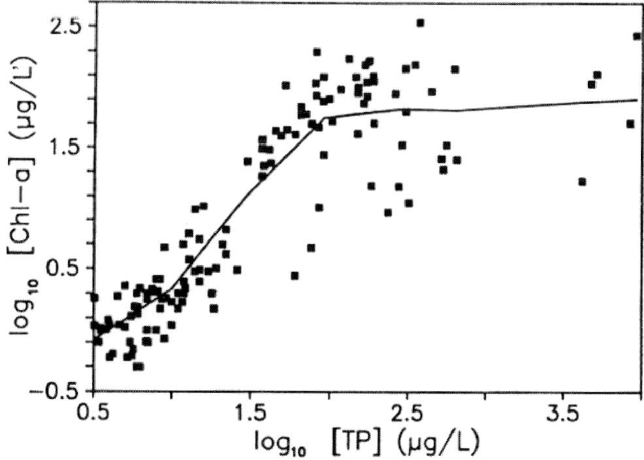

Figure 17-6. Algal biomass as a function of total phosphorus.[43]

Three factors make the true relation more complicated than the simple model (17-12): the absolute level of *TP*, average lake depth, and the abundance of large-bodied (>1 mm) zooplankton:

- *TP range*: A higher portion of the variance for the *TP–chl-a* relationship is explained with the sigmoid function shown in Figure 17-6.
- *Lake depth*: Lake depth is correlated with lake area and *TP* level. There is much scatter around the chl-a/TP relationship, A question is therefore how large a reduction in TP is required to observe a decrease in chl-a the following year with 80% probability. The percentage reductions required in TP-concentration is for deep lakes ΔTP >35% TP; for the average deep lake the reduction should be 14 mg m^{-3}. For shallow lakes it is ΔTP >10% TP; for the average shallow lake the reduction should be 30 mg m^{-3}.
- *Concentrations of large zooplankton*: Large zooplankton, like *Daphnia galeata* – a filter-feeding zooplankter, are effective grazers of phytoplankton, at least up to hypereutrophic concentrations of *TP*. In temperate lakes with a high density of large-bodied herbivores, the *chl-a* yield per unit of *TP* is half of what it is without grazers. This is consistent with the theory that large zooplankton can control phytoplankton and maintain them at a low level if phytoplankton concentration is not too high.

4.1.4 Loads of Conservative Compounds and Nutrients → Open Water Concentrations

Conservative compounds like metals and nutrients may be loaded into very large bodies of water with large surface areas, and extensive spatial heterogeneity in local pollutant concentrations may result. In some cases the immediate spreading of the compounds on the surface may be of particular interest (like some pesticides and oil). Oil will spread at 3% of the wind velocity measured about 1 m above the surface; for example the Exxon Valdes spill on the gulf of Alaska (marine waters) lasted for 56 days and contaminated 750 km of the coast.[44] One may as a first approximation assume that other substances that behave conservatively will spread in the same manner. For a particular case it is important to know if the main wind direction is along the shore or normal to the shore. For a lake with a long horizontal axis, spreading will most probably occur along this axis because wind will blow along it.

4.1.5 Lake Nutrient Concentrations → Lake Transparency

Since algal biomass reduces lake transparency, there is a relationship between total phosphorus and lake transparency. This property is often used in lake water quality assessment, because water transparency can be measured efficiently by a simple method using a Secchi disk. The **Secchi disc** is a black and white disc 20 cm in diameter attached to a metered string.

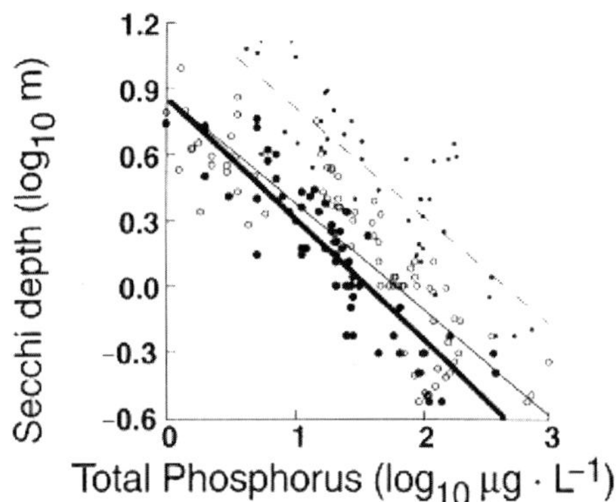

Figure 17-7. Transparency as a function of Total phosphorus. Relationship between total phosphorus and Secchi depth for temperate "small herbivore" (open circles) and "large herbivore" (dashed line and dots) lakes and subtropical "small herbivore" (bold line and dots) lakes.[45]

The disc is submerged into the water, and when the disk is no longer visible, the water depth is measured. The result of studies of TP-Secchi disc relationships in subtropical and temperate lakes is shown in Figure 17-7.

Model 17-13 was estimated for a sample of 85 temperate lakes without large zooplankton. *TP* explained 75% of the variation, and the significance level was less than 0.001. Subtropical lakes give approximately the same relationship.

$$\text{Log } Secchi = 0.86 \pm 0.05 - (0.48 \pm 0.03) \log TP \quad (17\text{-}13)$$

For temperate lakes with large zooplankton the equation is:

$$\text{Log } Secchi = 1.31 \pm 0.09 - (0.49 \pm 0.06) \log TP \quad (17\text{-}14)$$

40 lakes were in the sample, and *TP* explained 70% of the variations, with a significance level lower than 0.001.

In highly eutrophic lakes with $TP > 250$ mg m^{-3}, 50% of the reduction in transparency is related to non-plankton particles. Similarly, transparency in many reservoirs is dominated by non-organic suspended material.[46]

Exercise: Estimating TP concentration from transparency

- If you measure a Secchi disk depth of 2.4 m in a temperate, small herbivorous (SH) lake, and 6.6 m in a temperate "Daphnia" (LH) lake, what is the *TP* concentration of the two lakes?
- *Solution*: Using Equations 17-13 and 17-14, you will find *TP* concentrations of 10 mg m^{-3} in both lakes.

4.1.6 Lake Nutrient Concentration & Macrophyte Cover→ Lake Chl-a Concentration

Macrophyte cover increases and extends to deeper waters with increasing temperature and irradiation. The coverage ranges from 0% to 100% with typical densities between 0.2 and 90 kg m^{-2} in tropical or subtropical lakes, and from 0.2 to 3 kg m^{-2} in temperate lakes. Maximum depth of macrophyte colonization, Z_c (m), is approximately related to Secchi disk depth according to Equation1 7-15.[47]

$$Z_c = (0.53 \times \log Secchi + 1.51)^2 \qquad (17\text{-}15)$$

The model was based on a sample of 160, and explained 50% of the variations, with a significance level lower than 0.0001.

Also, littoral slopes should be less than 0.5%.[48] With high macrophyte coverage, >5 kg m^{-2}, dissolved oxygen concentrations may easily become lower than suitable for most fish species.[49] For very shallow lakes (<3 m) Canfield et al. (1984) developed a model which included the Percent Area Infested with macrophytes (%, *PVI)*

$$\text{Log } Chl\text{-}a = \log TN + 0.28 \log TP - 0.005\, PVI - 2.08;\ r^2 = 0.9 \qquad (17\text{-}16)$$

4.1.7 Lake Nutrient Concentration → Blue-Green Algae Concentration

Blue-green algae are particularly unwanted in lakes. The amount of blue-green algae is estimated to increase with nutrient concentration in deep lakes according to Equation 17-17.[50] The model is based on a sample of 14, taken on depths between 8 and 153 m. *BG* is blue-green algae biomass (g wet weight m^{-3}), and *TP* is total phosphorus (mg m^{-3}).

$$\text{Log } BG = 2.8 \log TP - 4.1;\ r^2 = 0.7 \qquad (17\text{-}17)$$

For shallow lakes, a corresponding estimate is:[51]

$$\text{Log } BG = 1.3 \log TP - 1.9; \; r^2 = 0.7; \text{ Depths: 3–11 m} \quad (17\text{-}18)$$

4.1.8 Lake Nutrient Concentration → Fish Abundance and Quality

Bays and Chrisman (1983) developed a set of models to estimate fish density in lakes. They used *Carlson's trophic state index*, *TSI*, as an independent variable, which converts *TP* onto a scale between 0 and 100.

$$TSI = 60\text{–}10 \log_2 (65/TP) \quad (17\text{-}19)$$

Relationships between lake trophic state measured by *TSI* and fish status are given in Equations 17-20 to 17-23. *TF* is total fish biomass (kg ha $^{-1}$); %SF is percent desirable sport fish; %RF is percent rough fish; and %FF is percent forage fish. The sample size was n = 39 for all models.[52]

$$\ln TF = 0.043\, TSI + 2.45; \quad r^2 = 0.4 \quad (17\text{-}20)$$

$$\%SF = -1.374\, TSI + 131.3; \quad r^2 = 0.5 \quad (17\text{-}21)$$

$$\%RF = 1.593\, TSI - 64.2; \; {}^2r \; = 0.6 \quad (17\text{-}22)$$

$$\%FF = -0.540\, TSI + 45.6; \quad r^2 = 0.2 \quad (17\text{-}23)$$

Table 17-7. Fish abundance and quality resonse to different nutrient loads.

TP (mg m^{-3})	3	6	12	24	48	96
TSI	20	30	40	50	60	70
Total Fish (kg ha^{-1})	27	42	64	97	149	227
Sports fish, %	104	90	76	63	49	35

Computed responses for different *TP* concentrations are shown in Table 17-7.

4.1.9 Lake Area → Fish Species Richness

Regression models for the number of fish species as a response to lake surface area and several other factors have been studied for subtropical and temperate lakes.[53] Surface area was the most powerful explanatory variable. The equations below were obtained for subtropical lakes (36 Florida lakes, size range 0.065–129.6 km^2), and temperate lakes (138 Wisconsin lakes, size range 0.004–15.66 km^2) respectively. N is Number of species, and A surface area (km^2).

$$\mathrm{Log}_{10}(N+1) = 0.16 \log_{10} A + 1.1;\ r^2 = 0.4 \quad (17\text{-}24)$$

$$\mathrm{Log}_{10}(N+1) = 0.37 \log_{10} A + 1.1;\ r^2 = 0.7 \quad (17\text{-}25)$$

4.1.10 Lake Area → Fish Size

Trout size (*Salvelinus namaycush*) has been found to be related to lake area for Ontario lakes.[54] The ultimate size of adult trout, L (cm), was related to lake area A (ha) as:

$$L = 37\,A^{0.071},\ r^2 = 0.227,\ p = 0.001,\ n = 51 \quad (17\text{-}26)$$

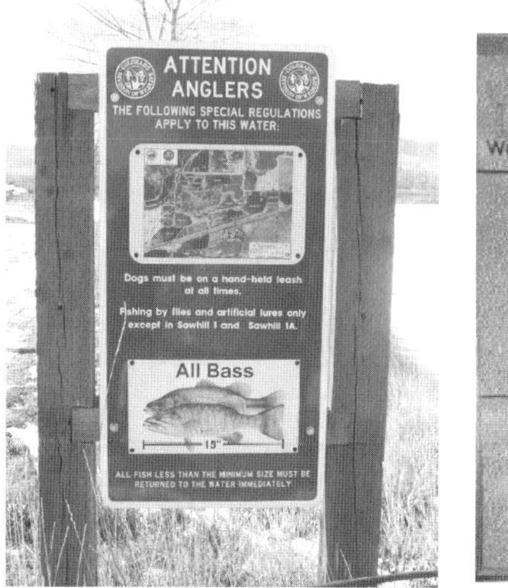

Figure 17-8 Instruction to anglers regarding fish size in Saw Hill ponds, Boulder, Colorado, 2005. Right: Water quality warning sign from a Hong Kong beach around 1995.

5. MITIGATION MEASURES

The simplest mitigation measure to avoid negative effects of poor water quality on people is to set up a warning sign advising people not to use the water if it is in poor condition. Figure 17-8 (right) shows a sign at a Hong Kong swimming beach. To maintain fishable lakes and ponds, signs are put up instructing anglers how to do their fishing. The sign in Figure 17-8 (left) asks anglers to discharge fish below a certain size.

There are several techniques available to improve water quality in lakes. Reducing the input of phosphorus and nitrogen appears to be the most effective and predictive abatement methods.

Other techniques that can be used in particular circumstances are i) diverting clean water into lakes with high nutrient concentrations, ii) removing nutrient rich bottom sediments in shallow lakes, iii) adding chemicals (alumina) or lime to the sediments so that they make phosphorus unavailable for phytoplankton growth.[55]

Removing nutrient rich sediments is expensive, and costs around US$0.15 million per km shoreline, and the effects tend to be of short duration (3–6 years).[56]

Biomanipulation may be used to restore aquatic plants and change a turbid water state into clear water, but if the water nutrient level is higher than 100 mg TP m^{-3}, it will normally not work.[57] Biomanipulation involves four measures or assumptions:

- Fish predating on herbivore zooplankton must be removed.
- Mysid shrimp (*Neomysis*) must have density lower than about 100 individuals per mL.
- Blue-green algae (*Cyanobacteria*) density must be below 100,000 individuals per mL.
- Carp grazing on aquatic plants must be removed.[58]

If water is an "open access" source that becomes over-used, mitigation measures may include extraction permits, transferable permits with a water volume cap, or withdrawal taxes.[59]

6. APPLICATION: LAKE REHABILITATION

The conflict over water

Lake water is a convenient medium: it can be used for navigation, cleansing and drinking. Most lakes are now too small – or the neighborhood communities too large – to accommodate all the various uses in demand. One therefore has to end many traditional activities, and lakes have to be renovated.

Lake management is often conducted in two phases: a survey phase where the lakes in a region are studied and their general conditions assessed, and a focus phase where efforts are directed towards particular lakes. To find the best trade-off between alternative uses of a lake, one has to find out how polluted it is, and determine the values of the competing uses.

The boundary of the ecosystems of a lake is the catchment area, and from historical sources one may get suggestions for what type of ecosystem it may support and thus how clean the lake may become after renovation. A reasonable time frame for bringing a lake back to pre-industrial conditions is around ten years.

6.1 Decision Analysis

Below, we propose a structure for rational decision analysis, with decision criteria and alternatives that could be applied to any lake. Then we look more closely at the consequence analysis by using a concrete case from lake Gjersjøen in Norway. This case gives us also an opportunity to show how diverse models in the book can be applied.

Decision makers: In most lake use cases, the decision makers are local authorities or the government

Stakeholders: The stakeholders consist of diverse interest groups. One group, including agriculture and industry, uses the lake as a *recipient.* Another group uses the water for fishing, which makes a small level of nutrients desirable. A third group uses it for drinking water; they want a low level of nutrients. A forth group uses it for recreation and wants the lake to be clean (use value) and a fifth group just wants it clean (existence value). See Chapter 3, section 2.3 on decision-making and Chapter 8, section 2 on the various categories of values of environmental resources.

Decision objectives: The overall goal of lake rehabilitation is to obtain a suitable cleanness that offers sustainable services at the least social and economic costs. Examples of sub-goals are:

- Maximize transparency (Secchi disk depth at least >2 m).
- Minimize toxic levels (at least below recommended standards, see Chapter 14 on toxicity).
- Maximize white fish catch; see Chapter 13, section 4 on harvesting and this chapter, section 4.
- Maximize supply of potable water; this Chapter, section 2.2.
- Maximize human activity along the shoreline, see Chapter 8 on recreational values and Chapter 10 on life quality.

- Maximize scenic beauty. See Chapter 11, section 6 on aesthetics.
- Minimize direct and indirect costs of lake management.

Alternatives: In the survey phase, one has to decide whether to maintain status quo or to consider rehabilitation measures, which may include:

- Reduction of the nutrient load to the lake.
- Biomanipulation.
- Chemical treatment.
- Dredging sediments (shallow lake).

Utility functions: In the case of lake management, primary utility functions would probably be judged risk-averse; that is, a small, but noticeable loss in water purity may not reduce its utility significantly.

Preferences: Preferences regarding performance criteria in lake rehabilitation studies are most often expressed by environmental authorities, which in turn should solicit the preferences of users and the general public. To do this, methods described in Chapter 7, section 3 are relevant. The results should be expressed as importance weights as in Table 3-8.

6.2 Consequence Analysis in Lake Gjersjøen

The objective of the survey phase is to obtain as much information as possible with as little effort as possible. One approach is to compare a few key observations with predicted properties of the lakes.

Morphological observations: The lake has an area of 2.68 km^2, an average depth of 23 m, and a maximum depth of 64 m. The water has a retention time of three years. The catchment area is 84 km^2. The observed spring concentrations in the lake water are about 30 mgTP m^{-3}. The phosphorus load to the lake is 3–6 ton yr^{-1} or 1–2 mgTP m^{-2} yr^{-1}. The TN:TP ratio in the water is about 10. Roach dominates the lake, but seven fish species have noticeable abundance.[60]

Question: What is the expected total phosphorus concentration and what is the "expected" lake area – assuming that you don't already know it – based on morphological observations only?

Solution: The lake volume is 2.68×10^6 $m^2 \times 23$ m = 61.2×10^6 m^3. From Equation 17-1 and 17-2 we find expected TP concentration as Log (TP) = $-1.13 \log 23 + 3.29$ which gives TP = 56 mg m^{-3} and Log (TP) = 0.38 log ($84\ 10^6/61.2\ 10^6$) + 1.57, which gives 41.8 mg m^{-3}, respectively. From Equation 17-3 we find Log(23) = $0.46 \times \log(A) + 0.728$ which gives A = 1.34 km^2. The "expected" lake concentration is about 20–60 mg m^{-3}. As a contrast, the real lake area was 2.68 km^2.

More questions:

1. Is phytoplankton in this lake limited by phosphorus or nitrogen?
2. What is the expected critical load of TP to this lake?
3. What are the concentrations of phytoplankton and blue-green algae in the water?
4. What is the density of fish in the lake?
5. How much would you have to decrease the TP load to this lake to have an 80% probability of observing an enhancement in lake water quality?

More answers:

1. The suggested TP concentration in this lake is around 30–50 mg m^{-3}. At concentrations below 200 mg m^{-3} TP concentrations contribute to phytoplankton limitation in the lakes.
2. The critical load – that is the load that would give about 10 mgTP m^{-3} in lake water – can be found from Equation 17-9 which gives $L_c = 50 \times 23^{0.6}$ = 328 mgTP m^{-2} yr^{-1}, that is, substantially greater than the actual load.
3. Equation 17-12 gives Log Chl-a = –0.60 + 0.96 log 40, which gives Chl-a equal to 9 mg chl-a m^{-3}, whereas the actual values are in the range 20 to 40 mgChl-a m^{-3}.
4. Fish densities can be calculated by Equation 17-20: Ln(TF) = 0.0425 × (TSI = 55) + 2.452 which gives about 120 kg ha^{-1}. The actual density is estimated to be maximum 300 kg ha^{-1}. The major species is roach (*Rutilus rutilus*) size 60–95 mm.[61]
5. The number of fish species can be estimated by Equation 17-24: Log(N+1) = 0.16 × Log(2.68) + 1.14 = 36, which is more species than observed. Using the text in section 4.1.3 in this chapter, the decrease should be 0.35 × 30 mgTP m^{-3} = 10 mgTP m^{-3}.

Final question: What is the aesthetic value of the lake?

Answering approach: Lakes and reservoirs are among the major contributors to the aesthetic value of a landscape; see Chapter 11, section 2.6. Land preferences for the actual lake could be estimated with the technique explained in the application section of Chapter 11.

6.3 What Goes On?

In developing countries in particular, large sums of money are currently used to renovate lakes, and many lakes have become swimmable that had become too dirty. The largest contribution to the renovation of lakes comes from the building of sewage treatment plants and restrictions on agricultural practice close to lakes. Regarding lakes in tropical or semitropical countries, there are few studies available to make educated guesses about possible recovery potentials.

---

[1] Pauly and Christensen (1995)
[2] Johnson et al. (2001)
[3] Howe (2005), Howe (2002) Charles Howe, Colorado University, personal communication
[4] Bulte and Kooten (1999)
[5] Cyr and Pace (1993)
[6] Currie et al. (1999); Prepas et al. (2001) Brabrand et al. (1990)
[7] Equations 1, 2, and 3 were developed or quoted in Seip et al. (1992: 65); Eq (2) is from Prepas et al. (2001) (both upland and lowlands, zero fens.) Hanna (1990) report relationship between lake area and lake depth (average and max) that obtain correlation coefficients, r = 0.74-0.76, n > 133
[8] Marshall and Peters (1989)
[9] Reynolds (1984)
[10] Imboden et al. (1983) p. 27
[11] Hanna (1990)
[12] Seip et al. (1991) p. 242
[13] Buerge et al. (2003)
[14] Skulberg et al. (1984), Reynolds (1984), Seip and Reynolds (1995), Kotak et al. (2000)
[15] As an example of aerial nutrient sources, the daily nutrient load to a lake in Canada from migrant Canadian geese was 1.57gN and 0.49 gP Manny et al. (1994) p. 122. For this lake, 70% of all phosphorus loading came from waterfowl excreta.
[16] See Chapter 12 on Ecology
[17] Li and Yang (1995)
[18] UN (2000)
[19] Confederation of European Paper Industries
[20] Data WRI (1994)
[21] Postel et al. (1996)
[22] Postel et al (1996) p. 787
[23] Ezcurra and Montana (1990)
[24] Holland et al. (1998)
[25] Wackernagel et al. (1999:612)
[26] Barlaz et al. (2002)
[27] Heyman et al. (1984)
[28] e.g., Smith et al.(1999) Smith (2004)
[29] Data: Kirchmann et al. (2002:404)
[30] see Edvardsen et al. (1990) on marine algae; Kotak et al. (2000) on freshwater algae
[31] Postel et al.(1996) p. 785
[32] Knowlton et al.(1984) p. 405
[33] Seip (1984)
[34] Lampert and Sommer (1997)
[35] Colborn et al. Holme and Dybing (1997)
[36] Edmondson (1991)
[37] Vollenweider (1976)

[38] Welch (1980) p. 127
[39] Downing and McCauley (1992)
[40] Seip (1994)
[41] Smith and Shapiro (1981)
[42] Prairie et al. (1989)
[43] Figure after Praire et al. (1989)
[44] Seip (1991)
[45] After Mazumder and Havens (1998:1659)
[46] Seip et al.(1992), Smith et al. (1999)
[47] Duarte and Kalff (1987)
[48] Duarte and Kalff (1986)
[49] Allen and Tugend (2004)
[50] Seip and Ibrekk (1987)
[51] Trimbee and Prepas (1987)
[52] Bays and Chrisman (1983)
[53] Keller and Crisman (1990)
[54] Shuter et al. (1998)
[55] Welch (1980)
[56] Allen and Tugend (2004)
[57] Li and Yang (1995), Moss (1990), Hosper and Meijer (1993)
[58] Li and Yang (1995)
[59] Charles Howe, Colorado University, personal communication
[60] Faafeng et al. (1990), Brabrand et al. (1990)
[61] Brabrand, (1990)

Chapter 18

RIVER

Water environment – rivers and wetlands

We describe the surface water environment of rivers and wetlands and discuss methods for monitoring, classification, and prediction of water quality. Human activity impacts surface water by withdrawing it for domestic use, irrigation, industrial process-water, and by discharging it polluted back into the surface water. Flood control by building revetments, dams and dykes along rivers will also change the habitat properties of the rivers. We especially discuss the the Streeter-Phelps model, which is the major model for river water quality predictions.

●

"Vitum capiunt, ni moveantur, aquæ" (Water becomes corrupt unless kept in motion.) Latin proverb

1. INTRODUCTION

Rivers were until the beginning of the 1900s used as open sewage channels in most developed countries and still are in many developing countries. Wastewater was discharged into rivers, while river water was used for domestic purposes. Figure 18-1 (left) shows an old factory built close to a river to obtain energy from a waterfall just upstream, and to discharge waste.

Historically, rivers are the veins of human society, but they also cause catastrophic floods. Thus river management involves compromises between conflicting interests, and proper management requires knowledge of what services rivers and wetlands provide, and their value.

Rivers contain much less of the worlds freshwater than lakes – 0.006% versus 0.26%, but are important parts of the environment and the cultural landscape. **Wetlands** are constantly flooded shallow areas along lakes and rivers. **Floodplains** are shallow areas that oscillate between terrestrial and aquatic phases, with alternating suitability for aquatic and terrestrial organisms.

Figure 18-1. River as recipient (left) and for navigation (right). Industrial complexes are often established at river banks, partly for hydropower, partly to discharge obsolete materials into the water. Right: The text for this map is "Transport routes in the Mediterranean" it shows rivers and ship routes, not roads. From Ambrosius (1923) p. 105.

1.1 Natural and Man-Made Impacts

Rivers are impacted by natural factors like floods and loads of erosion material. Dry weather may kill flora and fauna, and cold weather may freeze bottom sediments and kill fish. In Danish streams, the prime factors responsible for water ecosystem deterioration are: organic matter loading (46%), lack of habitat diversity (23%), high concentrations of dissolved iron and its oxides (11%), and seasonal drying up (9%).[1]

Withdrawal of water for domestic use and industrial purposes like cooling and manufacturing reduces river flow. This may cause lower flow than is required for efficient self-cleaning, leading to oxygen-depleted zones and riverbed destruction. In some regions, authorities issue water-withdrawal permits to guarantee a lower limit to the water flow. Water level regulation affects the river-margin vegetation. There are generally fewer species along a regulated basin than around an unregulated one, and in Sweden one has observed reduced vegetation after water level fluctuations lasting for more than 60 days per year.[2]

1.2 Pollution Status

The extent of water pollution in a region can be measured by quality indexes. River quality in most countries is generally poor, although it has improved recently, mainly in some developed countries. In England and Wales, the river stretches in good, fair, poor and bad conditions were in 1990 64%, 22%, 9% and 2%, respectively. The total river length is 38900 km. In Ireland, with a total river length of 2900 km, the percentage of river stretches in good condition decreased from 83% in 1971 to 65% in 1990.[3] The length of polluted river stretches that are poor or bad in England and Wales is 0.09 m per capita and in Ireland 0.11 m per capita (1990). In the USA, about 39% of the about 6 million km of rivers and streams are impaired for some uses and this corresponds to about 10 m per capita.[4] In China, 60% of the southern urban rivers and 89% of the northern urban rivers are in such condition that human contact with water is not advisable.[5]

2. BACKGROUND INFORMATION

In this section, we discuss habitat types in rivers, wetlands and flood-plains, and quote estimates of the value of services provided.

2.1 Value of Resources and Services Provided by Rivers and Wetlands

Rivers and wetlands offer services like habitat for fish and shellfish, organisms that can be harvested for food, and recreational value for residents and tourists. Estimates of the annual value are shown in Table 18-1. Data on commercial and recreational river fishing are uncertain. A recreational angler in Norway would typically catch 50–100 kg of fish per year. The commercial value is about €10 per kg, and adding leisure benefits increases the total value to around €50 per kg.[6]

Table 18-1. Ecosystem values. All values in € ha^{-1} yr^{-1}. Data Constanza et al. (1997), Kaly and Jones (1998):, Carpenter et al. (1998).

Habitat	Fish and Shellfish	Water purification	Water regulation [(4)]	Water supply
Rivers	41	665[(1)]	5446[(1)]	2117[(1)]
Wetlands	4177 [(2)]	500,000[(3)]	3800	–

[(1)] Water purification is the ability of rivers to remove nutrients and toxic substances, and to regenerate oxygen in water. Data-source includes the values for lakes.

[(2)] Calculated for an assimilation capacity of 1 g m^{-2} yr^{-1} of P and a value of €100 kg^{-1}.[7]

[(3)] Salt marshes as sea defense against flooding of low-lying land, 80 m wide, £30–60m^{-2}.[8]

[(4)] The ability of rivers and lakes to store and divert water to prevent overland flow.

Wetlands and mangrove areas are included in many "biodiversity hotspots" that should be kept in pristine state for conservation purposes.[9] Thus, wetlands have high conservation values.

2.2 Negative Externalities of Rivers

Negative externalities, or costs, of rivers are unwanted side effects for which there is no normal compensation. The most conspicuous costs associated with rivers are caused by floods, which sometimes cause extensive damage to life and property. In the USA about 20,000 communities in areas that are exposed to a 1% annual chance flood pay about US$ 2 billion annually in insurance premium through the National Flood Insurance Program (NFIP). The 1% chance flood has a 26% chance of occurring during 30 years and a 63.5% chance of occurring during 100 years.[10]

Rivers, wetlands and floodplains also take up valuable space, which could be used for other purposes. Figure 18-2 shows how the upper Rhine has changed course during a period of 80 years, partly by natural causes, partly by human interference.

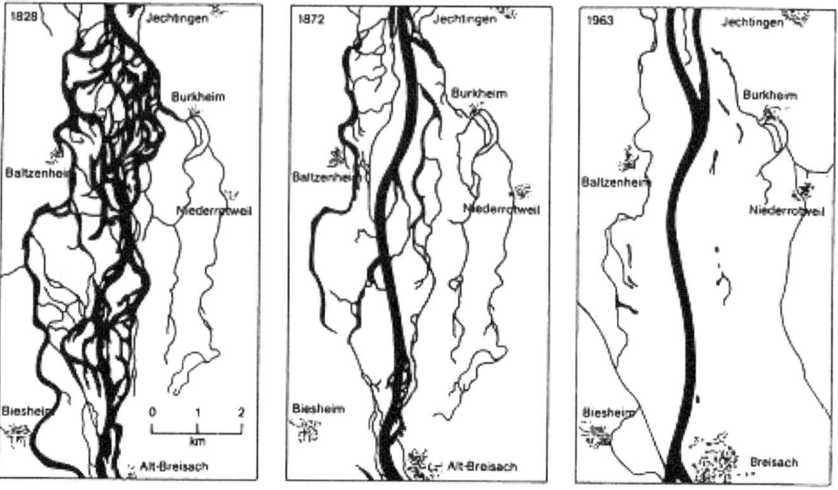

Figure 18-2. The Upper Rhine at Breisach: (a) In 1828, before river development; (b) in 1872, after the Tulla realignment; and (c) in 1963, after further canalization. Data Wieriks and Schulte-Wülwer-Leidig (1997).

2.3 River Morphology and Habitation

Basic morphological and physical properties of rivers and streams are discussed in Chapter 16 on hydrology. Here, we emphasize the relationship between morphology and the river as a habitat.

Water velocity is an important parameter for biological characterization of streams. Both average velocity, and the frequency of exceptionally large or small flows are important parameters. Water velocity may be inferred from the morphology, but is also easily measured. Velocities typically range from almost nil to about 2 m s^{-1}.

Associated with stream velocity are water flow characteristics such as tranquil flow and sluggish flow. The **Froude number**, *Fr*, summarizes flow characteristics in terms of the ratio between inertial forces – expressed by velocity – and gravitational forces. F_D is the density adjusted Froude number shown in Equation 18-1.

$$Fr = \frac{v}{\sqrt{gD}}; F_D = \frac{v}{\sqrt{\frac{\Delta\rho}{\rho_0} gD}} \tag{18-1}$$

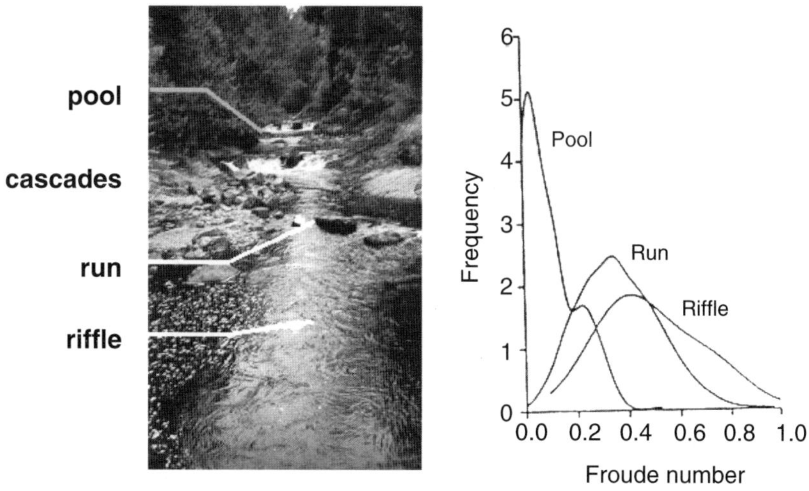

Figure 18-3. Left: A river stretch with cascades, pools, run and riffles. Right: Density function for Froude numbers in pool, run and riffle habitats. The Froude number is a property of water velocity and stream depth (See text). Right figure after Jowett (1993).

Here, v is flow velocity (m s^{-1}), g is gravity (9.8 m s^{-2}), D is depth (m), $\Delta\rho/\rho_0$ is water density change divided by average, or reference, density. A Froude number less than 1.0 means subcritical, slow, or tranquil flow. A number that is higher than 1.0 indicates fast and turbulent flow. Typical Froude numbers for pools, runs and riffles are shown in Figure 18-3. The Froude number is a better predictor of species abundance than water velocity or water depth alone.[11] If the density adjusted Froude number is larger than 0.32 there is no stratification, and if F_D is smaller than 0.01 the stream is strongly stratified. Also, if the difference in water density is large, $\Delta\rho$ > 0.01g m^{-3}, the stream is strongly stratified.

Resident trout are most abundant in steep streams, whereas salmon is most abundant in gently sloping streams with 1%–3% slopes and pool-riffle channels. Streams with slopes in the 1%–3% range may also be plane-bed channels with long distances between pools – more than seven channel widths. Wood debris that creates pools may enhance the abundance of salmon spawning sites.[12]

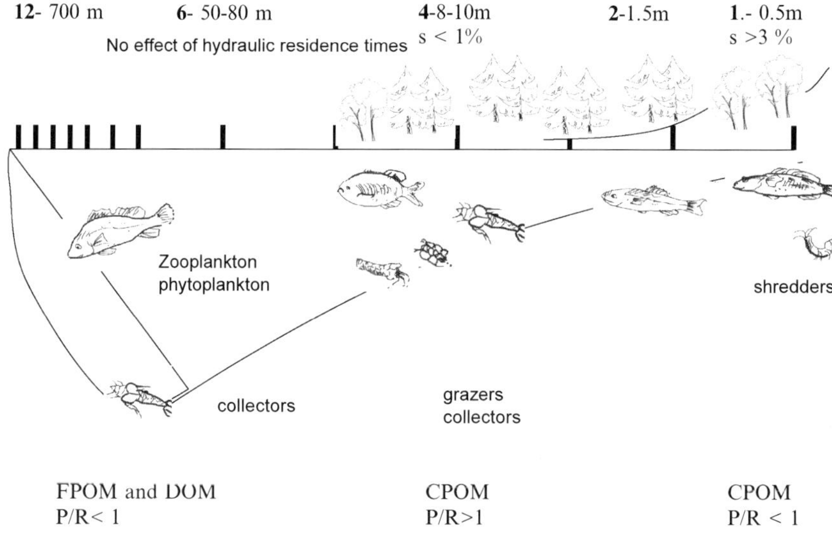

Figure 18-4. Relationship between stream order, (numbers in bold) and the progressive shift in structural and functional attributes of stream communities. Order 1 is about 0.5 m wide, 2 is 1–2 m wide, 3 is 4–6 m wide, 4 is about 10 m wide. *P/R* = production/respiration ratio. *s* = slope. FPOM and CPOM are acronyms for fine and coarse particulate organic matter. Idea from *from Vannote* et al. (1980).

2.3.1 Important Physical Attributes of Rivers and Streams

The main physical attributes determining abiotic and biotic characteristics are:

- Mean depth, mean width, and wetted cross section.
- Meandering characteristics: In natural lowland streams, meandering wavelength may be 10–14 times the channel width.
- Seasonal average and deviations in water velocity, often measured at the surface and 0.5–0.6 m from the streambed. The fastest point in the water column, v_{max}, will be at 30% of the total depth measured from the water surface.
- Mean water velocity, v_{mean}, is measured at 60% below the water surface and a multiplier of 0.85 is used to convert between the two velocities. Velocities larger than 2 ms^{-1} are exceptional.
- Dominant substrate often defined by grain size or reference to standard size fractions like sand, gravel, cobble or rock. Stones (e.g., larger than fist-sized) will often be recorded separately. Depth of substrate for burial of eggs is also important.

Table 18-2. River habitat types. D_{50} is the diameter of particles at the 50% percentile. FPOM and CPOM is fine and coarse particulate organic matter respectively. Data Raven et al. (1997), Kronvang et al. (1998), Montgomery et al. (1999).

Habitat type	Description	Water velocity
H_1 -Riffle	Areas along the stream edge and in the stream channel with high current velocity, shallow water depth and coarse substratum	Mean flow velocity: 0.50 m s^{-1} Gravel, $D_{50} \approx$ 10 mm. High density of macroinvertebrates and spawning fish. Riffle spacing 5–7 times channel width.
H_2 Riffle/gravel	Mid-stream, coarse substratum, mostly gravel	Mean flow velocity: 0.80 m s^{-1} Gravel, $D_{50} \approx$ 10 mm. River slopes 0–5%. High macroinvertebrate and spawning fish density
H_3 Pool	Deep areas with sluggish flow	Mean flow velocity: 0.30 m s^{-1} Typical depths 0.90 m in a stream with average depth 0.5m
H_4 Step-pool/ cascade/rapids	Steps and pools interchange	High water velocity >1.0 m s^{-1}. River slopes 2–10%
H_5 Glide/run	Mid-stream, deep water and fine substrate (mostly sand)	Mean flow velocity: 0.40 m s^{-1} Sand, $D_{50} \approx$ 0.20–0.30 mm
H_6 Edge	Sluggish flow and fine substrate along stream edge. Rich in organic matter, FPOM and CPOM	Mean flow velocity: 0.10 m s^{-1}
H_7 Undercut banks	Banks overhanging the river.	Shelter for large fish

- Aquatic macrophyte coverage and riparian vegetation, measured in % coverage, and ranging from 30% to 75% in lowland streams. Macrophyte vegetation in a lake was shown in Fig. 17-5.
- Woody debris abundance, size fractions, and sometimes including debris used in beaver dams.
- Overhanging banks or undercut banks.
- Weirs, structures made of stones or wood protruding out into the river.

2.3.2 Habitat Types

The relationship between stream size and typical changes in stream characteristics were shown in Figure 18-4. This relationship is often referred to as the **River continuum concept**, RCC. For example: the production respiration ratio, *P*/*R*, is greater than 1 in the middle reaches of the streams, but lower in higher and lower reaches. Table 18-2 specifies seven types of habitats. As the river channel widens and flow velocity decreases, the stream becomes more like a lake.

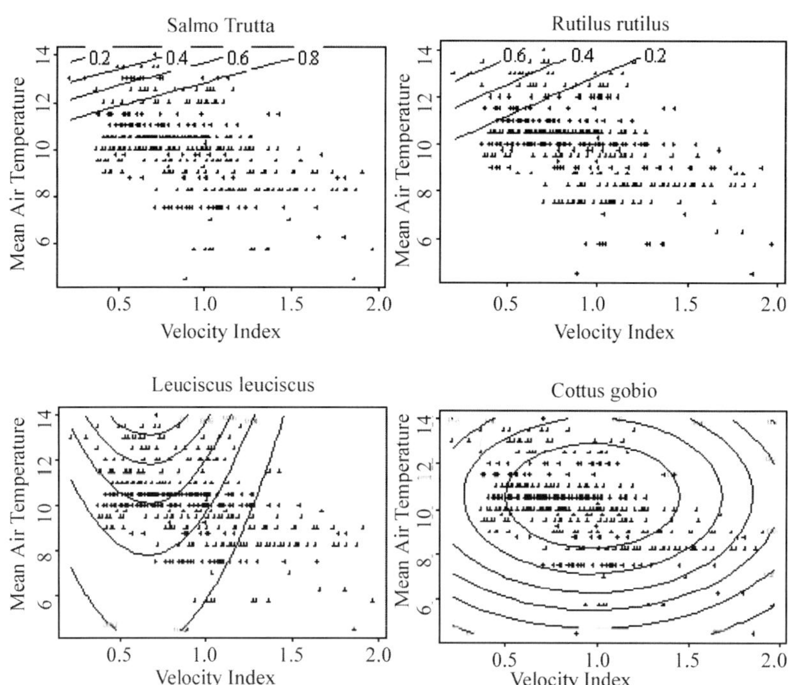

Figure 18-5. Niches for four species of freshwater fish. Dots show observation combinations; it is the same for all species. Data: Pont et al. (2000).

For streams of 5th order or larger, hydraulic residence times do not influence ecosystem properties and the ecosystem is more like that of a lake.[13] In the upper reaches, organic matter is supplied in the form of coarse particulate matter, which is leaves and branches, whereas soil erosion is the primary source for organic material in the lower reaches and thus supplies organic matter in the form of fine particulate organic matter.

Benthic fauna changes among shredders, grazers and collectors downstream are shown in Figure 18-4. Fish species and families have different requirements for physical and chemical environment.[14] The niches of a selection of freshwater fish species are shown in Figure 18-5. The figure shows relative suitability of fish habitats as a function of velocity and mean air temperature for salmon (ST, cold water species), roach (RR, cyprinid species), dace (LL, cyprinid species) and sculpin (CG, white fish, cold water). The lines represent a habitat suitability index from 0.0 to 1.0. The dots are observations. For example, a salmon prefers high velocity and low temperature whereas a roach has opposite preferences.

2.4 Wetland

Wetlands are shallow water-bodies, usually less than 0.5–1.5 m deep, allowing most of the wet area to be covered with macrophyte growth on the sediments. Wetlands are either natural or constructed, often formed at the mouth of rivers or along flat parts of the river course. Wetlands can be classified in seven groups as shown in Table 18-3.

Figure 18-6. Salt "mountains" in Camargue, France, 2000.

The largest wetlands in North America are in the Okefenokee Swamp in Georgia and the Everglades in Florida. Figure 18-6 shows part a wetland area in Camargue, France, which is used for production of salt

Wetland is an important part of natural ecosystems, cleaning fresh water and pumping nutrients from the soil into nutrient deficient coastal waters. It is habitat for a rich flora and fauna, especially birds. Wetlands are now a prime target for environmental protection. They are important nursing grounds for juvenile fish. Zooplankton abundance tends to increase with macrophyte cover,[15] and wetlands offer small fish protection against predators such as waterfowl and larger fish. Wetlands are sometimes created for the purpose of treating wastewater.[16] By the year 2000, there were more than 500 treatment wetlands in Europe and over 600 in North America. A possibly ethical question is whether constructed wetlands can be counted as "new natural habitats". Although nutrients disappear in wetlands, one is concerned about the fate of other pollutants such as heavy metals and pesticides. If these concentrations are high and detrimental to fauna, constructed wetlands may pose a double jeopardy to nature; first by attracting vulnerable species, and thereafter killing them.

3. WATER QUALITY STANDARDS

Water quality refers to physical, chemical and biological characteristics, and different water uses require different qualities.

Table 18-3. Freshwater wetlands. Data after Mitch (1983).

Type of wetland	Definition	Plants
Forested swamps	Frequently flooded wetlands with forest	Cypress Taxodium, tupelo Nyssa
Bottomland hardwood forest	Infrequently flooded floodplains with hardwood trees	Maple Acer, oak, Quercus, hickory, Caraya
Marshes and shallow ponds	Intermittently flooded wetlands with emergent or floating hydrophytes	Cattail Typha, reed grass Phragmites, bulrush, Scirpus
Bogs, peatlands and ferns	Soil is waterlogged and spongy, acid and nutrient poor	Mosses Sphagnum
Tundra	Soil alternatively freeze and taw, permafrost in ground	Lichens, willow, blueberry, cotton grass, mosses
Agricultural wetlands	Wetlands used for crop growth or for sewage treatment	Rice
Treatment wetlands	Wetlands used for sewage treatment	

Habitat quality includes quality of surroundings, like channel bed structure and vegetation along the shore. "Ecosystem health" is the ultimate criterion for habitat quality assessment, but it is difficult to define, and managers try to capture it with hydrological, physical, and chemical as well as biological parameters. Abundance, richness and quality of fish is often used, both as a habitat and water quality indicator. A second indicator system is based on parameters describing the assemblage of benthic and attached organisms on the riverbed.

3.1 Water Flow Quality Standards

In Montana, USA, recommended minimum flow to maintain fish habitats is in the order of 10% of mean annual flow for minimal habitat suitability, and 30% to provide good habitat for fish. It is recommended to allow high flows once every three years in the fall to inundate and maintain floodplain habitats.[17]

3.2 Chemical Water Quality Parameters

When biological material degrades or mineralizes, it uses oxygen, which is taken from the water. Since fish and other organisms require oxygen, high biological oxygen demand is negative. However, in real systems, organic matter that has not been mineralized within five days may be very hard to mineralize further, and is thus often not relevant as a pollution problem. Logs and branches mineralize over decades.[18] BOD_5 and *COD* are two common river water quality criteria related to oxygen and phosphorus, which measures oxygen demand:

- BOD_5 = biological oxygen demand after five days = the oxygen respired for oxidation of the biodegradable organic material present = oxygen used when organic materials decompose in a bottle for 5 days. BOD_5 is measured as mg oxygen per liter. Typical values are 0–3000 mg L^{-1}.
- *COD* = chemical oxygen demand = the oxygen equivalent of the chemically oxidable material present. *COD* is higher than BOD_5 by a factor of 2 to 12. *COD* accounts for all oxygen demand in the sample.

3.3 Water Nutrient Quality Criteria

Water quality parameters in rivers often refer to the macronutrients total phosphorus and total nitrogen. In Table 18-4 we compare macronutrients for some target organisms. A comparison of water quality criteria in rivers and lakes is shown in Table 18-5.

Table 18-4. Water habitat quality standards. TP = total phosphorus, IP = inorganic phosphorus, TN = total nitrogen, IN = inorganic nitrogen. Inorganic fractions are normally between ½ and ⅔ of total fraction. Data: summary by Smith et al. Smith, Tilman et al. (1999)185, nuisance levels of benthic algal biomass is 100 mg m^{-2} Dodds et al. (1997).

Protected species	Phosphorus	Nitrogen	Dissolved oxygen, DO	Sediment depth [1]
Benthic algal biomass	<30 mg TP m^{-3}	<350 mg TN m^{-3}	–	–
Salmonid species	<47 mg IP m^{-3}	–	>6 mg L^{-1}	>20 cm
Generic fish spp.	<60 mg IP m^{-3}	<610 mg IN m^{-3}	>4 mg L^{-1}	5–10 cm

(1) in spawning habitats

Table 18-5. Comparison of water quality criteria for lakes and rivers. After Smith (1982), Dodds et al. (1998); Smith et al. Smith, Tilman et al. (1999)180, and Biggs (1996).

Attribute	Oligotrophy	Mesotrophy
Lakes		
Total phosphorus, TP	TP <10 mgm^{-3}	10 mgm^{-3} < TP < 35 mgm^{-3}
Mean planktonic chl-a	Chl <2.5 mgm^{-3}	2.5 mgm^{-3} < chl < 8 mgm^{-3}
Mean planktonic chl-a [1]	Chl <4 mgm^{-3}	4 mgm^{-3} < chl < 6 mgm^{-3}
Mean Secchi	SD >6 m	3 m < SD < 6 m
Rivers		
Total phosphorus, TP	TP <25 mgm^{-3}	25 mgm^{-3} < TP < 75 mgm^{-3}
Mean suspended chl	Chl <10 mgm^{-3}	10 mgm^{-3} < chl < 30 mgm^{-3}
Mean benthic chl	Chl <20 mgm^{-3}	20 mgm^{-3} < ch < 70 mgm^{-3}
Mean benthic chl [1]	Chl <60 mgm^{-2}	60 mgm^{-2} < chl < 200 mgm^{-2}
Marine		
Total phosphorus	TP <10 mgm^{-3}	10 mgm^{-3} < TP < 30 mgm^{-3}
Total nitrogen	TP < 260 mgm^{-3}	260 mgm^{-3} < TP < 350 mgm^{-3}
Mean Secchi	SD > 6 m	3 m < SD < 6 m

(1) frequently flooded

Generally, rivers tolerate higher nutrient loads than lakes, and rivers that that are flooded often tolerate even higher loads. Oligotrophic to eutrophic water is called "blue water" and hyperutrophic water is "green water".[19]

3.4 Biotic Quality Indexes

The US Corps of Engineers index for river water quality, HQI, is based partly on stream morphological characteristics, partly on chemical properties, and partly on biological properties. The parameters and their weights are shown in Table 18-6. For comparison purposes, the lakes index is also included. Total dissolved solids, TDS, is a measure of the amount of soluble mineral and organic matter held by water in solution. Sinuosity is a measure of the stream's meandering pattern. Fish species association characterizes the river ecosystem. This sub-index is not easy to describe precisely, but depends on the river system. Parameters that include number of species, total number of fish, and fish health contribute to the measure.[20]

Table 18-6. Parametric description of stream and lake habitat quality. After U.S. Army corps of Engineers (1980).

Streams		Lakes	
Parameters	Weights	Parameters	Weights
1. Fish spp. Association	30	1. Total dissolved solids	30
2. Sinuosity index	20	2. Spring flood index	20
3. Total dissolved solids	20	3. Mean depth	15
4. Turbidity	10	4. Chemical type	15
5. Chemical type	10	5. Turbidity	15
6. Benthic diversity	10	6. Shoreline development index	5
		7. Total fish standing crop	(1)
		8. Sport fish standing crop	(1)

(1) If fish data are available, lake quality may be related directly to fish quality. Indexes are now being developed that address water quality directly in terms of fish abundances, much like the Chandler score for benthic fauna.[21]

4. DOSE – RESPONSE RELATIONSHIPS

We discuss some common models of the effects of nutrient inputs on rivers and wetlands, notably the important *Streeter-Phelps model* that predicts oxygen deficit when organic material is loaded into a stream.

4.1 River Models

The benthic community in a river will adapt to reasonable fluctuations in river flow, but become severely disturbed at extreme flows. Studies in County Cork, Ireland, showed that a catastrophic flood larger than 1.5 times full bank led to 95% reduction in invertebrate density and 30% reduction in the number of taxa.[22] During the flood, the main channel was heavily scoured, and mosses and gravel disappeared. Bank full floods recur every other year, and cause scouring of sediments sufficient to affect fish species that bury their eggs at shallow depths in the sediments.[23]

High flow in streams is defined as flows with a volume higher than three times the average. They disturb sediments and periphyton, but not fish, which take refuge in backwaters, or anchor themselves to fixed artifacts in the stream. Low flows are, on the other hand, detrimental to fish since part of the normal habitat dries up, and the oxygen concentration is low in the remaining water.

4.1.1 Nutrient Load & Time Between Flows → Periphyton Density

Nutrients cause buildup of periphyton biomass, but extreme floods flush the biomass off the sediments. From a management point of view this means that a greater load of nutrients can be allowed in streams that flood more frequently.[24] Equation 18-2 relates nutrient load in terms of inorganic dissolved nitrogen, or soluble reactive phosphorus, *SRP*, and the time, d_a (day), between flows that are three times the average flow to the biomass of periphyton, *P*. Typically, d_a is between 10 and 100 days.[25]

$$\log_{10}P = 0.5\log_{10}SRP + 4.7\log_{10}d_a - 1.1(\log_{10}d_a)^2 - 2.71 \quad (18\text{-}2)$$

If we define the oligotrophy-mesotrophy boundary at 60 mg *chl-a* m^{-2} and the mesotrophy-eutrophy boundary at 200 mg *chl-a* m^{-2}, we can construct the nomogram in Figure 18-7.

4.1.2 Mean Flow Velocity → Macrophyte Coverage

A model of summer macrophyte volume, *TMV*, as a function of stream velocity (m s^{-1}) was developed in New Zealand:[26]

$$TMV = e^{4.6 - 2.8\,V},\ r^2 = 0.84,\ p < 0.001 \quad (18\text{-}3)$$

TMV is measured in percent volume and defined as percent coverage times average plant height by average reach depth.

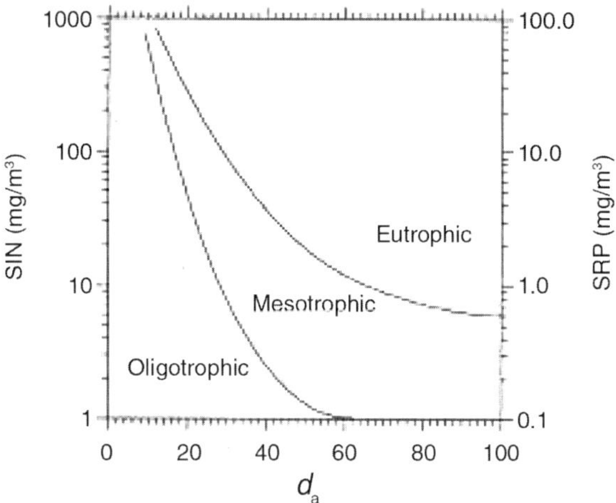

Figure 18-7. River pollution degree as a function of mean monthly soluble nutrient concentration and days between flows.[27] Accrual-days between flows, d_a, in gravel, cobble bed streams. SIN is soluble inorganic nitrogen; SRP is soluble reactive phosphorus and about 50% of total phosphorus. See text for further definitions.

Velocity ranged from 0.1 to 0.9 m s^{-1}. A 0.5 m s^{-1} velocity thus gives a macrophyte cover of 25%. Knowledge of such relations is potentially useful in river management projects where physical change of the river course will alter stream velocity.

4.1.3 Effluent Discharge into Rivers → Diluted BOD Strength

If a pollutant with BOD strength L_w (mg) is discharged at a flow rate Q_w (m^3 s^{-1}) into a river which has a flow rate Q_s and a BOD strength L_r, then the initial BOD strength L_o of the mixture in the well-mixed section just below the discharge point is:

$$L_o = (Q_w L_{w} + Q_s L_s)/(Q_w + Q_w) \qquad (18\text{-}4)$$

4.1.4 The Oxygen Sag Curve: Organic Load → Oxygen Concentration

Streeter and Phelps (1925) first formulated the relationship between organic load and oxygen concentration in a river.[28]

Assume that an industrial plant makes a sudden discharge of an effluent into a river, so that the BOD strength is L_0 just below the discharge point. As the water volume with enhanced BOD strength flows down the river, two processes take place.

1. A first order biological oxidation (degradation) of the organic material takes place, which decreases BOD strength and requires oxygen and contributes to an oxygen deficit $D(t)$. The deoxygenation rate K_1 depends upon the type of organic matter, as shown in Table 18-7. The deoxygenation process can be written $dD/dt = -L_0\exp(-K_1 t)$.
2. An opposing mechanism, called *reaeration*, mixes new oxygen from the air into the stream. The reaeration rate is proportional to the oxygen deficit, that is, the less oxygen the water already contains the more oxygen is mixed into it: $dD/dt = K_2 D(t)$. The reaeration coefficient, K_2, increases with the downstream velocity of the water and decreases with the depth of the water. Typical reaeration coefficients are shown in Table 18-8.

Table 18-7. The coefficient of deoxygenation. Fujiwara et al. (1986).

Water service	K_1 = coefficient of deoxygenation, day^{-1}
Untreated wastewater	0.35–0.70
Treated wastewater	0.10–0.25
Polluted river	0.10–0.25

Table 18-8. Reaeration coefficients, K_2. Data: Võrõsmarty et al. (1997) and citations therein, Statzner and Sperling (1993).

Environment	K_2 = coefficient of reaeration, day^{-1}
Reservoir	0.22
Small ponds and backwaters	0.10–0.23
Sluggish streams, large lakes	0.23–0.35
Low velocity rivers	0.35–0.46
Average velocity rivers	0.46–0.69
Fast rivers	0.69–1.15
River rapids	>1.15

Adding these two processes together and solving for the time development of oxygen deficiency, D_t, we get Equation 18-4:

$$D_t = \frac{K_1 L_0}{K_2 - K_1}\left(e^{-K_1 t} - e^{-K_2 t}\right) + D_a e^{-K_2 t} \qquad (18\text{-}5)$$

Here,

D_t = oxygen deficit at time t

K_1 = coefficient of deoxygenation, day^{-1}

K_2 = coefficient of reaeration, day^{-1}

L_0 = initial BOD in stream following mixing, mg $\times$ L^{-1}

D_a = oxygen deficit upstream of the discharge point, mg $\times$ L^{-1}

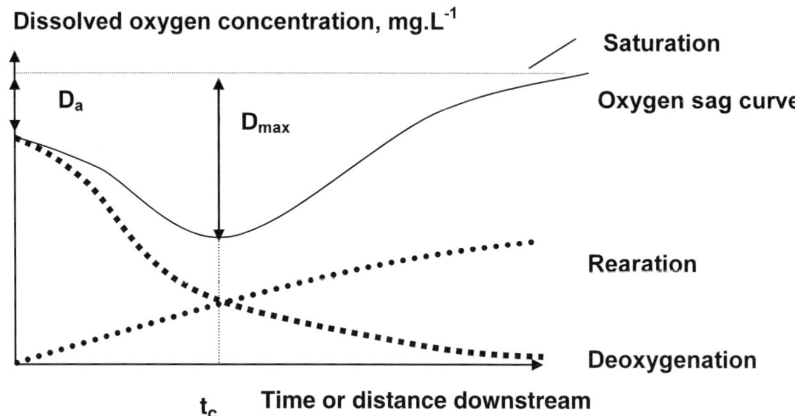

Figure 18-8. Dissolved oxygen concentration as a function of time or distance downstream, assuming that the water flows with a velocity $v = c$ (m s^{-1}), c = constant. D_{max} is the maximum oxygen deficit. t_c is time to minimum oxygen concentration.

The time and distance downstream to the minimum oxygen concentration, D, – corresponding to maximum oxygen deficit – can be found by differentiating $D(t)$ with respect to time, as shown in Equation 18-6.

$$t_c = \frac{1}{K_2 - K_1} \ln\left[\frac{K_2}{K_1}\left(1 - D_a \frac{K_2 - K_1}{K_1 L_0}\right)\right] \quad (18\text{-}6)$$

$$D_c = \frac{K_1}{K_2} L_0 e^{-K_1 t_c} \quad (18\text{-}7)$$

The distance downstream to this point can be found by multiplying by the water velocity. The maximum – or critical – oxygen deficit, D_c, is shown in Equation 18-7. If D_c is larger than the saturation concentration, so that the oxygen sag curve intersects the zero oxygen curve at two points, one has to return to Equation 18-5 and set $D_t = D_{sat}$ to find the intersection points. The minimum oxygen concentration in the river is $D_{min} = D_{sat} - D_c$.

4.1.5 Total Phosphorus (TP) → Chlorophyll (chl-a)

A linear relationship between *TP* and water column *chl-a* has been estimated for streams, mostly in North America.[29]

$$\log chl\text{-}a = 1.7 + 2.0 \log TP - 0.3(\log TP)^2,\ p < 0.001,\ n = 292 \quad (18\text{-}8)$$

The production of algae per unit of *TP* is often significantly lower in rivers than in lakes and reservoirs. There are at least three possible reasons:

- The washout rate is higher.
- Light levels are too low for algal growth. Low light levels are caused by suspended inorganic solids or shading by forest canopies.
- Light levels may be to high and cause light inhibition because the river is too shallow and phytoplankton cannot escape down into shade.[30]

In all three cases, the relationship between *TP* and *chl-a* may change downstream; suspended particles may settle, the river becomes wider and deeper so that canopies only shade part of it, and phytoplankton can sink to larger depths.

4.1.6 Stream Characteristics → Salmon Abundance

Stream salmonids select spawning sites based on substrate size, water depth, water velocity and stream width – in that order.

Table 18-9. Coefficient to compute the probability of finding salmonid spawning sites.

Variable	Coefficient x_i	Coefficient x_i^2	Best value	Range
x_1, Substrate size	1.52	–0.05	20 mm	0–100 mm
x_2, Water depth	0.63	–0.03	0 cm	0–140 mm
x_3, Water velocity	0.16	–0.002	50 cm^{-1}	0–120 cm s^{-1}
x_4, Ln(Stream width)	2.17	–	800 cm	150–800 cm
Constant	–29.44	–		

The probability of finding spawning sites is:

$$p = \frac{e^{\phi}}{1 - e^{\phi}} \tag{18-9}$$

With the coefficients of Table 18-9, we can calculate ϕ as:

$$\begin{aligned} \phi = 1.52x_1 - 0.05x_1^2 + 0.63x_2 - 0.03x_2^2 \\ + 0.16x_3 - 0.002x_3^2 + 2{,}17x_4 - 29{,}44 \end{aligned} \tag{18-10}$$

These four variables accounted for 59% of the variation in the observed number of spawning sites in a logistic regression model.[31] Substrate size, water velocity, and water depth have optimum values as suggested in Table 18-9, whereas spawning probability appears to increase with stream width. The number of spawning sites – or redds – per km is in the range 0 to 300.[32]

4.2 Wetland Model

Wetlands assimilate nutrients from runoffs from the landscape before they enter rivers, lakes or the sea. Wetlands have high storage capacity for some ions and a good ability to transform nutrients like nitrogen and phosphorus, and this makes them suitable and efficient instruments to treat the runoffs. Nutrient assimilative capacity is defined as the total mass per unit area that can be permanently absorbed by the wetland system with no significant ecosystem change and with no change in downstream output of nutrients.

Table 18-10. Removal rates in wetlands.[33]

Constituent	Removal rates in wetlands Kgha^{-1}day^{-1}	Concentration of outflow from storm water wetland mgL^{-1}
Total suspended solids, TSP	20	18
Biological oxygen demand, BOD	8	5.0
Total nitrogen, TN	2	1.0
Total phosphorus, TP	0.03-0.2	0.05

If constructing wetlands is an option for water purification, it has to be of a certain size to accommodate the average annual sediment load generated from the landscape. For the entire area examined by Tilley and Brown in the South Florida treatment area, requirements ranged from 0.2% to 4.5% of the basin area. For sub-basin sizes 100-1000 ha they estimate the area required as treatment wetlands as:

$$W\,(\%) = 0.0434\;U\,(\%) + 0.0042;\; r^2 = 0.94 \qquad (18\text{-}11)$$

Here W is percent of watershed required as treatment wetland, and U is percent of watershed land classified as urban.[34]

5. MITIGATION MEASURES

Efforts at river restoration have increased in recent years. Examples are:

- Re-meandering: If you look at the maps of the upper Rhine river in Figure 18-2, re-meandering would mean reversing the trend.
- Making groins (stream training structures) that produce invert habitats and have a pool at their tip.
- Addition of stone ripraps to existing groins across the stream channel to create weirs. The ripraps are most often constructed of stones, most of them weighing less than 40 kg. See Figure 18-9.
- Replanting of trees along the low-flow riverbank.
- Anchoring large woody vegetation to banks or bottom to create forced pools and thereby greater probability for reeds.

  Replacing wetlands by sewage treatment plants.

Figure 18-9. River mitigation measures. Left: stone ripraps; Right: tree trunks. From Boulder Creek, Colorado, 2005.

To restore wetlands, a landscape perspective is required, because water transport processes connect areas that are far apart and aquatic species may require different ecosystem types to complete their life cycle. Multipond systems may be used to control diffuse phosphorus pollution in a sustainable way. Multiponds are tiny ponds and ditches scattered around in agricultural fields. They may be used to store rainwater for irrigation, intercept runoff and decreasing erosion, and recycle phosphorus.[35]

---

[1] Windolf et al. (1997)
[2] Nilsson et al. (1997) p. 798, Nilsson and Keddy (1988)
[3] Kiely (1997) p. 271-4
[4] http://www.epa.gov/305b/2000report/chp2.pdf; (2004)
[5] World bank 1997 home page
[6] Estimated from Navrud (1995)
[7] Richardson and Quian (1999)
[8] King and Lester (1995)
[9] Myers et al. (2000) p. 854, Myers (2000)
[10] Search US National Flood Insurance program
[11] Gordon et al. (1992)
[12] Montgomery et al. (1999) p. 383-5
[13] Basu and Pick (1996)
[14] Useful information is given in : Oberdorff et al. (2000), Montgomery et al. (1999), Bult et al. (1999) p. 1299, Crisp (1996), Mann (1996) Summary by Smith et al.(1999) p. 185, Gerking (1978):production estimates, Matuszek and Beggs (1988): lake area and pH
[15] Grenouillet et al. (2002)
[16] Cole (1998), Tilley and Brown (1998)
[17] Allan (1995) p. 320 and Bonvechio and Allen (2005)
[18] Text based on Stum and Morgan (1996) and Kiely (1997)
[19] Ashton (2002)
[20] See Karr et al. (1987), Matuszek and Beggs (1988)
[21] Oberdorff et al. (2000), and Oberdorff, Guégan et al. (1995)
[22] Giller (1990), and Giller, Sangpradub et al. (1991)
[23] Montgomery et al. (1999)
[24] Biggs (2000)
[25] Biggs (2000) p. 21
[26] Henriques (1987)
[27] After Biggs (2000)
[28] Streeter and Phelps (1925)
[29] Nieuwenhuyse and Jones (1996)
[30] Heiberg and Seip (1994)
[31] Knapp and Preisler (1999)
[32] Montgomery, Beamer et al. (1999) p. 383
[33] After Tilley and Brown (1998), Richardson and Quian (1999)
[34] Tilley and Brown (1998)
[35] Yin and Shan (2001)

Chapter 19

TERRESTRIAL ENVIRONMENT

Habitat suitability in natural terrestrial environments

The natural terrestrial environment is affected by agricultural development, logging of forests, and human building activity – including infrastructures that increasingly fragment landscapes. We describe how to estimate natural areas required by wild animal populations, and how to develop habitat suitability indexes that will secure land for wildlife.

●

"Man, the cutting edge of terrestrial life, has no rational alternative but to expand the environmental and resource base beyond earth." Krafft A Ehricke

1. INTRODUCTION

In this chapter we discuss natural terrestrial environments – but exclude cultured areas like farmland and urban areas – and with reference to the motto above, also regions beyond earth.

Natural impacts on terrestrial ecosystems include natural fire, landslide, hurricanes and flood. Volcanoes are particularly devastating. On the morning of May 18th 1980, Mount Helena in Washington State erupted and ruined 600 km^2 of forested valleys and ridges. The consequences were severe; 10 million trees were uprooted, and fifty-seven people died. Worldwide, more than 500 million people live within striking range of potentially lethal volcanoes.[1]

However, also human activity impacts the terrestrial environment to a considerable degree. Examples are:

- Pollution.
- Harvest (timber, wildlife).
- Landfills.
- Introduction of new species to an area.
- Road construction.

Sometimes man's activities reinforce natural impacts and increase the vulnerability of terrestrial environments to natural forces like rain and erosion. Harvesting and plowing, for example, make soil vulnerable to erosion. Thus, disadvantages of reduced plowing and harvesting must be balanced against the benefits of lower erosion rates. If we assume that the ultimate ecological goal regarding terrestrial habitats is high diversity of species, the major question becomes how to achieve that in the best possible way. Corollaries to this question are:[2]

- What is the value of rare or endangered species?
- What is the value of old growth forest?
- What is the value of nature preserves or wilderness areas?

2. BACKGROUND INFORMATION

Terrestrial ecosystems provide many services. Vegetation makes surface soil less shifting and reduces erosion dramatically, typically to less than 1% of that without the vegetation. It also reduces evaporation, and provides shelter for animals and humans. Vegetation stores carbon and is thus a beneficial factor in climate regulation. The Northern Hemisphere forest regrowth absorbs 0.5 ± 0.5 billion ton carbon per year (GtC yr^{-1}), and other terrestrial sinks 1.3 ± 1.5 $GtCyr^{-1}$.[3] Forests provide material for building and carpentry. Round wood harvest in 1995 was 3.4×10^9 m^3 $year^{-1}$ and may soon reach the estimated worldwide maximum sustainable yield of around 5×10^9 m^3 $year^{-1}$.[4] The harvestable fraction of net aboveground primary production, **NAPP**, vary with crop, from about 25% of trees as timber, 50% of modern agricultural crops, and 60%–75% of grass as animal fodder.[5] Currently, between 40% and 95% of Earth's terrestrial surface has been cultivated, and the rest left as natural habitat.[6] Natural habitats have progressively become smaller and more fragmented. Even in the sparsely populated southern Norway, only 8% of the land area is more than 5 km from technical installations, giving 1 ha of wilderness area per capita.[7]

2.1 The Value of Terrestrial Resources and Services

Estimated values of the most important terrestrial resources and services are shown in Table 19-1. Wastewater treatment (87 US$ ha^{-1} yr^{-1}) and recreation (66 US$$ha^{-1}yr^{-1}$) are among the most important short-term services. If timber is logged according to "Living forest" criteria, a premium may be added to the sales-price of the order of €1 per m^3. In India, non-timber products such as pods and fuel-wood was found to account for about 10% of the forest benefit value.

Table 19-1. Summary of average global value of terrestrial ecosystem values. Global flow value is value per area unit and year times world area.[8]

Type	World area 10^6 ha	Main services (Partial values in parenthesis) US\$ ha^{-1} yr^{-1}	Value per area US\$ ha^{-1} yr^{-1}	Global flow value 10^9US\$ yr^{-1}
Total terrestrial area	15,323		804	12,319
Forests	4855	Climate regulation (141) Nutrient cycling (361) Raw materials (138)	969	4706
- Tropical	1900	Nutrient cycling (922) Raw materials (315) Climate regulation (223)	2007	3813
- Temperate/boreal	2955	Climate regulation (88) Waste treatment (87)	302	894
Grass/rangelands	3898	Waste treatment (87) Food production (67)	232	906

Mushrooms and wild plants constitute a substantial fraction of food intake in some societies. In Central Africa, mushroom consumption rates are 30 kg per year for villagers and 15 kg for townspeople.[9]

In the USA, the willingness to pay for preservation of endangered species, like Bighorn sheep and Bald Eagle, is US\$12 and US\$15 per household and year, respectively.[10]

2.2 Terrestrial Regions

The kind of species living in terrestrial communities reflects the geological and climatic conditions there. The most important parameters are steepness and fertility, which indicate organic matter content and moisture. Within the climate zones, land can be divided into smaller zones that are suitable for different land uses. A rough approximation is shown in Figure 19-1. *Agrosilviculture* – agriculture and forestry – requires 2% to 4% organic matter content and more than 500 mm yr^{-1} rainfall if irrigation is unavailable. Slopes as steep as 30% may be farmed for food if they are terraced. In some temperate regions more than 30% of the productive forest grows on slopes steeper than 50%.[11] Animals may utilize steep slopes; the mountain caribou in British Columbia is found on 70° slopes.[12] Terrestrial habitats can be classified in four biomes: tropical forest, temperate/boreal forest, grass/ rangeland, and wetland. One may also classify a region in vegetation zones according to the dominant vegetation. The main vegetation types are: tropical rain forests, sub-tropical rain forest, monsoon forest, temperate rain forest, summer green deciduous forest, needle leaved forest, evergreen hardwood forest, savannah woodland, thorn forest and shrub, savannah, steppe and semi desert, heath, dry desert, tundra and woodland, and cold desert.

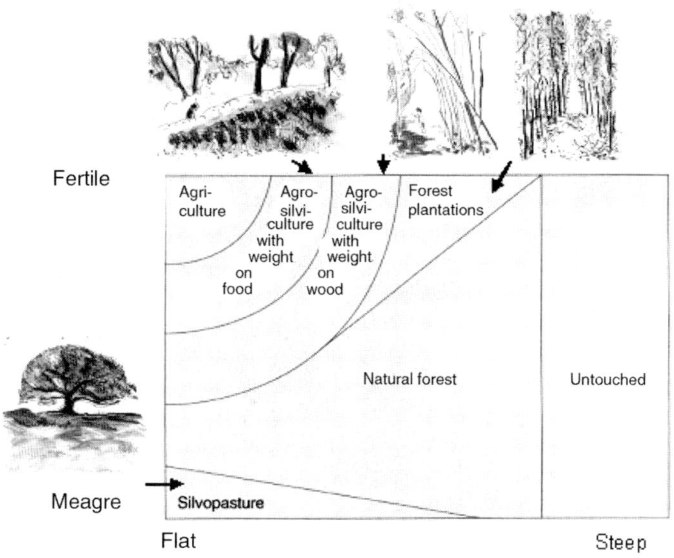

Figure 19-1. Suggested land use choices in the tropics according to land steepness and fertility.[13]

These categories are useful if one wants to value land for environmental decision making. For example, although tropical rain forests cover only 7% of total land area, they contain over 50% of all terrestrial plant and animal species.

2.3 Terrestrial Ecology

In this section we examine population parameters that are particular for terrestrial ecosystems, among them densities of vegetation and wildlife. Both are relevant for harvesting and preservation.

2.3.1 Population Density

It is difficult to determine the "natural" population density of species in terrestrial habitats. Many terrestrial species, in particular *microtine* populations, show annual or cyclic fluctuation, or variations that resemble chaotic behavior. In addition, dynamic variation of different species often interacts. For example, wolf (*Canis lupus*) density increases with the density of its prey, such as woodland caribou (*Rangifer tarandus*) and moose.[14] The density of plants, trees, and animals also depends on the quality of the habitat. It is therefore customary to quote density ranges instead of central estimates. Typical terrestrial vegetation density for grass in tundra and alpine regions is 1 to 30 ton ha^{-1}, and 2 to 50 ton ha^{-1} in temperate regions. The yield is 0.1 to 4 ton ha^{-1} yr^{-1} in alpine regions, and 2 to 15 ton ha^{-1} yr^{-1} in temperate regions. Savannah and brush vegetation may reach densities of 150 – 200 ton ha^{-1} with annual yields of about 10%. Softwood forests in temperate regions may reach 600 ton ha^{-1} and hardwood forests 2000 ton ha^{-1}. Tropical seasonal forests and rain forests may reach 600–800 ton ha^{-1}, but all forest types appear to have upper limit yields of 25–35 ton ha^{-1} yr^{-1}. The density – number per hectare – of animals generally decreases with the weight of the individual. There are 200–500 field mice and voles per hectare. Since they weigh only 20–50g per individual, the total mouse weight is 5 to 10 kg ha^{-1}. Wolves weigh around 50 kg, but you will only find 0.0001 to 0.001 wolves per hectare. Large birds, like owls that weigh 10–30 kg, have a low average density of 0.003 ha^{-1}.[15]

2.3.2 Plant Vegetation

Vegetation in its different forms is the ultimate food source. Vegetation varies, but only a small part is suitable for agriculture. Forests are natural or managed, but even in managed forests 30% may be left to itself because it is inaccessible or has too low quality.

The size, distribution and age of understory vegetation in a forest are important for the abundance and variety of wildlife. Decaying wood is habitat for owls and woodpeckers – the engineer that pecks the holes, as well as mammals like squirrels, martens, wood rats, and black bears. Forest wood densities are typically 1.1–4.5 ton ha^{-1} for open timber, grass and understory vegetation, 2.2–5.6 ton ha^{-1} for closed timber and litter with light dead fuels and 4.5–11.2 ton ha^{-1} for closed timber and litter with heavy dead fuels. Without logging, a forest is like a breathing necropolis. It contains dead standing trees that may number 75 to 100 per hectare and stand 40 years or more. Big fallen logs may last 300 years. In old conifer stands 400 linear meters of logs have been found per hectare.[16]

In temperate forests, trees may have diameters 150 cm at breast height, reach heights of 35 m and ages up to 650 years.[17] Bark thickness is an important factor determining the effects of forest fires, but is also an important parameter for bark and leaves as a source of food. Douglas fir bark thickness is 1.0–2.0 cm, whereas sub-alpine firs have much thinner bark, 0.3–0.6 cm. This is also reflected in the critical time for cambial kill, which is 3–13 minutes for Douglas-fir and only 0.3–1.0 minutes for subalpine fir.

2.3.3 Fauna

Vegetation can support a certain biomass of animals (grazers) over time; the maximum sustainable biomass corresponding to the maximum yield. Grazers that are able to keep the vegetation at maximum yield condition with their grazing mode will also have the potential greatest biomass.

Three methods are available to estimate faunal densities.

1. The amount of food required for each individual of a species can be compared to an estimate of the available net primary production.
2. One may use empirical regression models that show the relationship between species body size and species density.
3. One may use passive and active types of observations; passive such as examining litter, active such as tagging or radio collaring animals that mix with untagged oncs and are counted with catch-re-catch methods.

Animal biomass in tropical forests and temperate mountain ranges is inversely related to forest cover.[18] Evergreen closed canopy and moist forests have lower mammal biomass than open, seasonally deciduous forests. The total biomass of large mammals – over 1 kg adult body mass – in evergreen forests rarely exceeds 30 kg ha^{-1}. In more open forests with a mixture of forest and grassland, **standing biomass** can reach 150 kg ha^{-1}. Open grasslands can support standing biomasses over 200 kgha^{-1}. Overall, the standing biomass of mammals in tropical forests is an order of magnitude lower than in habitats that are more open, Table 19-2.

Table 19-2. Large Mammal biomass at tropical sites and in temperate mountain ranges.[19]

Habitat	Ungulates kg ha^{-1}	Domestic ungulates kg ha^{-1}	Primates kg ha^{-1}	Rodents kg ha^{-1}	Total kg ha^{-1}
Evergreen, closed forest	4.64	–	3.82	1.56	13.18
Deciduous open forest mosaic	50.0	39.0	1.43	1.65	54.0; 93.0 ⁽¹⁾
Grassland savannas	72.6	96	0	8.5	84.0; 200 ⁽¹⁾
Temperate mountain ranges	0.83	–	–	3.22 ⁽²⁾	4.1

(1) Including domestic ungulates

(2) Years with high density; in low density years biomass may be 0.001 kg ha^{-1}

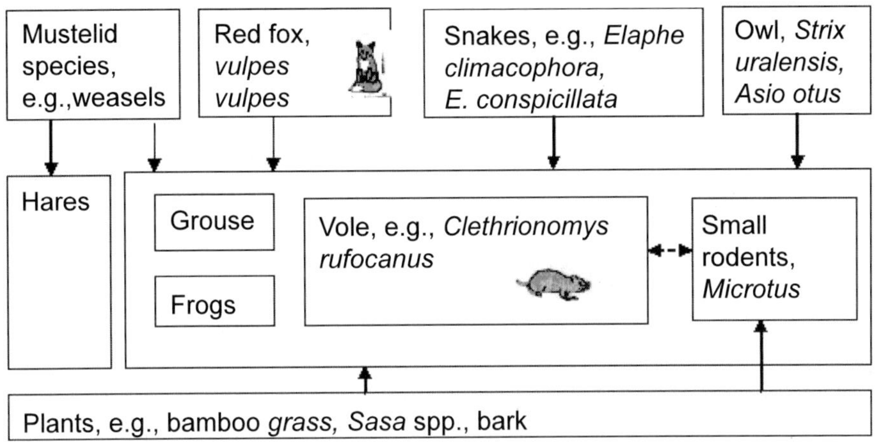

Figure 19-2. Part of food web in the forests of Hokkaido, Japan.[20]

2.4 Terrestrial Food Webs

A **food web** for a terrestrial habitat describes relationships among species that interact by eating each other. Since one species may prey on several other species, the food web will also often depict competing species. Predators with a large number of prey species, switching between different species according to taste and prey abundance can maintain a large population even if their preferred prey dwindle. Wolves in India prey on black buk (*Antelope cervicapra*), gazelle (*Gazella gazella*), hare (*Lapus nigricollis*), foxes (*Vulpus* spp.) and other small animals.[21] A food web from Hokkaido, Japan looks like Figure 19-2.

2.5 Ecosystem Services and Management Policies

In this section we discuss ecosystem services and management policies. We start with forest fires, since they pose particularly challenging management problems. Then we discuss four selected services that have considerable impact on the ecosystems that provide them: Logging, wildlife harvest, landfills, and hydropower dams.

2.5.1 Forest Fires

Forest fire, when moderate in extent, is usually regarded as favorable from an ecological point of view. It sweeps out old perennial species and gives new annual species a chance to establish themselves for a time.[22]

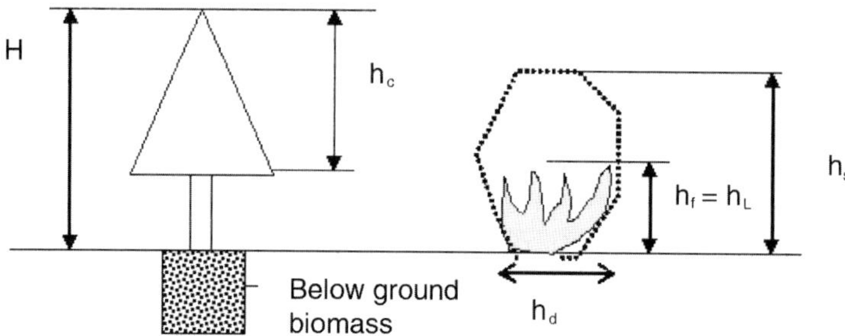

Figure 19-3. Forest and fire characteristics. Flaming depth h_d, is the width of the flames along the ground surface. With no wind, flame length, h_L is equal to flame height, h_f. Scorch height, h_s, that is the height at which burning takes place is somewhat higher. H is tree height and h_c is crown height.

Open terrestrial vegetation also supports a higher biomass of mammalian species than closed vegetation. Small and not too frequent fires – say every 50 years – also helps avoid large devastating fires.

The US Forest service recognizes two types of severe fires, those with high intensity – **crown fires**, and those of long duration – **Amazon fires**. The long duration of fires in the Amazon is due to high moisture levels, which give low flame heights, only 20 to 50 cm. When a fire starts in a tropical forest in the Amazon, it usually moves slowly along the ground with intensity like candlelight – but don't keep your finger above it.[23] See Figure 19-3. These fires consume little besides dry leaves, but kill some thin barked trees while large thicker barked trees survive. Models to predict the susceptibility of forest to fires are described in the section on dose-response.

In the Amazon and in Vancouver, Canada, fire return intervals of hundreds or thousands of years have been observed.[24] Fire-return intervals of 90 years or less may eliminate rain forest tree species, whereas intervals of less than 20 years may eradicate trees entirely. Some of the Amazon forest areas – paragominas and tailandia forests – are currently experiencing **fire return times** between 7 and 14 years. Previously burned forests are also more prone to burning, with calculated fire return times of less than 5 years. Forest fires are caused by humans or lightning. Although most Canadian forest fires are due to human activity, lightning is responsible for 85% of the burnt area.[25] Up to 8% of the fires in Europe are caused by lightning.[26]

2.5.2 Logging

Logged forests are also called industrial forests, managed forests or plantations. Logged forests often differ from native forest in tree composition.[27] Professional management of forests is called **siviculture**. As a rule-of-thumb, a forest area of 2×10^{-3} km^2 is required to give a long term harvest that will supply wood pulp for 1000 kg newsprint –3000 copies of a medium sized newspaper

2.5.3 Wildlife Harvest in Forests

Intensified logging operations has increased the access to the world's tropical forests and generated a significant harvest of wildlife resources, especially large animals with body mass above 1 kg, Table 19-3. Many tropical forest people rely on wild meat for over 50% of their protein intake.[28] The largely subsistence harvesting in the Brazilian Amazon is estimated to 67,000 to 164,000 tons of wild meat per year. This corresponds to between 9.6 and 23.5 million mammals, birds and reptiles. Average hunting rate of large mammals in tropical forests ranges from 6 to 17 animals km^{-2} yr^{-1}. With an average weight of 15 to 25 kg per animal this gives about 2.40 kg of wild meat live weight per hectare and year.

2.5.4 Landfills

Waste is generated by households, by industry, and by almost all other activities. Although waste is a household word, it is sometimes difficult to define what is waste and what is not. Paper is waste when it is delivered to a landfill, but not necessarily if it is burned for heat. Waste definitions may therefore be related to ownership, location, or to the state of materials.

Waste generation in the USA and Norway is illustrated in Table 19-4. The amount of waste in developing countries is increasing by about 5% per year. In order to dispose of one ton of waste, you need a landfill with an area of just 0.05 m^2 if the waste site has a depth of 50 m, which is the typical depth of waste sites. A waste depot not only occupies the land on which waste is filled, but also affects surrounding areas by leakage to soil and air. Leakages may follow streams and underground waterways in an area at least ten times the waste depot area. A landfill is called active until diffuse losses reach steady state, which may take 30 to 100 years. Stable leakage is obtained if the ratio of biologically to chemically available oxygen is less than 0.1 ($BOD/COD < 0.1$).[29]

Table 19-3. Wildlife harvests per year. Harvest of large animals in no-logging areas.[30]

Location	Number of individuals harvested, km^{-2} yr^{-1}	Biomass harvested kg km^{-2}
Southeast Asia	6	94
Africa	17.5	262
Latin America	8.1	121

Release rates from landfills have been reported to be 9 kg N ha^{-1} yr^{-1} and 0.04 kg P ha^{-1} yr^{-1}. This can be compared to surface runoff from residential and commercial developments, which is 8 to 22 kg N ha^{-1} yr^{-1} and 1.4 to 3.4 kg P ha^{-1} yr^{-1}.

2.5.5 Hydropower

Hydropower plants require a supply of water under pressure, and this can be tapped from waterfalls or piped from magazines such as natural lakes or man-made dams. Thus, reduced waterfalls and submerged territory are among the most significant direct consequences of hydropower development, which again has impacts on ecosystems and ecosystem services.

Norway – with a large number of rivers and waterfalls – has developed a master plan for 600 potential hydropower projects by identifying 11 end impacts as decision criteria: cost-efficiency of development and production, production capacity, forestry, agriculture, herding, wildlife hunting, fishing, clean water supply, flood prevention, cultural sites, and landscape. The Norwegian Parliament established an official Master Plan by ranking the 600 projects according to these criteria. The ranking list deviated considerably from a list just based on cost-efficiency. The difference corresponded to a willingness-to-pay to avoid the implied loss of resources and ecosystem services amounting to 25% of development and production costs. Thus, to internalize environmental costs of hydropower in Norway, one should impose a 25% environmental tax.[31]

Table 19-4. Waste mix in Norway[32] and USA[33]. The Norwegian data include municipal waste and other wastes. Norway has 4.4 million and USA 274 million inhabitants (1990).

Activity	Norway 10^3 ton	Norway kg per capita	USA 10^6 ton	USA kg per capita
Paper and paperboard	921	209	73.3	268
Glass	121	28	13.2	48
Metals	717	163	16.2	59
Plastics	321	73	16.2	59
Rubber and leather		–	4.6	17
Textiles	86	20	5.6	20
Wood	1101	250	12.3	45
Toxic waste	650	147	–	–
Other	–	–	3.2	12
Subtotal-materials in products	–	–	144.6	527
Other wastes	1078	245		
Food waste/wet organic	1556	354	13.3	48
Yard trimmings		–	35.0	128
Miss inorganic waste	–	–	2.9	11
Subtotal-other wastes		354	51.1	186
Total municipal solid waste	1355	308	195.7	714
Total waste	6551	1488		

3. TERRESTRIAL QUALITY

In this section we discuss what a "natural" habitat is and contrast natural habitats to managed and built land.

3.1 Terrestrial Quality Criteria

It is a challenge to define measurable criteria for terrestrial quality. First of all, what is considered quality will depend on the use of the land; whether it is meant to be in a natural state, is managed, or is built.

Natural habitats: In natural habitats, one wants to maximize abundance and diversity of vegetation as well as wildlife. Instead of trying to observe this directly, researchers have developed so-called **Habitat Suitability Indices** (*HIS*), which serve as indicators. These must be developed separately for different species, and are sometimes quite complicated. For illustration, we provide an example of *HIS* for the fox squirrel in Figure 19-4. When an average *HIS* for an area has been determined, multiplying *HIS* with the area indicates the total quality for that species. Alternatively, one may focus on endangered species and express damages to endangered species as reclassification of species in the IUCN classification: Near threatened, vulnerable, endangered and critically endangered, and possibly extinct. Another option is to use restitution time for endangered species populations. **Restitution time** is the time it takes for a population to regain 90% of its abundance by growth or immigration. Species listed as "threatened species" by national authorities are candidates for use under these criteria. The simplest, but probably the most unreliable indicator of habitat quality, is to identify the unsustainable areas (km^2) where ambient concentrations of pollutant are above guideline criteria or where the area is disturbed. If the area is exposed to several pollutants, the area corresponding to the pollutant that gives the smallest suitable area should be used. As far as we know there are no established guidelines for the two categories: managed and built land described below.

Managed land: Vegetation and wildlife is important in managed land, and the quality criteria will usually be different from those of natural habitats. In particular, quality criteria for tree plantations differ from natural, but harvested forests. In addition, we must include criteria that represent the value of resources and ecosystem services, such as water, wildlife harvest and wood extraction.

Built land: For built land, socio-economic values are dominating. The concept of multiple use area is gaining popularity, although there are mixed experiences.[34] Built land may require maintenance of parks and alleys and other semi-natural refuges.

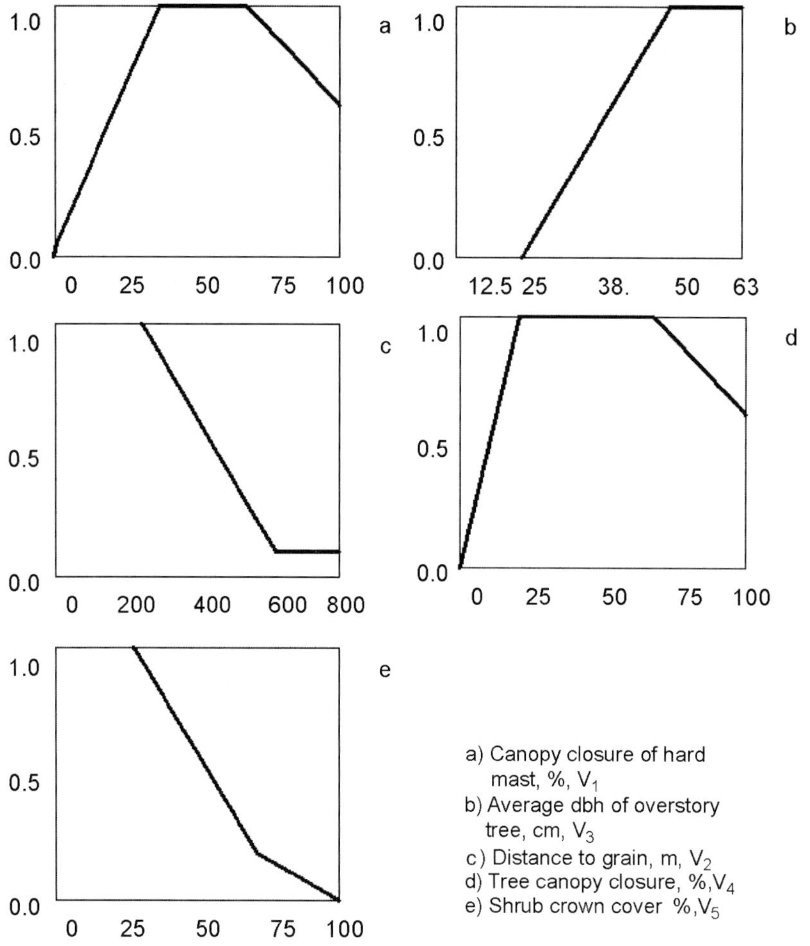

Figure 19-4. Habitat suitability index for the fox squirrel.[35]

3.2 Terrestrial Habitat Suitability Index

Habitat sustainability indices have been developed for some wildlife species. One intention is to help predict species abundance:

$$Abundance = HIS \times \text{Species interactions} \times \text{history and evolution} \quad (19\text{-}1)$$

As an example, we present a model for the fox squirrel.[36] The model has five habitat characteristics as shown in Figure 19-4.

Figure 19-5. Species succession. From top left: a) Forest fires releases nutrients and pioneer species invade; b) deciduous tree species start growing, which is good for ungulates, c) deciduous forest trees are becoming old, and create an important biotope; d) on nutrient rich soil, spruce returns, and shade for deciduous trees, after 200-300 years we have old growth forest with a high proportion of dead material.[37] Illustrasjon Pål Thomas Sundhell.

Two of the habitats in Figure 19-4 relate to winter food and three to cover and reproduction. Squirrels are most abundant in open forest stands with sparse understory vegetation such as in Figure 19-5b and Figure 19-7(left).

- $V_1 = f_1$(canopy closure of trees that produce hard mast (>25 cm dbh); %).
- $V_2 = f_2$(Distance to available grain, m).
- $V_3 = f_3$(Average diameter at breast height, dbh, of overstory trees, cm).
- $V_4 = f_4$(Tree canopy closure, %).
- $V_5 = f_5$(Shrub crown cover, %).

To compute the index, one must first measure the five variables and then find V_1 to V_5 by using the graphs in Figure 19-4. The index is then found by computing the winter food index, *WF*, and cover/ reproduction index, *CR*. *HSI* is the lower of these two sub-indices:

$$WF = (3V_1 + V_2) / 3 \qquad (19\text{-}2)$$

$$CR = (V_3 \times V_4 \times V_5)^{1/3} \quad (19\text{-}3)$$

$$HSI = \min(WF, CR) \quad (19\text{-}4)$$

Note that *WF* is zero only if both factors are zero, whereas *CR* is zero if any of the factors are zero. Since both *WF* and *CR* are required, the lowest scoring one is equal to *HSI*.

3.3 Quality Standards

Acceptable and non-acceptable conditions: Standards are defined for managerial purposes. The idea is to avoid poor habitats or by defining what conditions are acceptable and non-acceptable. If such critical requirements are not met, we call it *non-attainment conditions* Examples:

- Ozone levels should not be above a certain concentration value. For example, for a habitat to be in attainment conditions with respect to ozone, the daily maximum 1-hour average O_3 concentration should be below 0.12 parts per million all but three times during three years.[38]
- Noise and traffic should not be above certain levels. Logging cause nuisance if the noise level is above 45 dB.
- Dam water level fluctuations should not be above a certain level. For instance, hydropower dam regulations should be less than 2 meters.

In addition to critical levels like these, there are other criteria that depend upon vegetation and wildlife that needs protection. If all criteria are met at this lower limit, the habitat may be declared acceptable for the lowest habitat suitability class.

3.3.1 Standards for Critical Loads

Use of renewable and non-renewable energy sources will affect soil and ecology because of emissions during the burning. Nitrogen oxides, NO_x, will precipitate to the ground and act as a fertilizer. It may in some instances be damaging, in particularly for growth on thin soil. Sulfur oxides, SO_x contribute to acidification. The same concentration values for **No observable effect level**, NOEL, are often used for health and vegetation although there may be significant differences. The concept of "**critical loads**" has been suggested as a scientific base for negotiating pollutant discharges. Nitrogen has been shown to accumulate differently in forests of different ages, accumulation in fast growing young stands being largest. As a summary for critical loads for N one may assign 7 kgN ha^{-1} yr^{-1} for coniferous forests, 10 kgN ha^{-1} yr^{-1} for deciduous forests and bogs and 15 kgN ha^{-1} yr^{-1} for Mountain meadows. Critical loads to avoid acidification are about 9 kg ha^{-1} yr^{-1} (0.9gS m^{-2} yr^{-1}) for most Swedish ecosystems.[39]

3.3.2 Standards for Sustainable Forest Logging

There are several sets of standards for **sustainable forest logging**. Living Forest (1998) discusses forest planning for plots larger than 1000 ha[40] and prescribes:

- Closed logging should be used whenever natural generation of forest is likely.
- Protective forest at the timber line should be kept as old forest stands.
- No flat logging on moors, 5 to 10 trees per hectare among the oldest trees in the stand, that are also resistant to high wind, should not be removed; dead trunks older than 5 years should be left on site.
- On the border to lakes and streams, local or endemic vegetation should be preserved, in particular deciduous trees. Key biotopes should be maintained. These are restricted habitats that are managed to sustain or restore valuable biotopes that are not protected during normal forest operations. Biotope size should be 0.5 ha or 1% of productive forest area for properties larger than 50 ha. If forest fires destroy more than 0.5 ha, 0.5 ha should remain unmanaged for 10 years.

4. DOSE-RESPONSE MODELS

Dose-response models try to capture how the quality of the land supporting the ecosystem affects the quality and quantity of vegetation and wildlife.

4.1 Growth and Density Models

4.1.1 Temperature → Plant Growth

A plant biomass model for the biomass of crops is discussed in the chapter on agriculture and land use. Here we present an index for the vegetative growth period (also referred to as the "warmth index"). The index multiplies the number of days with temperature above a base temperature with the number of degrees above that base, $T - T_B$.

$$WI = \sum(T - T_B) \tag{19-5}$$

Thus, with a base temperature of 10°C (suitable for maize), 30 days at 20°C gives WI = 300. The larger the value of the index, the larger is the growing season.

The index can be helpful in calculations of the potential effects of global warming. In theory, a 1°C increase in mean annual temperature will move northern boundaries 150–200 km further north.[41]

4.1.2 Individual Animal Weight → Animal Numbers

Empirical relationships between individual body mass, W_g, (grams) and population density are shown in Table 19-5.

Table 19-5. Mammalian species density.[42]

Species	Number of mammals, N, per km^2
Tropical herbivores	$650\ W_g^{-0.57}$
Tropical carnivores	$11000\ W_g^{-1.02}$
Temperate herbivores	$45000\ W_g^{-0.66}$
Temperate carnivores	$95000\ W_g^{-1.14}$

4.1.3 Individual Animal Weight → Home Range

The **home range** of an animal is the area it routinely traverses. It is large enough to meet the energy need, but large animals share parts of it with neighbors. An animal weighing 1 kg uses 31% of the resource supply within its home range, while a 100 kg animal only uses 7%; the rest is shared with others.[43] Home range increases from tropical areas and towards the north. The home range H (ha) can be related empirically to body weight W (gram) by the formula $H = AW^k$ where k is a constant that depends on what the animal eats.

Large herbivores – mammals that eat less than 10% animal matter:[44]

$$H = 0.002 \times W^{1.02},\ r^2 = 0.74 \tag{19-6}$$

Omnivores – animals that ingest 10% to 90% animal matter:

$$H = 0.059 \times W^{0.92},\ r^2 = 0.90 \tag{19-7}$$

Carnivores – animals that ingest more than 90% animal matter:

$$H = 0.11 \times W^{1.36},\ r^2 = 0.81 \tag{19-8}$$

Note that the home range of grazers and omnivores increase about linearly with body weight, but greater than the metabolic requirement proportion[45] $W^{0.75}$, whereas the home range of carnivores increases faster than linearly. Wolves (*Canis lupus*) are reported to have a home range of 20,000 ha,[46] and reported densities for roe deer are in the range 5 deer km^{-2} to 25 deer km^{-2}, or an average of 6 ha per deer.[47]

4.2 Response to Species Introduction

There are several examples of species introduced intentionally or accidentally[48] into new areas that alter the ecosystem to a great extent. A famous example is the introduction of rabbits to Australia. Likewise, the Golden Apple snail has a great impact in Asia, and so has the Spanish slug (*Arion lusitanicus*) in Europe. Snails eat fresh plants and leave holes in the leaves, and by their numerous offspring they multiply in large numbers and cause extensive damage. The Spanish slug is an especially devastating snail in European gardens. From its former home on the Iberian Peninsula, it has extended its area of distribution far east and north into Europe. Its offspring can count up to 400 young, so the slug outnumbers all other local slugs.

Another well-known example is the killer bee, or the Africanized bee, which is a strain of bees that originated in Brazil in the 1950s as a cross between an aggressive African bee and a honeybee. It has now spread as far north as Texas, and has become a terrible nuisance.

Invasion of alien species are among the primary causes of extinction of endemic or native species. The introduction of a new species into an ecosystem may be considered as a *dose*, and the *response* is the reaction to the introduced species. It would probably not be meaningful to use "number of new species" as the unit on the dose axis. More relevant is the strength of the new species' colonizing ability, but to quantify this property is not easy. A successful invasion of any exotic species depends both on the vulnerability of the ecosystem to invasions and on the characteristics of the invader as defined by its population dynamics and dispersal activity. It appears that the number of species present at the site of introduction is not a predictor for success.[49] The response variable in these dose-response relationships can be defined as the *dominance index* for the ecosystem, i.e., $(x_1 + x_2)/\Sigma x_i$, where x_1 and x_2 are the biomasses of the two most abundant species.

4.2.1 A Portrait of an Invader

The following list shows some of the most important factors that will enhance the possibility of a successful invader:

- Simplified receiving ecosystem (human-altered) with lower number of species and fewer prey-predation relationships than in a similar natural or undisturbed system.
- Similar ecosystem; the native ecosystem and the receiving ecosystems are similar with respect to climate and biology.
- High reproductive output.
- Short juvenile period (to avoid being eaten).
- High rate of dispersal.
- Dormant escape stage (to survive non-optimal transient periods).

Based on these characteristics, it should be possible to assess the probability of a successful introduction of an alien species, and whether the invasion would be too successful in the sense that it becomes dominant and out-competes native species

With respcct to the golden apple snail that was introduced from South America to Asia, it should have been possible to estimate with 50% probability that the golden apple snail would become a nuisance, and in hindsight with more than 90% probability, given that virtually all reported invasions of Asian rice-ecosystems have been successful.[50]

4.3 Land Fragmentation

Statistical data show that wilderness land – defined as land more than 1 km from encroachment – decreases with real Gross Domestic Product (GDP) and population density. In Norway, with 1% annual GDP growth, 0.2–0.5% of existing wilderness land has disappeared annually.[51]

4.3.1 Patchiness

Land fragmentation is an important factor in terrestrial management and deforestation. It causes the circumference to be disproportionately larger than the core area and can be measured as *patchiness*, which is the perimeter to area ratio (*P*/*A*). Generally, patches with large *P*/*A* ratios (>15m ha^{-1}) tend to disappear. In a dammed area in China, small islands (<1 ha) lost 75% of their species in 15 years, and all islands lost their top predators within 4 years.[52]

Several studies have shown that vegetation and fauna does less well on the fringe. Forests in the Amazon lost up to 35% of their biomass in the first 10 to 17 years after fragmentation.[53] Also, seed germination suffers on the edge.[54] Other studies suggest that edge effects occur 300–500 m from the circumference.[55] However, edges may be valuable biotopes by themselves, supporting other type of species than in the core area. Figure 19-6 shows typical processes on a forest fringe.

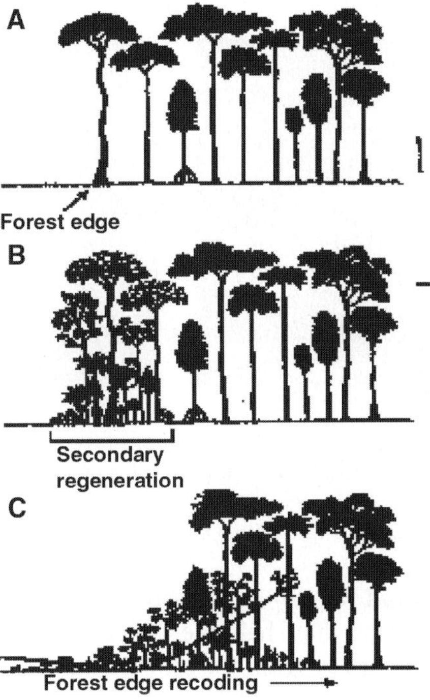

Figure 19-6. Forest edge reclining. Picture: Gascon et al. (2000).

4.3.2 Forest Fire Models

Tree mortality is a function of damage to the crown and damages to the tree stem or the cambium of the bark. Since fires are of limited duration, trees with very thick bark, say >4 cm, will usually resist a fire. For the fire to cause damage, the cambial temperature has to be in the range 60°C–70°C.[56]

The following data can be used to predict the severity and character of fires for different forest types:

- c_k: Fraction of the crown that is killed.
- t_c: Critical time for cambial kill.
- t_L: Duration of lethal heat (range 0.1 to 10 minutes).

The probability of tree mortality can be calculated as.[57]

$$P_m = c_k^{\,t_c/t_L - 0.5} \tag{19-9}$$

Figure 19-7. Left. Logs are left in the forest as habitat for wildlife. The animals in the background are goats brought in to graze down vegetation. Right: Forest fire warning signal. Logs increase the fuel value of the forest. Both pictures from Boulder, Colorado

The equation is valid for t_c/t_L ratios between 0.5 and 4.5. If the critical time for tree kill is less than half the duration of the heat, the tree is completely girdled and dead even if the crown is not killed.

First-time fires can be controlled and put out manually with minimal equipment. After the first fire, fuel levels rise substantially because combustible fuels of all sizes fall from the standing dead trees, and the open canopy allows greater solar heating and dryer forest fuel. Second-time fires spread faster and are much more intense. Large trees have little survival advantage during these more intense fires, and more than 95% of the trees are susceptible. More than 30% of observed fires in previously burned forests have fire line intensities that are beyond the limits of manual control. Combustion may remove approximately 15, 90, and 140 ton wood per hectare in the first, second and recurrent burns. Figure 19-7 illustrates a dilemma between leaving logs for wildlife and at the same time increasing the fuel value of the ground.

5. MITIGATION MEASURES

Landscapes designated for wildlife can be managed under different schemes. A large corridor project is undertaken by eight countries in Central America to provide passages between protected areas for wildlife, like jaguars and monkeys. However, construction of corridors in densely populated countries causes conflicts with several other uses of land, like tree harvesting or creation of farmland.[58]

5.1 Protected Landscapes

The most common landscape types designated for wildlife are:[59]

- *National parks*: where the objective is to protect plants and wildlife.
- *Protected landscapes*: land areas dominated by private holdings interspersed with publicly owned sites. They do not exclude local residents. The moors of Great Britain are mixtures of grassland, upland bogs, and wooded valleys maintained by low-intensity livestock grazing.
- *Multiple use areas*: land areas where limited preservation takes place. This approach to conservation depends upon two assumptions. First, it requires that local communities have sufficient cultural traditions and political, legal and economic power, as well as the support of regional institutions for sustainable management. Second, the resources available must sustain the existing population.
- *Zoological gardens*: refuge for species whose survival in the wild is in immediate peril. Captive-breeding animals in zoos, growing plant specimens in botanical gardens, and storing seeds and other plant components in gene banks are called *ex situ*, or offsite, conservation measures.

5.2 Patchiness of Protected Areas

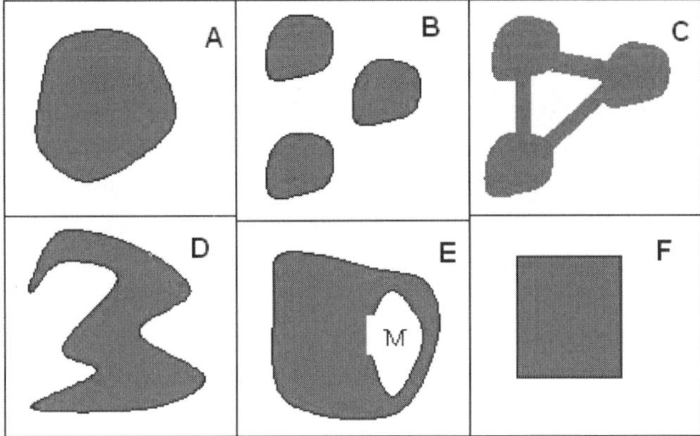

Figure 19-8. Land fragmentation.

Look at Figure 19-8. For the same effort, which landscape type is best for protection of species richness?

A. A normal protected landscape.
B. *Several smaller areas instead of one large one.*[60] How large must the sum of the smaller areas be to offer the same protection as a larger, continuous one? Fragmentation may have positive effects, like limiting dispersal of diseases and securing uncontaminated sources.
C. *Corridors.* If corridors of some size are established as step-stones between the patches, could the size of each patch be decreased?[61]
D. *Special shape.* Is there any other shape than the circular one that offers better protection per unit area? (The edge to area ratio is larger with any other shape than a circle).
E. *Multiple use areas*: These will try to respect commercial, ecological and socio-economic goals. Old growth pockets only tens of hectares in size within a larger area may shelter arthropods and microflora, and fish refuges may act in the same manner.[62] How large would such an area have to be to offer the same species protection?
F. *Zoological gardens.* Management practice is important for managed systems like E and F.

5.2.1 Case Study: Habitat Fragmentation

To solve the case, you need information in Chapter 12 on ecology.

Problem. An approximately square wilderness area, *A*, is 400 km^2 in size and imbedded within a greater rural landscape. Potential evapotranspiration for this area is 500 mm per year. With reference to Figure 19-9 and Table 19-6.

1. What is the species richness this area? If the area is divided into two or four smaller, equally sized areas as shown in Figure 19-9, what would be the species diversity of each area? And of the two/ four areas combined? (You have to make some additional assumptions here).
2. What type of species could not survive in the new habitats, but would survive in the old, connected habitat?
3. Would the edge effect be important?

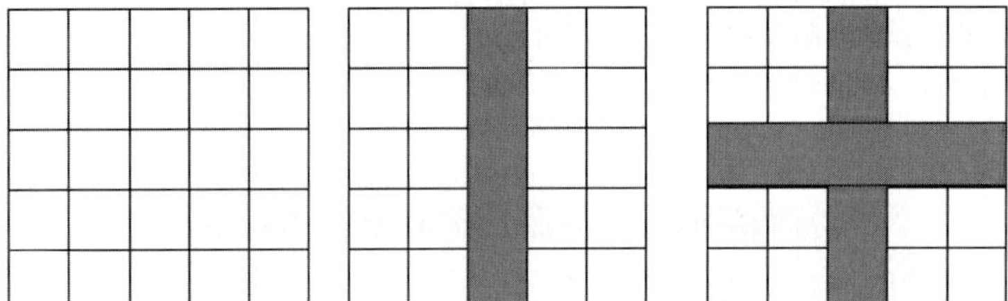

Figure 19-9. Land fragmentation. Subdivision of land into two and four habitats. The shaded areas are barriers.

Table 19-6. Land fragmentation. The figures in the parentheses in column 6 are expected number of species and assume that the number of species is lower in areas with low energy. The column for total number of species assumes that 95% of the species will be the same in all areas.

No	Total area, km^2	Single area, km^2	Number of areas	P/A ratio km km^{-2}	#spp, single area	Total # spp.
1	400	400	1	0.20	6.12 (3.06)	3.06
2	360	160	2	0.35	4.55 (2.28)	2.51
3	256	64	4	0.50	3.38 (1.69)	1.94

Solution:

1. We use the species area equations in Chapter 12 on Ecology with coefficients for mammalian species on mountaintops (US data, > 1000 mm PET): $N = 0.194 \times 40000^{0.326} = 6$ for the whole connected area. The number of species within each single area is shown in Table 19-6. Since PET is smaller than 1000 mm, the number of species may be lower than average. The site corresponds approximately to a location in North Canada. To see the expected number, we scale with Equation 12-13 in Ecology and get $N' = 0.5 \times 6 = 3$. Neighboring "areas" would typically have 95% of the species in common, thus total species numbers is as shown in Table 19-6.
2. We use Tables 12-17 and 12-18 in the Ecology chapter with additional information on habitat variability and herbivore/carnivore distinctions. The smallest area on which a population with 3000 kg herbivores can survive is 40 km^2 and the smallest area on which a population of 100 kg carnivore could survive is 150 to 210 km^2, thus large carnivores may not survive if the area is subdivided. We do not know if the estimated "home range" is meaningful, and we do not know if corridors could mitigate the subdivision for the large carnivores.
3. Assuming that the interior has to be 0.5 km from the edge, then the outer periphery is 25% of the single patch area for the smallest area. The

ecosystem of the total patch is now probably endangered. The *P/A* ratio is in the range 0.20 to 0.50; thus, no area has a ratio that would suggest that it has an increased susceptibility to become built land.

Figure 19-10. It is easier to raise preservation funds for animals that look charismatic than other animals. Linked to the ferret recovery program home page, there is a kid's page.

6. APPLICATION: THE CASE OF THE BLACK-FOOTED FERRET RECOVERY[63]

The black-footed ferret, *Mustela nigripes*, is a nocturnal carnivore indigenous to the western United States. It is a slender, wiry animal with a black facemask, black feet, and a black-tipped tail, Figure 19-10. It feasts on prairie dogs, which have been decimated to improve range-land for cattle. Therefore, by 1970 this apex carnivore was believed to be extinct, declining in concert with prairies dogs. A remnant population of 24 ferrets were found and captured in Wyoming in 1981-85, of these six died from infection.

6.1 Decision Analysis

This is an actual decision-making case, which we re-structure according to the rules of a rational decision-making scheme.

Problem: How should one save the black-footed ferret? The region of concern is the western United States and the time frame is several hundred years, which is the normal time frame for assessing species extinctions.

Decision makers and stakeholders: The *decision makers* were federal agencies responsible for species preservation. Stakeholders are wildlife and land management agencies, conservation groups, private and public land owners, and native American groups. For a discussion of decision makers and stakeholders see Chapter 3, section 2.3.

Decision objectives: The goal is to preserve the black-footed ferret as a species in the Unites States at as small social and economic cost to the society as possible.

Among the sub-goals were:

1. Preserve sustainable populations.
2. Maximize peoples enjoyment of the ferret.
3. Minimize restriction on land that is habitat for the ferret.
4. Minimize economic costs.

Structuring objectives and goal hierarchies are discussed further in Chapter 3, section 3.

Decision alternatives: Among the decision alternatives were:

1. Status quo, that is, hope that the ferrets would survive in their habitat in Wyoming.
2. Bring the ferret population to a captive breeding facility or to a zoo and maintain it there;
3. Bring the ferrets to a breeding facility and then reintroduce the ferrets into the wild, either in Wyoming or at other sites where they would survive.

Consequence analysis: The ferret's habitat is the grasslands of the prairie ecosystem. However, prairie ecosystems are few and separated by great expanses of cropland and built land. Thus, the habitats available for the ferret in present day US are few, but they have to be large, since the ferret is a carnivore, see Chapter 12 on ecology. It has been estimated that by establishing 10 or more separate, self-sustained wild populations, the black-footed ferret can be downgraded from endangered to threatened. One reason there has to be separate populations is that the ferret is prone to natural disasters or diseases. To support a viable subpopulation of 10 breeding ferrets, one needs prairie dog populations of 10,000 individuals. The ferret is elusive and difficult to get sight of, but it has charismatic facial characteristics. With very few individuals surviving, breeding programs that include assisted breeding are important to ensure genetic diversity. Such programs are costly.

Preferences: The U.S. Fish and Wildlife Services determined preferences for end-impacts for this species, amid strong engagement from the public that probably helped in bending the preferences towards preservation. A willingness to pay (WTP) study was designed on a severity scale from "extinction" to "endangered" to "threatened" to "abundant". The results of the consequence and preference analyses are supposed to result in a consequence table looking like Table 3-5 and a utility table with weights looking like Table 3-9.

6.2 What Happened?

Alternative 3 was chosen. A captive breeding program had by 2001 produced 3000 ferrets that are being reintroduced to the wild. One foresees that the ferrets will be declassified from "endangered" to "threatened" and subsequently to "abundant".

---

[1] Newhall (2000) p. 1181
[2] Burton et al. (1992)
[3] Bach (1998) p. 501
[4] Wackernagel et al. (1999)
[5] Haberl et al. (2001) p. 42
[6] Summarized by Rosenzweig (1999) p. 277 and (2001)
[7] Skonhoft and Solem (2001) p. 293
[8] After Constanza et al. (1997)
[9] Non-timber forest benefit: Ninan and Lakshmikanthamma (2001); mushroms: Dijk, Onguene et al. (2003)
[10] Bulte and Kooten (1999)
[11] Tomter (1994) p. 36
[12] Caribou occur on elevations from 500 to 3000m but most frequently at 1200m Johnson et al. (2004)
[13] Figure after Seip (1996)
[14] Hayes and Harestad (2000) Johnson, Seip et al. (2004)
[15] Data from Hokkaido, Japan: Stenseth and Saitoh (1998), generic values: India Rajpurohit (1999); Madhusudan and Karanth (2002), Canada: Hayes and Harestad (2000); European grasslands, Hector (1999), Morell (1997). For comparison purpose, hens under ecological production conditions count 4000 hens ha^{-1}; hens in cages have a density of 140 000 hens ha^{-1}
[16] Krajick (2001)
[17] Krebs (1995) p. 490
[18] Robinson and Bennett (1999)
[19] Data: Robinson and Bennett (1999). Primates typically monkeys (Cebus olivaceus)

---

[20] Drawn after information in Stenseth et al. (1998)
[21] Rajpurohit (1999)
[22] Science (285:1836)
[23] Cochrane et al. (1999), Saaty and Vargas: (1994) p. 1832
[24] Cochrane and Schulze (1999)
[25] Weber and Stocks (1998)
[26] Winiwarter et al. (1999)
[27] Shugart (1998) p. 307
[28] Robinson et al. (1999) p. 595
[29] Barlaz et al. (2002)
[30] Data: Robinson et al. (1999) p. 595
[31] Carlsen et al. (1993)
[32] NS (1999)
[33] Callan and Thomas (1996) p. 580
[34] McRae (1997), Rice et al. (1997)
[35] After Allen (1982)
[36] The US Forest Service's Wildlife and Fish habitat relationships (WFHR) inventory system has been developed for 800 species, Atkinson (1985)
[37] Aanderaa et al. (Eds.)(1996)
[38] Chameides et al. (1999)
[39] Pelley (2000) p. 248 A
[40] Living Forest by Anonymous (1998)
[41] Graves and Reavey (1996)
[42] After Peters and Raelson (1984)
[43] Jetz et al. (2004)
[44] Harestad and Brunell (1979)
[45] for plants the exponent is 1.0; Hedin (2006), Reich et al. (2006)
[46] Harestad and Brunell (1979) p. 391
[47] Cederlund et al. (1998)
[48] Gittenberger et al. (2006)
[49] Blackburn and Duncan (2001)
[50] Naylor et al. (2001), (1996)
[51] Skonhoft and Solem (2001)
[52] Wu et al. (2003)
[53] Laurance et al. (1997)
[54] Bruna (1999) p. 139
[55] Laurance et al. (2000)
[56] Peterson and Ryan (1986) p. 98
[57] After Peterson and Ryan (1986)
[58] Kaiser (2001)
[59] WRI (1994)
[60] See Crooks and Soulé (1999), Gascon et al. (2000), WRI (1994)
[61] See Hale et al. (2001)
[62] Burton et al. (1992)
[63] Dobson and Lyles (2000) p. 985, www.blackfootedferret.org; Black-footed Ferret Recovery Implementation- Kits & Kids

Chapter 20

AGRICULTURE AND LAND USE

Agriculture and land use with emphasis on developing countries

We describe how agriculture and other land-uses conflict with environmental goals, and examine how techniques for land-use management and social organization can support sustainable land management. We also present models of loss of soil and nutrients. We present the most important models for agriculture and land use, which are mass balance equations for nutrients and pesticides in soil, and dose/response models for soil loss.

"No man made the land. It is the original inheritance of the whole species. Its appropriation is wholly a question of general expediency When private property in land is not expedient, it is unjust." John Stuart Mill

1. INTRODUCTION

The concept *land-use* includes agriculture and all other activities that transform land from a pristine state. Forestry was discussed in Chapter 19 in connection with terrestrial environment. This chapter concentrates on agriculture and puts emphasis on developing countries, partly because this is where the biodiversity "hotspots" are, and partly because there is a wide range of land management options in these countries.[1] Management goals are challenged by food shortages, which are early warning signals of conflict and violence.[2] Another reason is that indigenous agricultural practices may act as an idea-bank for how sustainable agriculture can be developed from a technical, social and economic point of view.

The farmer, or the herder, is an important decision maker for his own farming or herding system. Here, the consequences depend to a great extent upon weather and climate changes. A typical herder in Kenya must make a decision on where to move up to ten times a year. He must consider food and water availability, the possibility of disease, encountering predators, raiders that steal animals or even kill people, and also in an informal way short- and long-term life insurance measures.[3]

1.1 Statistics

About 11% of the earth's surface is farmland.[4] People compete for land and the areas allocated to different land-use reflect the outcome of this competition.

The world's land area is about 13 thousand million ha, or 29% of the total land surface. There is about 3.5×10^9 ha of pasture available worldwide. Forest land for production of timber, pulp and woods occupy about 0.57 ha per capita, arable land about 0.25 ha per capita, and pasture land used for dairy and cattle about 0.57 ha per capita. Built-up land refers to areas for human settlement and roads and is about 0.15 ha per capita worldwide.

1.2 Natural and Man-Made Impacts

To increase agricultural production, farmers in almost all agricultural systems have to increase soil fertility, remove weeds, and apply pesticides and water, Figure 20-1. This clearly has impacts on the environment. The single most important drawback of agricultural use of land is that soils become more exposed to erosion. In addition, agriculture affects the environment in four major ways:

It entails abstraction of soil material and nutrients. Removal of soil nutrients by harvesting crops without replacing the nutrients is called **soil mining** and has several consequences:

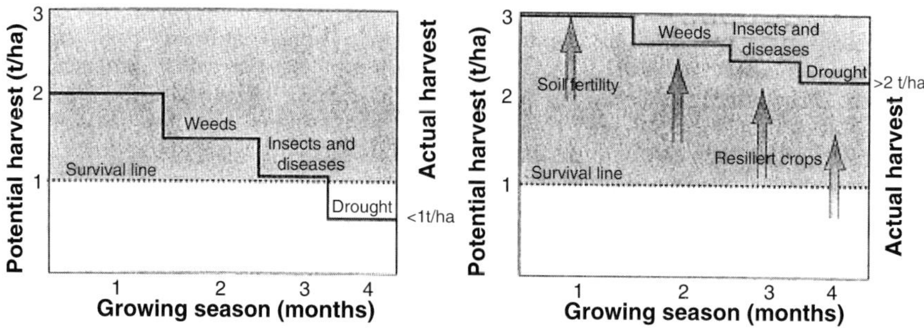

Figure 20-1. Farms in Africa that will support (right), and not support (left) a family on one hectare farmland. A farmer needs one ton of mixed crop just to survive and two tons to generate a modest income. The figure suggests which factors reduce the potential income from one hectare of land. Data: Conway and Toenniessen (2003).

- It causes pollution by pesticides and other chemicals.
- Modern agriculture affects large landscape areas by leveling fields and changing surface structure and soil structure by heavy machinery.
- It changes land areas from natural habitats to cultivated land. Man-made impacts on soil are often precursors to natural impacts. Land-use for agriculture and farming may conflict with land as protected reserves and land used for recreation.

At country level, decisions have to be made on how much and what type of land to be set aside as environmentally protected or as multiple use areas.

Sustainable agriculture can be discussed from three positions: The first position holds that even small regions, like a country, or even catchments, should be self-sustained with food and natural waste handling areas.[5] The second position holds that it is sufficient that the world as a unit is self-sustained. A third position holds that a self-sustained world, as it can be imagined today, is not any longer an option – with wild tigers roaming the wilderness. Thus, the world may be self-sustained in the future, but at a yet unknown equilibrium point.[6]

2. BACKGROUND INFORMATION

Agriculture is systematic cultivation of crops and livestock. It varies in intensity from greenhouse cultivation to ranching. Hunting and gathering are normally not considered agricultural activities. Across the world's countries, between 1% and 65% of GNP comes from agriculture.[7] Agricultural producers are often among the poorest people in the country where they live, their subsistence is more vulnerable to chance events than in any other profession.

Agricultural land has been predicted to be the limiting factor for the size of the human population. In 2005, about 14% of the world's population was chronically or acutely malnourished, mostly in Asia, but also in sub-Saharan Africa. These areas have hunger hotspots where more than 40% of children are underweight.[8]

2.1 The Value of Agriculture Services

Agricultural land provides several services to humans. In terms of energy intake, agriculture has to supply about 2600 kcal – 11,000 kJ – per day for the 6×10^9 people on the earth.[9] Since the human organism needs energy in different forms, agriculture has to deliver it in the right proportions of proteins, carbohydrates, other essential elements, and fat.[10]

Table 20-1. Summary of annual average global value of agricultural services. After Constanza et al. (1997).

Type	World area 10^6 ha	Main services US\$ ha^{-1} yr^{-1}	Value US\$ ha^{-1} yr^{-1}	Total global flow value 10^9 US\$ yr^{-1}
Crop land	1.4	Food production (54) Biological control (24) Pollination (14)	92	$1.4 \times 92 = 129$
Grass / rangeland	3.9	Waste treatment (87) Food production (67) Erosion control (29)	232	$3.9 \times 232 = 905$
Sum	5.3		324	1034

However, also the productive sea contributes to the food supply of the people of the world.

The most important services are listed in Table 20-1, which is based on a more complete list by Constanza (1997).[11] This is 16 percent of the total services offered by the environment. The values in the table are grand world averages, and currently the best grand estimates of agricultural services in economic terms. Among the countries with highest incomes from agricultural land are the Netherlands (\$1700 ha^{-1} yr^{-1}) and Japan (\$1427 ha^{-1} yr^{-1}). Farmers in the United Kingdom, France, and Hungary earn 400–600 \$ ha^{-1} yr^{-1}. Farmers in Tanzania earn about 20 \$ ha^{-1} yr^{-1}, and rural people in the Peruvian Amazon flood plains 13 \$ ha^{-1} yr^{-1}.[12]

2.2 Agricultural Negative Externalities

Farmers contribute to increasing soil erosion and run-off, which pollutes lakes and rivers. Since livestock is susceptible to predation by wild animals, many of the large predators – like tigers and wolves – are hunted almost to extinction in some agricultural regions.

Non-sustainable farming practices influence future soil quality, and may cause permanent soil loss and desertification. Typical soil loss rates in the USA are 17 ton $ha^{-1}yr^{-1}$. Loss accompanying the erosion is 75 mm of water yr^{-1}, 2 ton ha^{-1} yr^{-1}organic matter, and 15 kg $ha^{-1}yr^{-1}$ available nitrogen. This translates into an overall reduction in crop productivity of 8%, assuming that water and nutrients are not replaced. Replacement costs for nutrients and water amount to US\$ 196 $ha^{-1}yr^{-1}$. The economic cost of soil losses caused by water erosion in the USA has been estimated at $\$7.410 \times 10^9$ annually. Assuming that soil loss mainly occurs from cropland and pastures –176×10^6 ha in USA – this corresponds to \$44 ha^{-1} yr^{-1} or \$34 $capita^{-1}$ yr^{-1}, but these estimates are uncertain.[13] Costs in Sweden are summarized in Table 20-2.

Since farming needs a large share of the landscape and more water than most other uses, agriculture must take large responsibility for water degradation.

Table 20-2. Marginal costs of reducing loads of macronutrients to the coast around 1990, € kg^{-1}. Measures for reducing loads include runoff reducing measures, emission controls for industry, construction of wetlands as nutrient sinks, etc.. Data Turner et al. (1999).

Macronutrient	Agriculture	Sewage treatment plants	Atmospheric deposition	Wetlands
Nitrogen	1.5–30	0.9–9	17–1130	1.5–8
Phosphorus	14–825	2.5–12.5	–	68–2280

In some states in the USA, 75% of the land area and 80% of the total water is used for agricultural production.[14] In many areas of the world soil is in short supply, and in almost all regions water is in short supply. For pastoralists, conflicts regularly occur during movements to feed herds, and there are reports about resource availability provoking violence.[15]

2.3 Energy and Protein in Food

The size of crops and husbandry is given as **wet-weight**, fresh weight, or live weight; crops sometimes as **dry-weight**. Some typical wet-weight/dry-weight relationships as well as energy and protein contents of various crops are given in Table 20-3.

Table 20-3. Dry-weight/wet-weight ratios, energy and protein in food. Data Ruthenberg (1980:336), Lunden (1984:42) and others. 1 kcal = 4.2 kJ. ww = wet-weight, dw = dry-weight

Product	Dry-weight as % of wet-weight	Energy kcal per kg ww	Proteins gram per kgww
Soil	15–25	–	–
Wood	60	2900–4300	
Fish	30	2000–3000	170–200
Herring (1)	44	2930	170
Whale, aquatic animal	32	1440	213
Meat, cow (2)	30–35	2363–3175	145–200
Meat, camel	20	985	100
Meat, goat	20	125–724	78
Meat, pig	45	3222	159
Meat, elk	27	1290	21
Meat, bird (3)	27	1110	233
Milk, cow	–	831	34
Milk, goat	–	1200	67
Salad	4	140	10
Cereals	85	3480	114
Root crops	8–20	310	12
Legumes	26	1020	7
Grass	20	500	100

Edible part 60%; (2) Edible part: 76%; (3) Edible part 65%

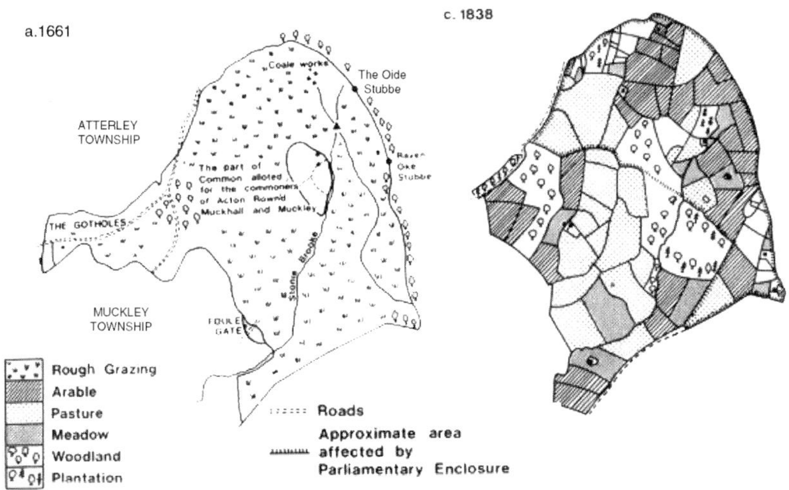

Figure 20-2. Development of land use in Shropshire, England 1661–1838. Most of the land was common land around 1661. Then the land became cultured, in part by squatters. In the lowland areas, enclosed land was hedged and fenced. From Briggs and Courtney (1994) p. 27.

2.4 Farming

The **footprint** of farms has evolved over time. The earliest farm practice depended heavily on **common land**, that is, land that was used by a pool of people under some informal rules. No one had the title to the land. However, with time, techniques and social development, part of the common land got owners and became hedged or fenced, Figure 20-2.

Farming systems may be classified according to the intensity of cultivation, and intensity may be defined in three ways:

- Intensity as external inputs to the agricultural system, like kg fertilizers per hectare.
- Intensity as yield per hectare.
- Intensity as "rotation" or shifts, between crop growth and fallow periods.

Longer fallow periods decrease erosion and increase soil quality. It is therefore considered a mitigation measure, which we discussed in section 5. The normal farm type in the developed world can be classified as intensive both in terms of external input and in terms of yield. One family would normally obtain sufficient economic support by selling products from 5–15 hectare of land. However, in spite of strong requirements for fertilizers, 40% of Ethiopian peasants had around 1995 never used fertilizers, only 1% is irrigated, and nearly 40% of the farms are less than ½ hectare in size.[16]

Table 20-4. Types of indigenous stock keeping in the tropics. Note that water available is precipitation minus evaporation. Data: Ruthenberg (1980) p. 323.

Rainfall, mm per year	Predominant type of farming	Animal mainly kept
<50	Only occasional nomadic herds	Camels
50–200	Nomadism with long migrations and supplementary arable farming	Camels
400–600	Semi-nomadism, arable farming	Cattle, goats, sheep
600–1000	Partial nomadism	Cattle
>1000	Partial nomadism and permanent stock-keeping	Cattle

In the tropics, agricultural practices are closely related to rainfall, as shown in Table 20-4. When there is less than 50 mm average rainfall per year, meaning that many years are without rain, there are only occasional visits by nomadic herds.

With rainfall in the range 50 mm to 200 mm there is usually systematic nomadic grazing with herds. Above 1000 mm permanent herding is the normal husbandry system. Tropical rainforests normally have more than 2000 mm yr^{-1} of precipitation.

Pastoralists and **ranchers** may use vast stretches of low potential and scarcely populated land. Such systems are particularly common in Latin America and in Africa. **Total nomadism** means that the animal owners do not have a permanent place of residence. They do not practice regular cultivation, and their families move with their herds.[17] A typical farm in Tanzania around 1967 is shown in Figure 20-3. In tropical Africa, permanent gardens with fruit trees and perennial crops like bananas and papayas are often found in the immediate vicinity of the living quarters. Fallows arc particularly widespread in tropical Africa, and intensive fallows may be found in concentric circles around the permanently cultivated areas. The fields are often used for growing staple food and cash crops.

2.5 Nutrient Cycles

Nutrient cycles describe how nutrients circulate between soil, water and air. It is important to understand these cycles because they explain how desired nutrient balances can be obtained within each stage, in particular in soil. The properties of the soil are important both for nutrient cycling and hydrology.

Agricultural practice influences the nutrient balance of agricultural soils, for example by application of fertilizer or manure. In areas with intensive husbandry, application of manure may be so large that runoff and leaching from the soil enriches waters above their tolerance limit.[18]

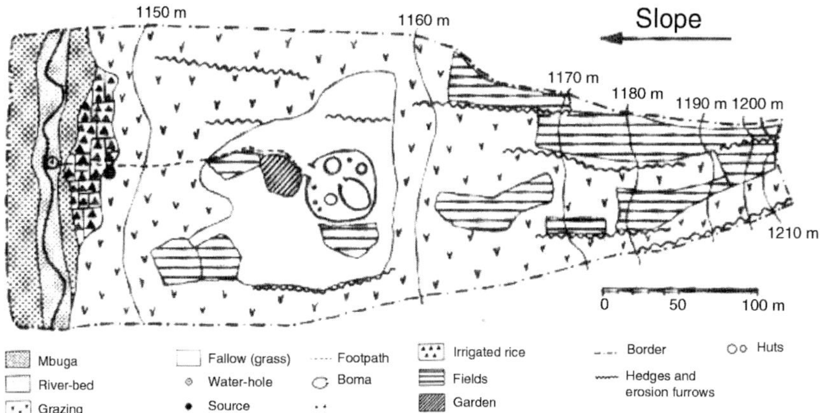

Figure 20-3. Land use in a Sukuma holding, Tanzania. Irrigated rice fields are close to the river. The gardens are just outside the living quarters (Boma). Some fields are closed to the living quarters, but most are located on the slopes uphill the quarters. Figure quoted in Ruthenberg (1980) p. 98.

The three most important cycles in relation to soil management and soil sustainability are the cycles of nitrogen N, phosphorus P, and organic matter OM. The first two cycles relate to agriculture, but also to nutrient enrichment of rivers, lakes and forests. The third also relates to the soil as a "fixed film reactor" for treatment of organic matter.

Three plant nutrients are critical for the growth of plants: nitrogen, phosphorus, and potassium K. On average, plants contain the elements N, P, and K in the ratio N/P/K = 2/0.44/0.83. Plant nutrient content may also be expressed as elemental N and the oxides of P and K. Conversion factors on a weight basis are: $P \times 2.29 = P_2O_5$; $K \times 1.2 = K_2O$. This gives $N/P_2O_5/K_2O$ = 2/1/1. Nitrogen in the forms of NO_3^- and NH_4^+ are important cell compounds and occur in proteins, chlorophyll and genes. Phosphorus, $H_2PO_4^-$, HPO_4^- occurs in genes and has a major role in plant energy transfer and protein metabolism.

The nitrogen cycle differs from that of phosphorus in at least three respects:

- Nitrogen occurs in many different oxidation states.
- Reactive nitrogen is constantly being exchanged with the large pool of atmospheric N_2.
- The bulk of nitrogen does not leak as easily as phosphorus from the soil matrix.

Thus, the phosphorus cycle is much simpler than the nitrogen cycle. Here we discuss only the nitrogen cycle.

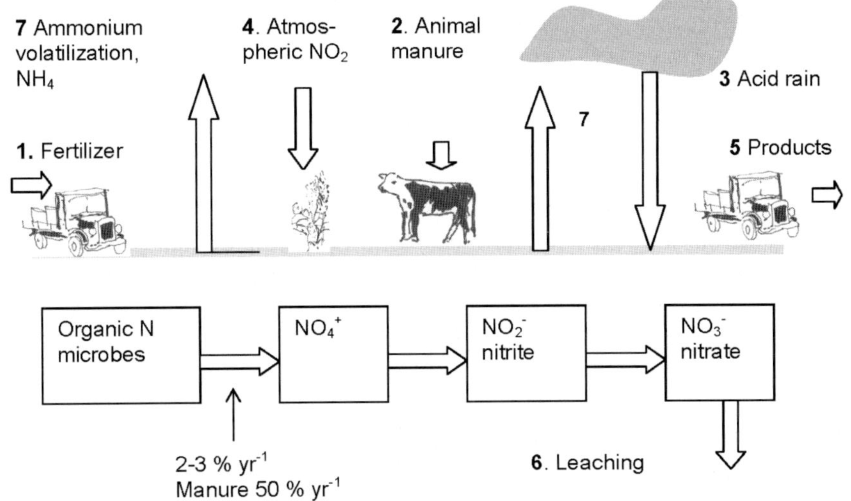

Figure 20-4. The nitrogen cycle. Main compartments and transfers.

2.5.1 Input and Output of Nitrogen

Inputs and outputs of nitrogen to and from agricultural soil may be calculated for a farm, region or country. The generic nitrogen flow of a typical farm is depicted graphically in Figure 20-4. Organic nitrogen is found in organic material. It is transformed to inorganic nitrogen by bacteria in the soil. Inorganic nitrogen has nine oxidation states and two of the forms, ammonium NH_4^+N, and nitrate NO_3^-N, can be used by plants. The numbers in the figure and in the text below refer to corresponding numbers in tables 20-5 and 20-6. Table 20-5 shows *input* quanta of nitrogen. Figure 20-4 shows how mineral fertilizers, or N-containing organic material (1) is added to agricultural land.

Table 20-5. Nitrogen budgets from Germany (1986, west), Tanzania (1996), and worldwide. Reference numbers in column 2 refer to Figure 20-4 and to figures quoted in the text. After Smaling et al. (1996) and Pimentel (1996).

Inputs ($kgha^{-1}$)	Reference	Germany	Tanzania	Worldwide [(1)]
Mineral fertilizer	1	126	2	53
Animal manure	2	47	1	246
Wet and dry deposits	3	30	3	Not available
Biological N fixation	4	12	4	66
Others		3	0	–
Total		218	10	370

(1) Recalculated after (Pimentel 1996), (2) Animal manure and human waste.

Table 20-6. Outputs.

Outputs (kg ha^{-1})	Reference	Germany	Tanzania	Worldwide
Plant and animal production	5	51	17	Not available
Leaching, runoff, erosion	6	51	15	
Gaseous losses	7	69	5	
Total		171	37	

In areas with high levels of husbandry, manure from farm livestock grazing within or outside the farm can contribute a significant portion of the organic material (2). However, the proportions of nitrogen, phosphorus and potassium in manure often differ from the requirements of the plants. Wet and dry deposits (3) also add to the nutrient input to the soil, particularly in areas with acid precipitation such as industrialized regions and areas with high animal density. Biological N-fixation (4) occurs through legume plants. Nitrogen fixation can be substantial in tropical systems and may amount to 75 kg ha^{-1} yr^{-1} of nitrogen in a secondary rain forest, 45 kg ha^{-1} yr^{-1} of nitrogen in a highland forest, and 15 kg ha^{-1} yr^{-1} of nitrogen in a savannah forest. This is more than is suggested for nitrogen fixation in agricultural soil in Table 20-2. As long as crop removal is about 50% of what is applied to cereal crops, which is the normal efficiency at economic optimum, leaching is hardly affected by fertilizer input.[19]

Sedimentation because of irrigation, natural flooding (the main source of new soil material in the old Nile delta), erosion material from upslope sites, and subsoil exploitation by trees and perennial crops may also contribute substantially to nutrient input for some farms.

Table 20-6 shows *output* quanta of nitrogen. Macronutrients are removed from the soil system by harvested or grazed crops; and from the farm system by crops and animal products such as meat and milk exported out of the system. In addition, crop residues – for example for heating – and manure may leave the farm (5), by leaking through the soil system and by erosion as overland transport of organic material (6) and by gaseous loss (7). These losses are of three types: volatilization of ammonia, NH_3^-N, to N_2, denitrification in which heterotrophic bacteria under anaerobic conditions convert NO_3^-N into N_2, and through losses resulting from fires.[20] The category "others" in Table 20-5 include wind erosion which may be substantial, for example, in many African countries.

2.5.2 Nutrient Budgets

The nitrogen budgets (kg·ha^{-1}) for Germany and Tanzania (1982 – 1984) in Table 20-5 show that the total nitrogen input in Germany (218 kg ha^{-1}) is higher than the output, whereas the input is lower than the output in Tanzania. Thus, the soil in Tanzania is depleted for nutrients, making it increasingly susceptible to erosion forces.[21] In general, fertilizer use is highest in China and Europe (126 and 110 kg ha^{-1} respectively), medium in North America, the former USSR, and India (around 50 kg ha^{-1}) and lowest in Africa and Latin America (11 and 22 kg ha^{-1} respectively.) However, application rates easily change over time as fertilizers become available.

With respect to nutrient budget, there are three agro-ecosystem classes:

- *Class I: N in > N out*: This leads to net accumulation of nutrients, which is common in countries where mineral fertilizers and imported foodstuffs are extensively used, and where atmospheric deposition is high due to air pollution. Floodplains may receive nutrients from the upper reaches of river basins.[22]
- *Class II: N in = N out*: This long-term equilibrium is typical of more or less closed systems, such as tropical rain forests and unperturbed savannas. Slight disturbances do not necessarily disrupt equilibrium, as in the case of traditional long fallow. Perennial agro-ecosystems dominated by high-input cash crops such as tea, coffee, and palm oil may be close to equilibrium when input levels are high and losses low, because a high proportion of crop residues is left.
- *Class III: N in < N out*: Many agricultural areas in Sub- Saharan Africa belong to this group. Fertilizer use is very low – less than 9 kg ha^{-1}. In Tanzania, nitrogen outputs in crop and animal products are about 17 kg ha^{-1}, whereas the sum of all inputs is about 10 kg ha^{-1}.

According to the *footprint ideology* a population should only depend upon their own region. Even small regions should be self-sustained with food and natural waste handling systems; see Figure 20-5.[23]

2.6 Soil Erosion

Over-harvesting and over-grazing lower soil fertility. Modern plant varieties also often have less ability to develop deep root systems that reduce erosion. Over-harvesting causes a chain reaction:[24]

> Over-harvesting → low soil fertility → low efficiency in use of available water → low vegetative biomass production → a decrease in biological activity as a consequence of low availability of energetic carbon required to support heterotrophic organisms → poor soil cover, surface crusting and compacting → exposure to runoff and to large surface wind velocities → soil erosion by water and wind → land degradation → creation of desert land.

Figure 20-5. Left. Agricultural areas in the USA supporting large populations far from the crop production site. Right: Traditional ploughing, Swat valley, Pakistan. A double ox plough would use 50 hours to plough 1 ha of land.[25]

Water and wind erosion are the acting degrading forces, but they are allowed to work because of man-induced effects that have occurred in many places in advance. However, as we shall see below, there are options for agricultural practice that reduce erosion. Worldwide, human-induced soil degradation has affected 24% of the inhabited land area. About 1.5×10^9 ha land is cultivated. Of this area about 12×10^6 ha or 0.8% is destroyed and abandoned every year because of non-sustainable farming practice and natural erosion.[26] On a per capita basis this is about 0.1 ha $capita^{-1}$ yr^{-1}. Water erosion is responsible for 2/3 of the erosion whereas wind erosion is responsible for 1/3. The various types of soil degradation are listed in Table 20-7.

In terms of soil weight, 7.5–9.3×10^9 tons per year is eroded worldwide by wind and water. This corresponds to about 5 ton ha^{-1} yr^{-1} on average. Expected level of erosion in the year 2040 are 45–60 10^9 ton yr^{-1}, and an 85% reduction is desired.[27]

Table 20-7. Major types of soil degradation in the world. Million hectares per year. The world's inhabited area is 8200 million hectares. After UNI/ISPRI (1991).

Cause	Light	Moderate	Strong & extreme	Total
Water erosion	343	527	224	1094
Wind erosion	269	254	26	549
Chemical degradation	93	103	43	239
-Loss in nutrients	52	63	20	135
-Salinazation	35	20	21	76
-Pollution	4	17	1	22
-Acidification	2	3	1	6
Physical degradation	44	27	12	83
Total	749	911	305	1965

In the USA soil loss is larger than the world average, approximately 17 ton ha^{-1} yr^{-1}, but these large figures are disputed.[28] Soil erosion also causes loss of phosphorus and nitrogen. Eroded soil typically contains three times more nutrients than the soil left behind.[29] The US estimate suggests that loss of phosphorus and nitrogen is 6 kg $capita^{-1}$ yr^{-1} and 23 kg $capita^{-1}$ yr^{-1} respectively. Total losses as P and N are 1.16×10^6 ton yr^{-1} and 4.65×10^6 ton yr^{-1} respectively.[30] In terms of losses per area unit, 6 kg P ha^{-1} yr^{-1} and 26 kg N ha^{-1} yr^{-1} are lost. In the section on dose-response models we will discuss two models for calculating of soil erosion.

2.7 Pesticides

Pesticides are used to protect crops from pests and pathogens. In 1996, about 3.0 million tons of pesticides were used at a cost of $\$30 \times 10^9$ every year, that is \$5 per capita.[31] Amounts per hectare depend on crop and pesticide type. Typical application rates are 5.7 kg ha^{-1} in the former USSR, 2.1 kg ha^{-1} in North America, and 0.1–0.2 kg ha^{-1} in Africa and Latin America.[32] There is a trend away from persistent pesticides (organochlorines) to more acute toxic pesticides (organophosphates), which are broken down more easily in the soil, persisting only for days or weeks in the environment, instead of years. Sublethal effects, however, are not yet well researched.

Despite pesticide use, about 40% of the world crop production is lost to pests every year. Insect pests cause about 13% crop loss, plant pathogens 12%, and weed about 10%. See Figure 20-1. Pesticides may reach unintended sites by leakage from storage tanks, loss during spraying, mixing and diluting, run-off, by aerial spraying and by uptake of pesticides through water and food. When pesticides are applied, generally 50% to 70% of the volume misses the target.[33] A severe consequence of pesticide use is that pesticides beget pesticides. The more one uses, the more must be used the next time because organisms tolerate higher doses and because species that formerly controlled weeds or insects start to disappear.

2.7.1 Damage to Humans

Side effects and misuse of pesticides cause severe damages. Worldwide cases of pesticide poisonings have been estimated at 26 million each year, with about 20,000 deaths. Problems with mood and memory and other neurological damages have also been reported, children and pregnant woman being especially vulnerable.

2.7.2 Damage to Ecosystems

Highly toxic pesticides, although rapidly degrading in the environment, may cause severe harm to birds. For instance, the pesticide *Carbofuran* has an LC_{50} for birds of less than 1 mg kg^{-1}. However, it is distributed as granules and one granule of the pesticide is enough to kill a small-sized songbird. The bird is also doomed if it eats an earthworm contaminated with carbofuran applied at the lowest rate in the cornfields.[34]

More subtle effects may also influence ecosystem functioning. Parasitic insects often have a complex search and attack behavior, which sublethal insecticide dosages may alter, disrupting biological control mechanisms. In England, organochlorine poisoning of zooplankton (Cladosera) in the 1950s and 1960s may have been responsible for the switching of lakes from a desirable stable state dominated by macrophyte-plants to an undesirable state dominated by phytoplankton. The biological mechanism behind this is supposed to be increased frequency of mismatch incidents in food supply and food requirement.[35] In the US, 0.5% of animal illness and 0.04% of all animal deaths have been attributed to pesticides.

2.8 Crop and Husbandry Yields

The most abundant crops on a world-wide basis are the cereals wheat, maize and rice. In 2005 the global production was 60–70 million tons of wet-weight for each. Cassava, sugar beets and potatoes are produced in quantities of 20–330 million tons of wet-weight. This gives 30 kg cereals and 15 kg root crops per capita per year on a world basis. An average of 20–30% of the crops is removed from the fields as usable product.

Over time, yields have increased for almost all crops. Partly because of new agricultural techniques, and also because new species or varieties of traditional crops have been developed. The potential yield for grain crops is about 20 tons of wet-weight per hectare. Scientists have also developed plants with a higher **harvest index**, which is the ratio between the harvestable weight and the total weight of the plant. The painting in Figure 20-6 shows the size of grain stalks compared to human size around the 15th century. At the beginning of the 19th century, the harvest index was typically 0.25; now many grains are approaching unstressed harvest indexes of 0.6. Globally, yields of wheat, rye, oats, barley and corn have risen by a factor of 2.5. Austrian yields rose by a factor of 4.3 from 1926 to 1995.[36]

Crop yields in tons per hectare for major crop types are: Root crops 20–60, tree crops 2–5, serials 1–8, brush crops 2. Meat yields are much lower, 0.005–0.050 ton per hectare.

Figure 20-6. "The Harvest" by P. Bruegel (section). The wheat stalks are almost as tall as the farmers at the time of Bruegel's painting (1526–1569).

Wildlife harvesting in forests is significant for some indigenous people, but the yields are much smaller compared to agriculture. For example, the average extraction area from the Amazonian flood plain forest is 113 ha per household, whereas 1–3 ha would suffice for agriculture.[37] See Table 20-8.

3. LAND-USE QUALITY STANDARDS

Soil and land-use quality standards can be defined with respect to:

- Crop yield.
- Erosion quality.
- Pesticide residues.
- Pristine habitats.

Quality standards have two measures, one defining a maximal standard, and one defining a sub-standard below which land is in non-attainable condition.

Table 20-8. Husbandry, hunting, gathering, and fishing yields. Data: Ruthenberg (1980) p. 339, Robinson et al. (1999) p. 595.

Husbandry	Standing stock	Yield, live weight kg ha^{-1} yr^{-1}	Comments
Camels	0.014 animals ha^{-1}	2.56	100–300 mm rainfall
Cattle, large	0.007 animals ha^{-1}		
Small stock	0.06 animals ha^{-1}		
Cattle	0.5–1.4 animals ha^{-1}	5.3–56	750–1000 mm rainfall
Hunting (mammals > 1 kg)	0.02–50 kg ha^{-1}	2.5	Forest >1000 mm rainfall
Gathering	1–10 kg ha^{-1}	Not available	
Fishing	50 kg ha^{-1}	5–20	Medium nutrient rich lake

Global soil maps show that poor soils dominate the tropical latitudes, whereas the most fertile soils are largely limited to certain areas of the temperate zone.[38] Standards for agricultural soils differ among soil types and crops. **Topsoil**, or plough depth soil, is the soil down to 15–20 cm depth.[39] Soil quality can be summarized by the amount of living and dead plant biomass, soil drainage conditions, soil water holding capacity, and the relative amounts of P, K, Mg, Fe, and Ca.[40]

3.1 Erosion Quality Standards

We know of no quality standards for soils with reference to resistance against erosion. However, soil on gentle slopes grown with dense vegetation is resistant to erosion. The term "**poor soil structure**" characterizes soils that are not easy to cultivate, and which have increased susceptibility to water and wind erosion. Such soils are common in many countries in Africa and in parts of South America.[41]

3.2 Pesticide and Heavy Metal Quality Standards

There are standards for pesticide concentrations in soil, crops and food, as well as for recommended maximal intake through food and water. For example, food concentration of DDT should be lower than 0.007 ppm. Daily intake should be less than 300 μg kg^{-1} bodyweight This contrasts with actual concentrations in many regions of the word. There are similar criteria for heavy metals. Further information can be found in databases on the Internet, for example the IRIS database published by the US EPA.

4. DOSE-RESPONSE FUNCTIONS

Dose-response models are of three types: i) Quantification of how agriculture detracts from a pristine state ii) Estimation of crop yields iii) Estimation of soil loss caused by agriculture. Long time harvest can be measured in several ways. One way is to calculate the annual yield in tons per hectare. Other possibilities are monetary units for a market economy (US$ ha^{-1} yr^{-1}), and nutritional yield in subsistence and non-monetary societies (kcal ha^{-1} yr^{-1}). One may also define a unit, for example the amount of food required to support 10 people for one year.[42] However, although agriculture supplies food for mankind, it subtracts land from the natural environment.

4.1 Agriculture → Loss of Pristine Land

Agriculture and land-use reduces the amount of land in a pristine ecosystem. On the average, each human being requires a footprint of 2 to 7 hectares as discussed in Chapter 13. Footprints can be calculated in different ways and under different assumptions. One example is to use the area required to feed livestock. If 1 cow requires 2 hectares of land, a "unit dose" for husbandry is 1 cow, and the response is "2 hectares of land removed from the pristine state".

4.2 Crop Yield

The ultimate response parameter in agricultural dose/response models is long-term harvest of crops or other food items, or long-term suitability of land as pristine habitat.

4.2.1 Soil Properties and Agricultural Practice → Long Term Harvest

Long-term harvest is a function of agricultural practice as well as soil properties, moisture and nutrients, which again are a function of topsoil thickness, concentration of nutrients and other essential elements, and absence of toxic elements. For example, the yield of barley increases with available rooting depth, potential summer soil moisture, available water capacity to 60 cm and accumulated temperature during the growing season. Yield decreases with a rising percentage of sand. Whereas heavy tillage may cause soil compaction and reduced crops, leaving large plant residues may reduce short-term harvest, but enhance long-term harvest because runoff and erosion are reduced.[43]

4.2.2 Precipitation → Net Primary Production

Net primary productivity of an agricultural area is closely related to precipitation when nutrients are not a limiting factor. For a general crop, the wet-weight and above ground net primary productivity, Y can be measured in grams per square meter per year (g m^{-2} yr^{-1}). The relation to evapotranspiration has been estimated for a southern Ontario agro-ecosystem:[44]

$$Y = 2400\,(1 - e^{-0.0009695\,(E-20)}),\ r^2 = 0.42,\ n \approx 40 \tag{20-1}$$

E is total evapotranspiration (mm). The formulae subtract a certain yield from a maximum yield of 24 ton ha^{-1} yr^{-1} (2400 gm^{-2} yr^{-1} $\approx$ 24 ton ha^{-1} yr^{-1}).

This occurs when evapotranspiration is higher than 2000 mm. An evapotranspiration of 20 mm gives zero yield (the exponent of the function is zero). When evapotranspiration is 600 mm, the yield is about 8 ton ha^{-1} yr^{-1}.

In general, agricultural productivity correlates with factors that vary with latitude. For example, the production of root and tubular crops, (ton wet-weight ha^{-1}) can be written:

$$c = 0.7L, \quad r^2 \approx 0.6, \quad p < 0.05 \tag{20-2}$$

L is latitude (degrees north or south, range 0°–60°). The crop, c, was within the range 0.3–40 ton wet weight ha^{-1}.[45] Different crops require different amounts of water. For example requires sugar canes 1300 mm of precipitation, whereas wheat only requires 400 mm.[46]

4.2.3 Air Pollutants → Yield Loss

Crops are reduced with elevated concentration of pollutants in the air, like SO_2, ozone, O_3, and acid depositions. Crops may either increase or decrease with deposition of macronutrients like nitrogen and phosphorus. Among the major crops that will be affected by depositions are wheat, barley, rye, oats, potato, and sugar beet. With present day concentrations of SO_2 and ozone, losses are of the order 0.05%.[47] To calculate losses one may either use the air concentration directly, or use data that summarizes the number of days during the growth period that the crop is exposed to concentrations above a certain level during the growth period.

$$I = \Sigma\,(+\,1 \times \textit{number of days}) \ \text{ if } c_{\text{ambient}} > c_{\text{crit}}; \tag{20-3}$$

4.3 Natural and Human Landscape Impacts → Erosion

The probably most common equation that describes soil erosion mechanisms, and calculates the erosion rate is The Universal Soil Loss Equation, USLE.[48]

$$A = R \times K \times LS \times C \times P \tag{20-4}$$

The legends are shown in Table 20-9 together with typical ranges for each parameter. The equation includes the effects of a series of factors affecting erosion. Note that the equation is of the multiplicative type. The factors may decrease erosion with orders of magnitude. For example, whereas bare soil gives a multiplication factor of 1.0, forest and dense shrubs give 0.01.

Table 20-9. Legends and parameter ranges for Equation 20-4.

Legend	Parameter	Typical range
A =	Computed storm loss for a given storm, ton ha^{-1}	1–20
R =	Rainfall energy factor, J m^{-2} mm hr^{-1}	400–2,000
K =	Soil erodibility factor, ton ha^{-1} per unit R	0.01–1.0
LS =	Slope length factor	0–2.0
C =	Vegetation factor	0.001–1.0
P =	Erosion control practice factor	0.01–1.0

5. MITIGATION

There are two main avenues to improve agricultural yield: to improve the properties of the plants and to improve external factors. Probably the most controversial measure is genetic modification, GM. Some people argue that GM is just an extension of spontaneous mutations; others argue that GM may cause dangerous species to develop beyond human control.[49]

There is an abundance of mitigation measures to prevent loss of agricultural soil, and among them planting of rapidly growing vegetation seems most efficient.[50]

5.1 Riparian Buffer Strips

USLE (20-3) describes soil loss from one field. However, an agricultural area will consist of a patchwork of several fields bordering each other. Use of riparian vegetation along waterways may decrease the amount of runoff, and the transport of nutrients to the water. An intercepting riparian zone will decrease transport of nutrients from agricultural land to a waterway if the runoff occurs as sheet, or overland flow. If it occurs as rill or gully erosion, the effects will be much smaller. A 15 m wide gently sloping riparian grass strip will reduce the amount of PO_4-P reaching the watershed by almost 95%, with about half of the reduction occurring within the first 5 meters. The corresponding numbers for nitrogen NO_3-N is about 10% after 5 m and 50% after 15 m. Thus, a riparian zone reduces phosphorus runoff much more efficiently than nitrogen. Increasing the width of the riparian zone to more than 20 m does not help reduce transport of nutrients appreciably.[51]

5.2 Crop Rotation

Crop rotation is an important management practice for sustaining agricultural land. Farming systems may be classified according to the intensity of rotation between cultivation and fallow periods. The longer the fallow period, the more time the system has to recover from cultivation impacts. Also, the longer time is there for ecosystem evolution through natural development and succession. The rotation factor R is defined as:

$$R = 100\frac{T_C}{T_C + T_f} \approx 100\frac{A_C}{A_C + A_f} \tag{20-5}$$

Here, T_c is cultivated time and T_f is fallow time in one cycle, and A_c and A_f is area being cultivated or lying fallow on the total area of a farm unit. The most intensely cultivated systems have $R > 66$. The term *ley* systems means farmlands where grass is planted, or establishes itself, on land that has carried crops for some years. Systems where R is between 33 and 66 are called *fallow*. The fallow may be bushes, savanna systems or grass systems. Savanna is a mixture of fire-resistant trees and grasses where grasses are ecologically dominant. The fallow may be used for pastures. *Shifting cultivation systems* are systems where the fallow period is very long. For example, two years of cultivation may be followed by 18 years of fallow in rainforests. Land may also be divided into paddocks and rotated between animal use and forage growth.

5.3 Other Measures

Terracing is an old and often used measure. See Figure 20-7. Several other soil loss abatement measures are implied in the Universal Soil Loss Equation described above. The equation only applies to one field, but use of sediment-catch basins is also useful where agricultural land is molded so that there is a slight uphill slope close to the waterways, forming a basin along it.

Figure 20-7. Terracing is an effective method for erosion control. Rice fields terracing in Western Sumatra, Indonesia.[52] Narrow lines of trees act as windbreaks to help slow down the wind and prevent sand storms and wind erosion, Agadir Morocco 2002.

6. APPLICATION: SPRAYING PESTICIDE

Side effects from trying to combat Nature

In 1962 Rachel Carson published her book *Silent Spring*, warning that pesticides may harm and kill birds. Thus, a silent spring would be the first sign that pesticide spraying jeopardizes the environment.

6.1 Decision Analysis

Below we suggest a framework for a rational decision analysis.

Decision problem: The *decision problem* is how to strike the optimal balance between use of pesticides and environmental effects. The positive effects last one year, whereas the negative effects may last dozens of years. The problem is common all over the world.

Decisionmakers and stakeholders: The *decision maker* is usually a farmer or landowner. Stakeholders are farmers who apply pesticides, consumers that buy and eat farm products, and conservationists who want to maintain healthy ecosystems.

Objectives: The overall *goal* is to obtain maximum quality and quantity of crop at the least social and economic cost. The government will be strongly involved in regulating pesticide management, and it is reasonable to assume that governmental objectives have long-term horizon – at least several decades. Sub-goals are:

1. Maximize consumer satisfaction with the product (crop quality).
2. Minimize damage to the environment.
3. Minimize health impact on people living in the neighborhood of the farm.
4. Maximize equity; winners and losers should be the same people.
5. Maximize profit for farmers.
6. Maximize long term profit for the country.

Alternatives: In general, we have three main *alternatives*:

1. Enforce no spraying of pesticide.
2. No initiative from the government, spraying is uncritical – sounds weird, but is unfortunately common.
3. Enforce best management practice to pesticide spraying.

In future decision situations one needs to develop alternative 3, and probably split it on several sub-alternatives.

Consequence analysis: Increased use of pesticides will usually increase crop quality and quantity, but pesticide residues make the product less attractive. Pesticide spraying has many side effects on the environment, but the effects on birds are probably most important. In some cases, spraying has decimated the number of birds and threatened species existence. In many countries, handling of pesticides is rather casual, causing illness and even death to farm workers. People in the neighborhood of the farm may be exposed to the above critical concentrations of pesticide residues both in water and food. For the government, to change pesticide practice may be costly. They have to educate and fund agricultural advisors, to control enforcements, and to get acceptance for their views both among farmers and among consumers. To get acceptance for no spraying – because pesticides beget pesticides – may be very difficult. In Chapter 21, section 7, we discuss the related problem of birds susceptibility to oil pollution.

Preferences: Preferences could be expressed both by farmers, producers of pesticides, as well as agricultural experts and Nature conservation groups. The result could be several sets of importance weights, like the weight in Table 3-8 or Table 7-4. To elicit weights, methods outlined in Chapter 7 could be used. If the analysis results in the same ranking of decision alternatives with all sets of weights, then the ranking is probably robust.

6.2 What Happens?

Pesticide management varies dramatically among sites; everything from alternative 1 to 3 is seen in practice. Pesticides are often given as aid to developing countries, but without adequate advice on its use. In the UK, bird species extinction and human nuisances were identified as the two major negative effects of pesticide spraying and it was found that people were willing to add about £13 to the current pesticide price of £20 per kg to mitigate the side effects.[53]

Whether £33 per kg is the "real" social cost of pesticides depends on whether people actually included all the detrimental effects when they responded.

[1] Hotspots: Myers et al. (2000) , Balmford et al. (2001), Balmford et al. (2001)
[2] Prendergast (1997) p. 165, McCabe (2004)
[3] McCabe (2004)
[4] Lægreid et al. (1999) p. 35
[5] Wackernagel et al. (1999)
[6] Marsh (1857)
[7] Huston (1993)
[8] Sanchez and Swaminathan (2005)
[9] That is 6.5 10^{16} kJ year^{-1} or 65 PJ or 18 TWh
[10] Warren-Rhodes and Koenig (2001)
[11] Constanza (1997)
[12] Huston (1993) Gram et al. (2001)
[13] Pimentel et al. (1995) p. 1119, Pimentel and Skidmore (1999) p. 1477, Trimble (1999)
[14] Cooper and Lipe (1992)
[15] McCabe (2004)
[16] Prendergast (1997) p. 166
[17] Ruthenberg (1980) p. 322
[18] Lægreid et al. (1999)
[19] Kirchmann et al. (2002)
[20] McEachern et al. (2000)
[21] Smaling et al. (1996)
[22] Lægreid et al. (1999)
[23] See Wackernagel et al. (1999)
[24] Modified after Syers et al. (1996)
[25] Lægreid et al. (1999) p. 44, 197
[26] Pimentel (1996)
[27] Rennings and Wiggering (1997)
[28] Pimentel and Skidmore (1999) p. 1477, (1999) p. 1478:1478
[29] Pimentel et al. (1995)
[30] Lægreid et al. (1999) p. 235, Cooper and Lipe (1992) p. 221, Pimentel and Skidmore (1999) p. 1477
[31] Personal communication, David Pimentel
[32] Finkelman (1996) p. 12, Pimentel (1996) p. 87, 94
[33] Pimentel (1996)
[34] Hui et al. (2003) p. 79
[35] Moss (1990) p. 374
[36] Haberl et al. (2001) p. 38
[37] Gram et al. (2001)
[38] Huston (1993)
[39] Montagnini et al. (1997), Lægreid et al. (1999) p. 97
[40] Summarized by Huston (1993)
[41] Ruthenberg (1980) p. 23:23 and Fearnside (1996)
[42] Monetary units: Huston (1993) p. 1677
[43] Briggs et al. (1994) p. 241,44
[44] Briggs et al. (1994) p. 242-3: 242-243

[45] Huston (1993) p. 1678
[46] Charles Howe, Colorado University, personal communication
[47] Krewitt et al. (1998) p. 178
[48] Wischmeier et al. (1978), Morgan (1995)
[49] Fedorff (2005)
[50] Mann et al. (2000)
[51] Vought et al. (1994)
[52] Lægreid et al. (1999) p. 108
[53] Mourato et al. (2000)

Chapter 21

AIR

Emission, dispersion and effects of airborne pollution

We describe the most important air pollutants, and their impact pathway from emission and dispersion to effects on ecosystems and human health. We look more closely at three models of local dispersion of pollutants. Finally, we give a short description of bird communities and describe a succeptibility index for pollution.

"Don't trust an air you can't see." Woody Allen

1. INTRODUCTION

The air is our largest "global **commons**". A common is a property that is shared by a population, but without legal ownership rights. Air cannot be divided into slots and assigned private property rights. **Air pollution** is the presence in the outdoor atmosphere of one or more contaminants – pollutants – in sufficient concentration and duration to make it harmful to humans, plants, or animal life. Air is also a vehicle for substances that expose humans to bad smells and winter fog, and it transmits noise. Air transports corrosive gases that slowly damage buildings, monuments, and cultural artifacts, causing serious concern about the future of world heritage monuments.

Cars, factories, and power plants are the most important sources of air pollutants, and city people are most exposed to it since the major sources are near urban areas. It is estimated that 81 million Latin Americans live in cities that exceed WHO-recommended thresholds for air pollution. This is 26% of the total urban population in the region.[1]

Natural sources also discharge toxic elements. For example, forests are a major source (34%) of non-methane volatile organic carbon, NMVOC, and animal husbandry is the major source (64%) of ammonia, NH_3.

Measures to reduce car traffic, the need for air conditioning, and discharges of pollution into air are being implemented on a continuous basis to improve air quality. However, some measures are expensive, and some measures – like reduction in car use – unpopular.

2. BACKGROUND INFORMATION

The year 1990 is used as a reference year for defining emission targets of air pollution and climate gas in the Kyoto protocol. The situation in Europe in 1990 can be characterized by the following observations:

- The critical level of no observable effect (NOEL) of SO_2 was exceeded in one third of Europe.
- NOEL of NO_2 was exceeded in one third of the whole area of Germany.
- NOEL of monthly O_3 average was exceeded in the whole of Germany.[2]
- Concentrations in the Nordic countries were normally well below NOEL, both with respect to human health, flora and fauna.

However, emissions and concentrations have been reduced considerably since 1990. By 2005 emissions of SO_2 have been reduced by about 50% and emissions of NO_2 by about 30%. The fate of air pollutants is shown schematically in Figure 21-1.

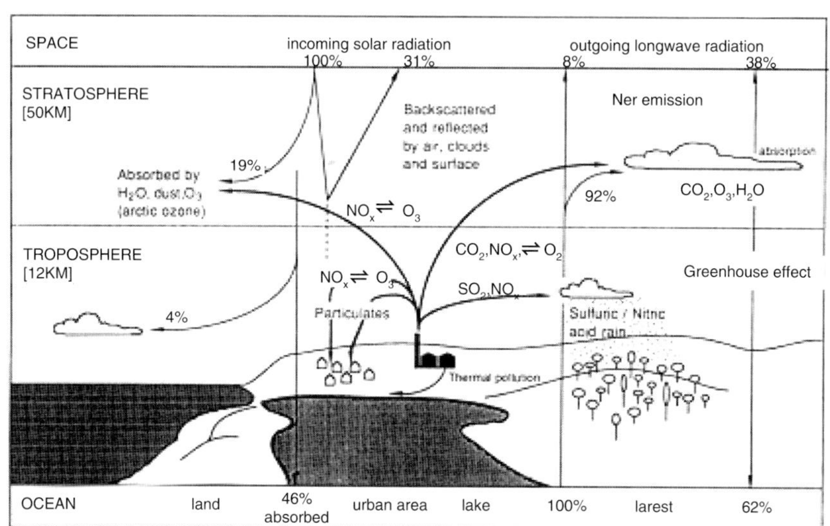

Figure 21-1. Emissions of air pollutants. Solar fluxes including ultraviolet radiation (thin lines) on the left, long wave fluxes (thermal IR) on the right. NO_x (thick lines are transport) may form ozone, O_3, in the troposphere and stratosphere. NO_x may also indirectly contribute to destruction of arctic ozone, which shields against ultraviolet radiation. CO_2 contributes to the greenhouse effect; SO_2 and NO_x contribute to acid rain. Data Seip et al. (1991)

2.1 Air Pollution, Units of Measurement

At least four different units of measurement are used for air pollutant concentration:

- ppm = parts per million: 1: 10^6 (volume or mass).
- ppb = parts per billion: 1:10^9 (volume or mass).
- Milligram per cubic meter: mg m^{-3}, or microgram per cubic meter: μg m^{-3}.
- Milligram per normal cubic meter: mg Nm^{-3} (Nm^3 = one m^3 normal dry air at 0°C and pressure 1013 mbar).

One mole of gas occupies 22.4 liter, or 22.4×10^{-3} m^3 at standard temperature and pressure. Thus: 30 mgCO $m^{-3} = 30 \times 10^{-3} \times 1/(12+16) \times 22.4 \times 10^{-3}$ $m^3 \times$ mole$^{-1} \times$ g mole^{-1} = 24 10^{-6} = 24 ppm.

The most important atomic masses are C:12, O:16, N:14, H:1, S :32. This gives the following conversion factors:

1 ppm CO = 1.16 mg m^{-3}
1 ppb SO_x = 2.66 μg m^{-3}
1 ppb NO_x = 1.91 μg m^{-3}
1 ppb O_3 = 2.00 μg m^{-3}

2.2 Air Pollutants

The following chemical components are among the most important air pollutants: carbon monoxide CO, nitrogen oxides NO_x, methane CH_4, sulphuroxides SO_x. Other important pollutants are volatile organic compounds VOC, particulates, dust PM_{10} (particulate material less than 10 microns in diameter), ground level ozone, O_3, photochemical oxidants, fluorides, and heavy metals. Carbon dioxide, CO_2, is not a pollutant in the conventional sense, but it is the most important contributor to the greenhouse effect. Six attributes are particularly relevant for predicting the effects of these pollutants: background concentration, residence time in air, major sources and sinks, toxic levels, and typical exposure times for resources that are susceptible to them.

Sulfur dioxide: SO_2 is a constituent of fossil fuels. It oxidizes in the atmosphere to sulfate. Inorganic and organic radicals enhance the oxidation. These occur naturally in the atmosphere, but their concentrations increase substantially with photochemical activity. SO_2 also transforms from gas to aerosol, which is suspended liquid particles. The most important sources of SO_2 are combustion and metal production. High levels of sulfur increase health risks to bronchi and lungs. Sulfur also contributes to acidification of water and corrosion of materials. From a reference emission of 100 in the year 1990 the EU goal for SO_2 emissions was 60 in 2000, and is now set to 16 in 2010. SO_2 still poses severe problems in developing countries.

Nitrogen: N – constituting 78% of the air volume – is fixed, mainly by oxidation, and a building block of organic matter. It is released into the atmosphere in the form of oxides from decaying organic matter, and as ammonia NH_3 from animal urine and fertilizers. There are three nitrogen oxides NO_x, NO and NO_2; 90% of NO_x in emissions consists of NO. It impacts the ozone layer in the troposphere. Fossil fuel contains NO_x, but NO_x is also extracted from air during combustion at very high temperatures with the presence of oxygen. All nitrogen in the atmosphere is actually of biological origin, but the average recycling time is very high, several million years.[3] Other sources of nitrogen are agriculture, production of fertilizers, industrial combustion and automobiles. NH_3 neutralizes precipitation, but contributes to acidification of soil and water through nitrification. NO_2 contributes to the greenhouse effect. NO_x increases health risk to bronchi and lungs and contributes to urban ozone O_3. The largest nitrogen sink is the atmosphere. With a per capita reference emission of 100 kg NO_x in 1990 (per capita emission is of the order 100 kg for NO_x and SO_2), the EU goal for NO_x emissions was 70 units NO_x in 2000, and is now set at 45 units for 2010.

Oxygen: O – constituting 21% of the air volume – is utilized by vegetation and animals, which re-emit it as CO_2 or water vapor H_2O. *Tropospheric ozone* O_3, or urban ozone, is a strong respiratory irritant and plant pathogen. Ozone in the upper atmosphere is beneficial because it blocks the sun's harmful ultraviolet rays. It is formed by photolysis of NO_2. Photolysis and oxidation of reactive non-methane organic compounds NMOC, provide a major pathway leading to oxidation of NO without destroying ozone. There is general consensus that controlling NMOC emissions in urban areas will reduce ozone, whereas controlling NO_x may – or may not – reduce ozone, depending upon the NMOC/NO_x ratio.

Carbon monoxide: CO is a colorless, odorless and tasteless gas. It is the product of incomplete combustion. It is deadly poisonous since it can replace oxygen in the blood.

Carbon dioxide: CO_2 – constituting 0.004% of the air volume – is a greenhouse gas with small annual and local variations and increasing from about 315 ppm in 1950 to 360 ppm in 1996. Recycling time is roughly a century, but there is also a continuous transfer from the atmosphere to marine life, and slow sequestering in coral reefs on the seabed, forests and the upper soil layer. The main source of CO_2 is combustion of fossil fuel, and deforestation. Trees sequester carbon. The oceans contain about 50 times more CO_2 than the atmosphere. Increased concentrations of CO_2 increase the greenhouse effects.

Water vapor: H_2O is present in the atmosphere in variable amounts, from 0.05% to 5.0%, depending on temperature. It is recycled rapidly by storms.

Methane: CH_4 is a greenhouse gas. It is present in about 1.8 ppm in the atmosphere, is central to atmospheric oxidation chemistry, and is the ultimate source of stratospheric water vapor. The most important sources of methane are agriculture, waste depots, and production and use of fossil fuel. Methane contributes to the formation of urban ozone.

Hydrogen sulfide: H_2S is very toxic, but smells so bad even well below toxic concentrations that it is not a threat to humans.

Lead: Pb is poisonous. It used to be added to petrol, but it is now less of a threat, at least in countries where leaded fuel is of little use.

Volatile organic compounds: *VOC* is emitted through oil and gas-production. Other sources are road traffic and the use of solvents. VOC may contain carcinogenic compounds, and contributes to the formation of urban ozone. With a reference emission of 100 units VOC per capita in 1990, the EU goal for NMVOC emissions was 70 units in 1999.

Particulate matter: PM is usually considered separately as PM_{10} and $PM_{2.5}$, where PM_{10} are particles less than 10 μm in diameter, and $PM_{2.5}$ less than 2.5 μm in diameter. The smallest particles pose the largest danger to human health. About 80% of ash from oil and coal is particulates, mostly large particles. The most important sources of particulates close to the earth surface are road traffic and wood burning. Particulates increase health risks to bronchus and lungs. Studies suggest that every 10-microgram increase per cubic meter of fine particles in the air increases the risk of death from cardiopulmonary disease by 6%, and 8% from lung cancer.[4]

2.3 Air Pollution Costs

Costs of air pollution include reduced health and welfare, deterioration of buildings, monuments and materials, and damage to ecosystems and natural resources. If abatement costs were lower, it would pay off to reduce emissions, but PM_{10} and chemical compounds containing SO – SO_x – are among the most expensive air pollutants to eliminate, costing about €5 per kg.

Overall economic damages of air pollutants have been estimated for several countries. Damages to materials, agriculture and forestry have been estimated for Germany, and general damages for The Netherlands. Estimates of total damages, and damages split on subcategories show that total air pollution costs per capita and year in Germany is in the range US$320–$360, and in the Netherlands US$30–$50. Loss of welfare is the largest cost item in Germany, about US$260 capita^{-1} yr^{-1}, whereas recreation, forestry loss, material damages and respiratory diseases, all are in the range US$13–$30.[5] Table 21-1 shows estimated costs of one unit of emission of five important pollutants. Such data should be used with great care, however, since uncertainties and regional variation are considerable.

Table 21-1. Social damage estimates (US$1992) of emissions to air. US$ ton^{-1}. Data: summary by Mathews and Lave (2000).

Pollutant	No. of studies	min	median	mean	max
Carbon monoxide (CO)	2	1	520	520	1050
Nitrogen oxides (NO_x)	9	20	1060	2800	9500
Sulfur oxides (SO_2)	10	770	1800	2000	4700
Particulate matter (PM_{10})	12	950	2800	4300	16,200
Volatile organic compounds (VOC)	5	160	1400	1600	4400

2.4 Emission From Transportation

Transportation is a major source of emission of pollutants to the atmosphere. Table 21-2 compares different carriers.

Table 21-2. Air pollutant emission factors from transportation in gram per passenger kilometer if not otherwise stated. Data compiled by Canter (1996:159), MacLean and Lave (1998), cars in Mexico and USA: Schifter et al. (2000).

Transport mode	Carbon dioxide	Organic compounds	Carbon monoxide	Nitrogen oxides	Sulfur dioxide
Truck	437	2.0	17.0	1.3	0.2
Car (<1990)	316	1.6	12.7	1.0	0.1
Car (1990–2000)			1–3.2	0.2–0.9	
Car production, kg	26,000		11	33	6
Car life-time use, kg	57,000		2460	215	7
Bus	110.0	0.2	0.8	1.1	–
Rail	121	0.7	0.4	0.6	0.3
Ship, kg yr^{-1}	–			30,000	42,000
Airplane	161	0.3	0.3	0.7	0.1

2.5 Emission From Energy Production

Table 21-3. Distribution of emissions from different fuels in electricity production. % Weight.

Pollutant	Wood (peat)	Coal	Oil	Natural gas
Carbon,%	27. 3	77	82.5	CO_2: 0.1%
Oxygen, %	16.6	2.1	–	
Hydrogen, %	2.6	3.0	11.3	
Nitrogen, %	–	1.3	–	N_2: 0.6%
Sulfur, %	0.3	1.0	2.8	0
Particulates, %	2	6.2	0.03	0
Heavy metals, %	–	–	–	–
Moisture, %	50	8	0.3	–
Calorific value MJ kg^{-1}	12–18.5	8.3–35.2	40.5–44	40–46
Efficiency %	20–33	20–40	40	40–60

Table 21-4. Annual emissions from a 200MWe power plant based on coal, natural gas (conventional plant) or fuel cells, and 7000 operating hours per year. A coprocessor is assumed to be used for heat with Solid Oxide Fuel Cell technology (SOFC). Based on Seip et al. (1991). MWe is Mega Watt energy in the form of electricity.

Chemical	Unit	Coal	Natural gas	Fuel cell SOFC
Efficiency	%	20	40	60
Fuel required	10^9MJ yr^{-1}	25.2	12.6	8.4
Fuel required	10^6 ton yr^{-1}	1.26	0.32	0.21
CO_2	Ton yr^{-1}	1400 10^3	695	460
NO_x	Ton yr^{-1}	1500	325	0.8
SO_x	Ton yr^{-1}	5300	10	7
SOx	g sec^{-1}	210	0.4	0.28
Particulates	Ton yr^{-1}	500	3	2
Heavy metals	Ton yr^{-1}	3	0	0
Heath loss	GWh yr^{-1}	5600	2100	930
Calorific value	MJ kg^{-1}	20	40	40

Energy production is another major source of emission of pollutants to the atmosphere. Table 21-3 shows percent distribution of pollutants.

Wood, coal, oil, and gas have different contents of chemical elements, as well as different calorific values. Emissions from natural gas depend on the technology used to convert gas to electricity. Table 21-4 shows total annual emissions from a power plant with a given production, depending on the source of energy.

2.6 Emissions Per Capita

As a first approximation, pollution is proportional to the size of the population, but political and technological factors also play a significant role. This can be demonstrated with a regression model based on SO_2 emission data[6] for European countries in 1990:

$$SO_2 = 20P + 852E - 65, n = 27, r^2 = 0.54, p < 0.0001 \quad (21\text{-}1)$$

SO_2 is measured in 10^6 kg, and population P in million people. E is a dummy variable, where E = 1.0 for earlier East Block countries, 0 for the others. p = 0.0006 for both factors. Neither GNP nor GNP per capita contributed significantly. The former Soviet Union was not included.

Emissions per capita for a selection of chemicals are listed in Table 21-5.

2.7 Corrosion

Corrosion of materials affects buildings and outdoor sculptures, bridges and other installations made of corrosive materials.

Table 21-5. Discharges per capita of selected air pollutants in 1995 five years after 1990, which frequently is used as a reference year for emissions. Data for USA (U), The Netherlands (Ne), Norway (N), Ireland (I), and West Germany (WG). The acronyms are used in the second column to indicate which country has the largest emissions.

Pollutant	Emission kg person^{-1} year^{-1}	Main source
CO	80–244 (U)	Mobile sources (70%)
CO_2	5000–11,300 (WG)	Fossil combustion, forest clearing
SO_2	15–90 (U)	Fossil combustion
NO_2	37–80 (U)	
CH_4	114 (N)	Inert, anaerobic decay of organic material
Pb	0.02–0.048 (I)	Industry, traffic
PM_{10}	6.2 (N)	
VOC	30–100 (WG)	Vehicles, solvent industry
HC	30–80 (WG)	

Concrete is the most common building material, and least susceptible. Galvanized steel and marble are the most susceptible materials. One usually distinguishes between damage to general constructions and damage to cultural or historic monuments, where serious deterioration occurs many places in the world.[7] Corrosion is usually a slow process. Visible decay may require as much as 70 years at an air concentration of 1.4 ppm SO_2.[8] Air pollution in large cities, like Rome, varies with season and day. It is mainly due to traffic, but also to domestic heating in the winter season.

Figure 21-2. The Ara Pacis monument in Rome. The picture shows the bas-relief blackened by soot in 1995 after being cleaned in 1984.[9]

This is particularly problematic because of greater atmospheric stability with reduced diffusion potential.

The daily cycles of NO_x and SO_2 show a bi-modal distribution, with the first maximum peaking at about 9 in the morning and the second at 19 o'clock in the evening. As an illustration, Figure 21-2 shows damage to the Ara Pacis monument in Rome. The Ara Pacis is a sacrificial altar inside a marble enclosure. It dates back to the Roman emperor August (9 BC). The inside surfaces are decorated with ox skulls and garland motives. In spite of it being sheltered by a glass house, the white marble monument is visibly blackened after a period of only two years.[10]

3. QUALITY STANDARDS FOR AIR POLLUTION

Quality control with regard to air pollutants can be performed at any of the four stages in the impact pathway:

1. The source or *End-of-pipe* (kg per hour).
2. Ambient concentration in the air ($\mu g\ m^{-3}$).
3. Concentration in the target medium (mg per g in the soil).
4. End impacts (impacts on humans, ecosystems, buildings, etc.).

Since this chapter concentrates on air pollution, we shall discuss stage 1, 2 and 4 below.

3.1 End-of-Pipe Standards

As a rule of thumb, concentrations at the discharge points are restricted to 30 times the ambient concentration. The German Air Standards separates substances into three classes, and sets different concentration limits as shown in Table 21-6. The mass rate ($g\ hr^{-1}$) is supposed to be measured during the most unfavorable hours.

Table 21-6. German end-of-pipe pollutant emission classes and concentration limits. (c) means carcinogen. Source http://www.umweltbundesamt.de/luft/emissionen/emissionen.htm

Class	Emission volume, $g\ hr^{-1}$	Concentration in emission $mg\ m^{-3}$	Pollutant example
I	0.5	0.1	Asbestos (c), Chloromethane
II	5	1	Arsenic (c), Chromium (c), Cobalt (c), Naphthalene
III	25	5	Benzene (c), Vinyl chloride (c), Parafin HC

3.2 Ambient Conditions

Much work has been done to establish guideline values or "**critical levels**" for air pollutants, which can be tolerated without harm. Tables 21-7 and 21-8 provide information on pollutant background values, that is, normal concentrations of the pollutants found in presumably uncontaminated areas, and guidelines for **safe concentrations**, respectively. National ambient air quality standards may differ among countries. However, many quality standards are based on standards set by WHO.

Table 21-7. Background values for pollutants. Normal concentrations in unpolluted air.

Pollutant	Background	Residence time	Pollutant characteristic
CO	120 ppb	30–60 days	Carbon monoxide, incomplete combustion, greenhouse effect
CO_2	315,000–380,000 ppb [(1)]	5–15 years	Carbon dioxide, combustion, greenhouse gas
SO_2	0.2–10 ppb	1–5 days	Sulfur dioxide, colorless, combustion, damages humans, flora and fauna, and causes material corrosion
NO	0.01–1.0 ppb	1–5 days	Nitrogen monoxide
NO_2	0.1–1.0 ppb	1–5 days	Brown-orange, combustion, acid rain
O_3	20–80 ppb troposphere average	3–5 weeks	Ozone, highly reactive, damages humans, flora, fauna and materials
NH_3	1–6 ppb	20 days	Nitrate
CH_4	1500 ppb	8–10 days	Methane
H_2S	0.2 ppb	–	Hydrogen sulfide, anaerobe degradation of organic materials, toxic
PM_{10}	<100 gm^{-3}	–	Particulate matter <10 microns diameter, black smoke [(2)]
Pb	$5\ 10^{-3}$ mgm^{-3}	–	Lead, heavy metal, leaded petrol, damages humans and fauna

[(1)] Fossil fuel combustion has increased the concentration of CO_2 from the lower number in 1960 to the higher number in 2005. Residence time is problematic in this context. Turnover time is estimated at 5 years. The time a pulse of CO_2 has been adjusted to $1/e \approx 1/\ 2.7$ has been estimated at 5–200 yrs by the international Panel of Climate Change, IPCC.

[(2)] Particles greater than about 10 mm will settle within a few km from the source, particles in the range 0.1–1 mm may settle more than 1000 km from the source.

3.3 End-Impact Criteria

The **end-impacts** of air pollution are damages to fundamental values as appreciated by humans. The values are systematized in Figure 21-3. They can be organized into four main categories: Damage to humans, damage to buildings and monuments, deterioration of local weather conditions, and a category that includes winter fog and clouds from discharge of water vapor. Air also transports pollutants to soil and water, which affects flora and fauna.

Table 21-8. National ambient air quality standards. Data: WHO air quality criteria.[11]

Pollutant	End impact	Type of average	Critical concentration
CO	Health	8-hour	10 mg m^{-3}
		1-hour	30 mg m^{-3}
CO_2	Climate	–	none
SO_2	Health	Annual mean	80 µg m^{-3}
SO_2	Corrosion	24h mean	3600µg μ^{-3}
	Health	24 - hour	125 µg m^{-3}
	Corrosion	24 hr	3600 µg m^{-3}
NO	–	–	–
NO_2	Health	1 yr	40 µg m^{-3}
O_3	Health	Maximum daily hourly average	235 µg m^{-3}
	Vegetation	24 hr	65 µg m^{-3}
Fluorides	Health (1)	24 hr	25 µg m^{-3}
	Vegetation	24 hr	1 µg m^{-3}
Pb		Maximum quarterly average	1.5 1 µg m^{-3}
PM_{10}		Annual mean	50 µg m^{-3} (2)
		24 -hour	150 µg m^{-3}

(1) Grass should contain less than 30 mg fluoride per kg dry matter, or a fluoride concentration below 0.4 µgF m^{-3}. (2) Should probably be zero, is steadily adjusted to smaller concentrations

This fourth category is discussed in other chapters. Figure 21-3 shows a goal hierarchy for air pollution management regarding the first three categories. For air pollution quality standards there are goals for the optimum quality and for "safe" quality standards.

The end impacts can be measured in several ways:

Lethal effects from toxic substances in the air: Excess angina attacks per year can be used as a proxy.

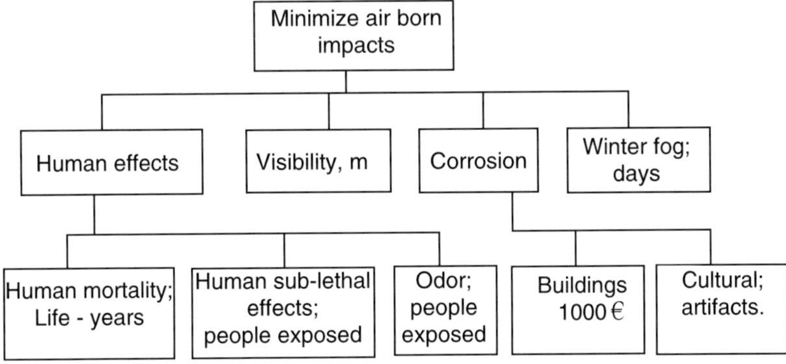

Figure 21-3. A goal hierarchy for reduction of air pollutants.

- Sublethal effects: Increased coughing days in children, increased prevalence of colds.
- Exposure to noise and odors: Number of people exposed above a critical level.
- Corrosion damage to buildings: Annual excess maintenance or rebuilding costs in Euros per year.
- Damages to cultural monuments: Number of monuments visibly affected.
- Local climate change represents the nuisance effects of foggy winter days and can be measured as days with excess fog. Visibility reduction is both a nuisance and may be dangerous for road traffic.

4. TRANSPORT AND DISPERSION MODELS

In this section we consider only a part of the impact pathway, namely models of transport and dispersion of airborne pollution. End impacts on ecosystems and health are covered in Chapter 14 on toxicology. There are long and short distance models of transportation. Since long distance models are complicated computer models and outside the scope of this book, we only discuss a few rules-of-thumb. We then present three simple models of short distance dispersion, and conclude the section by discussing impacts of airborne pollution on corrosion and visibility.

4.1 Long Distance Transport

Air pollution is not only a local problem; it is also transported across national borders. In the 19th century only a fool would paint his white house on the southern coast of Norway while the wind was blowing southwesterly; the house would become grey with air-born coal particles from industrial Britain. This situation is much better nowadays, but we still need to know what distance a pollutant will be transported.

A **long distance transport** of 4000 km is a normal range for pollutants like SO_x and NO_x.[12] Models show that with an average wind of 7 m s^{-1} – common in the US – such transport would take approximately 7–10 days. For a chemical with a half-life in the air of two days, only $(½)^4 = {}^1/_{16}$ would remain after eight days. Equation (21-2) predicts approximately the transport distance D (km) of a chemical with a half-life in air $T_{1/2}$ (hours) before the concentration has become halved.[13]

$$\log D = \log T_{1/2} + 2.3 \qquad (21\text{-}2)$$

Thus, chemicals with 100 hours half-life can travel about 20,000 km.

Involuntary importation of airborne pollutants is a source of international tensions. To quantify the problem, estimates have been made of the fraction of pollutants that is of indigenous origin in 12 European countries. It ranges from 5% to 87% for sulfur, from 6% to 46% for oxidized of nitrogen, and from 28% to 86% for reduced nitrogen.[14] Dust from volcanoes may reside in the atmosphere for five years.[15]

4.2 Local Dispersion

Local transport and dilution of airborne pollutants can be modeled mathematically, and monitored through observations of actual concentrations. Monitored data tend to be more reliable than model estimates, but they are often unavailable. Pollutants from industry and power plants are usually emitted from chimneys, and disperse, dilute and precipitate to the ground in a more or less oval shape around the chimney with the long axis of the oval in the prevailing wind direction. A central concept is the **dispersion coefficient**, which predicts how much a pollutant will spread, depending on meteorological conditions. They are estimated from empirical observations.

4.2.1 Atmospheric Stability

Air turbulence determines how airborne chemicals spread; the more turbulent the air, the more the pollutant will spread.

Turbulence is related to the wind speed we find 10 to 100 m above ground and the temperature gradient, which is the temperature change with thc height above ground. Air moving upwards becomes about 1°C cooler per 100 m if there is no heat exchange. The cooling effect is due to reduced barometric pressure. The cooling effect is called **dry adiabatic decrease** and the curve describing it is called “dry adiabatic lapse rate” DALR. If the temperature change deviates from DALR, the air will be unstable and create turbulence. How often this happens depends also on the pattern of solar radiation and nocturnal clouds, and sites can be classified as more or less stable according to these parameters.

Table 21-9. Pasquill atmospheric stability classes, A, B, C and D. A is the most unstable. Data Gifford (1976), Pasquill (1961).

Surface wind speed $m\ s^{-1}$ at 10 m height	Solar radiation, day			Cloud cover, night Percent	
	Strong	Moderate	Slight	>50	>50
>2	A	A–B	B	–	–
2–3	A–B	B	C	E	F
3–5	B	B–C	C	D	E
5–6	C	C–D	D	D	D
>6	C	D	D	D	D

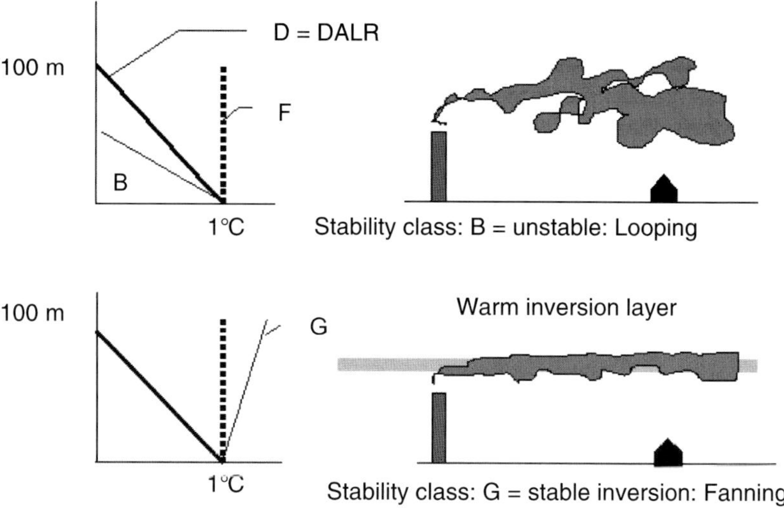

Figure 21-4. Plumes from chimneystacks for two stability classes. DARL = Dry Adiabatic Lapse Rate. Thin lines, ambient lapse rate. Looping means unstable, and fanning very stable. If inversion G starts from a point above the chimney opening, discharges are fumigating in the corridor between the warm inversion layer and the ground.

Such stability classes are used in atmospheric dispersion models. The most popular classification scheme is due to Pasquill, and shown in Table 21-9.

For classes A, B, and C the temperature decreases more than 1°C per 100 m, while Class D is neutral. Class E is fairly stable and F is very stable, with temperature unchanging with altitude. Inversion means that the temperature increases with height, which allows for a final, very stable class G that is not included in Pasquill's scheme. Figure 21-4 shows the situation for two different stability classes. The stability classes A and B are very unstable, occurring 9% of the time – during little wind and on sunny days; class C is unstable and occurs 9% of the time, class D is neutral and occurs 44.5% of the time – during overcast and strong winds, class E is stable and occurs 27% of the time. Class F is very stable and occurs 11% of the time. The prevalence does depend on locality, but these figures are often good approximations.

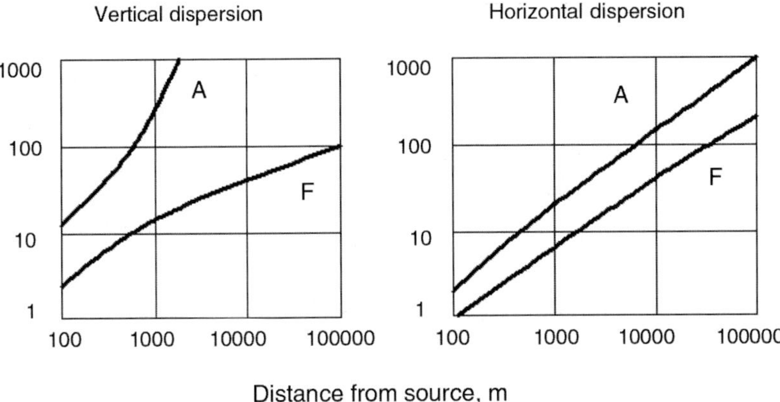

Figure 21-5. Vertical and horizontal dispersion coefficients (m) as a function of distance from source, and the Pasquill stability class. Equations for the other stability classes can be found in Hanna et al. (1982).

4.2.2 Vertical and Horizontal Dispersion

Pollutants disperse vertically as well as horizontally in the atmosphere. The dispersion coefficients describe the dispersion and have different values depending on distance from source and stability class, as well as landscape.

A gross picture of the coefficients is shown in Figure 21-5, but they are also given in equation form in Table 21-10 for vertical coefficients and Table 21-11 for horizontal coefficients.

4.2.3 Plume Rise and Effective Stack Height

The height of the stack has strong influence on the spreading of pollutants, but it is the so-called **effective stack height** that counts, which is the sum of the height of the stack and the **plume rise**, that is, the extra height the plume rises before it starts to spread out. A **chimney** is typically 20–150 m and up to 300 m, which is an extremely high chimney.

It is most relevant to calculate the effective stack height for situations where the plume spreads easily, and for the most frequent stability situations, which are the unstable and neutral conditions. The plume rise is related to two factors: the velocity of the gas out of the stack opening relative to ambient wind velocities, and the temperature differences between the gas and the ambient temperature.

Table 21-10. Vertical dispersion coefficients. D is distance in meters.

Stability class	Open country [1]	Urban [1]	Low sources [2]	From figures [3]
A	$0.20D$	$0.24D(1+0.001D)$		$0{,}005D^{1.7}$
B	–	$0.24D(1+0.001D)$	$0.07D^{1.02}$	–
D	–	–	$0.1D^{0.8}$	–
E	–	$0.08D(1+0.00015D)^{-1/2}$	$0.22D^{0.61}$	–
F	$0.016D(1+0.0003\,D)^{-1}$	$0.08D(1+0.00015D)^{-1/2}$	$0.26D^{0.5}$	$0.168D^{0.64}$

[1]Briggs (1973); [2]Bøhler (1987); [3] By fitting curves to the original nomograms for vertical and horizontal dispersion coefficients, cf. Figure 21-5.

Table 21-11. Horizontal dispersion coefficients. Legends as for Table 21-10.

Stability class	Open country	Urban [1]	Low sources [2]	From figures [3]
A	$0.22D(1+0.0001D)^{-1/2}$	$0.32D(1+0.0004\,D)^{-1/2}$	–	$60.8+0.167D$
B	–	$0.32D(1+0.0004\,D)^{-1/2}$	$0.31D^{0.89}$	–
D	–	–	$0.22D^{0.8}$	–
E	–	$0.11D(1+0.0004\,D)^{-1/2}$	$0.24D^{0.69}$	–
F	$0.04D(1+0.0001D)^{-1/2}$	$0.11D(1+0.0004\,D)^{-1/2}$	$0.27D^{0.59}$	$8.33+0.031D$

The plume rise caused by excess velocity can be estimated as.[16]

$$\Delta H_p = 3 \times D \times w_s/U \quad \text{for} \quad w_s/U > 4 \tag{21-3}$$

Here, w_s is exit gas velocity (typical values 10–20 m s^{-1}). U is wind velocity (typically 2–5 m s^{-1}); D is inside stack-top diameter (typically 0.5 to 4 m; stack heights 80–50 m).

With U = 2m s^{-1}, w_s = 20m s^{-1}, and D = 0.6 m, we get $\Delta H_p = 3 \times 0.6 \times 20/2$ m = 18 m. The plume rise is of the same order as the stack height under favorable conditions. If there are buildings or forest around the chimney, the chimney should be built that much higher, and at least 2.5 times the height of close tall buildings.

4.3 Local Dispersion Models

We will discuss three models: a simple box model, a "rule-of-thumb" model[17] and the classical Gaussian plume model. Even if the last model is a bit complicated, one can perform the calculations rather easily in a spreadsheet. The dispersion coefficients can be read from the graphs in Figure 21-5 or calculated from the formulas in Tables 21-10 and 21-11.

4.3.1 The Box Model

Given a chimney with effective stack height H, and a pollution emission rate Q, the box model estimates the area A that may become exposed to pollution concentrations above a critical level C_{cr}. The idea is to estimate the volume, V, of a box with height H and base area A that may contain the emissions at a stable concentration C_{cr}. This is done by multiplying the emission rate by an assumed residence time T, and dividing by C_{cr}.

$$V = QT/C_{cr} \qquad (21\text{-}4)$$

T should correspond to a typically long residence time of the air in the vicinity of the emission site, depending on the pollutant and the residence time of the pollution in the area. Shifting wind directions implies higher residence time than a stable direction. Twenty minutes to one hour have been suggested as typical residence times. Critical concentrations with regard to odors are often very low, and episodes with concentrations above critical level need only to be of short duration, say ten times repeated episodes of each 0.5–1 second, to cause complaints. The affected area depends on effective stack height; the higher the stack, the better: $A = V/H$. However, actual mixing of air depends upon air stability as well. Therefore, one must also divide by the stability class dilution factor S in Table 21-12.

$$A = V/(HS) \qquad (21\text{-}5)$$

Table 21-12. Dilution factors for air stability classes. Estimated by the Gaussian plume model.

Class	A	B	C	D	F
Factor	1000	500	100	50	10

4.3.2 A Rule-of-Thumb Model

This model predicts the maximum ground level concentration, c_{max}, and the distance from the source where the maximum occurs. The model can be used to predict the effect of chimney height for a given dilution requirement. The source is the chimney outlet, but since the plume bends in the direction of the wind, the cone describing the plume appears to come from a source at a higher level than the actual chimney opening, and slightly displaced upwind. See Figure 21-6. The total height is H. The angle defining the plume width is α. The first impact point B is then given by:

$$B = H/\text{tg}\alpha \qquad (21\text{-}6)$$

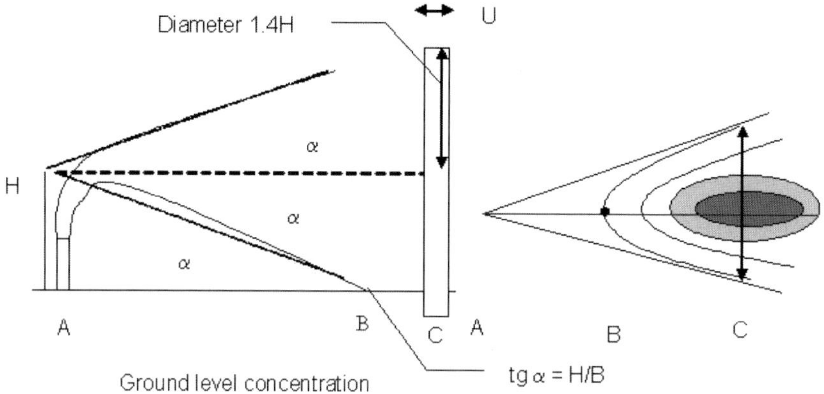

Figure 21-6. Dispersion from a chimney (vertical and horizontal sections) Adapted after Scorer (1990) p. 60.

Rules-of-thumb:

- The point of maximum impact *C*, is roughly 1.4 times as far from the source as the first impact *B*.
- If the average concentration across the plume is c_{av}, the concentration at the center of the plume above *C* is about 2.5 × c_{av}.
- The ground level concentration at *C* is about 2.0 × c_{av}.
- The cross section of the plume above *C* is a circle of radius about 1.4*H*, without the part below ground. Its area is $5.7H^2$.

If we draw a plan of the impact of the plume on the ground, the contour will appear something like in Figure 21-6 right. Now, if *Q* is the rate of emission ($m^3\ s^{-1}$), and *U* is wind speed ($m\ s^{-1}$), then Equation 21-7 gives a best guess estimate for the emission volume:

$$Q = 5.7UH^2 C_{max} / 2 \qquad (21\text{-}7)$$

This gives

$$C_{max} = \frac{2Q}{5.7UH^2} \approx \frac{Q}{3UH^2} \qquad (21\text{-}8)$$

The effective stack height for odor emissions with a concentration *C* can be estimated with Eq. 21-9:

$$H = \sqrt{\frac{2.3QD}{U}} \tag{21-9}$$

Here $D = C/C_o$ is the dilution factor to obtain threshold level for odor, C_o. These formulas are very approximate, but still useful. Experience suggests that a correspondence within a factor of two between prediction and observations is tolerably good, but a factor of more than three starts to become troublesome.[18]

4.3.3 The Classical Gaussian Plume Model

The classical **Gaussian plume model** can be used to analyze air quality. It predicts ground level concentrations of gas as a function of the distance from the source of emission and its elevation, emission rate, wind speed, and dispersion coefficients. It is based on the Gaussian distribution and gives for ground level concentrations, $z = 0$:

$$C_{x,y,\sigma} = \frac{Q}{\pi\sigma_y\sigma_z u} e^{-\left(\frac{H^2}{2\sigma_z^2}+\frac{y^2}{2\sigma_y^2}\right)} \tag{21-10}$$

Here,

$C_{x,y,\sigma}$(mg m^{-3})	Ground level concentration of gas or particulates (size < 20 μm), at a distance x (m) downwind from source, and a distance y (m) crosswind.
Q (μg sec^{-1})	Release rate from an elevated point source
σ_y (x)	Horizontal dispersion coefficient representing the amount of plume in crosswind direction at a distance $x = D$ downwind from the source under a given atmospheric-stability class
σ_z (x)	Vertical dispersion coefficient representing the amount of plume in vertical direction at a distance $x = D$ downwind from the source under a given atmospheric-stability class
U (m sec^{-1})	Mean wind speed
H (m)	Effective stack height (actual physical height plus any rise of plume as it leaves the stack)

The maximum ground level concentration occurs at the plume's center line, at $y = 0$. The horizontal dispersion coefficient increases linearly with distance for all stability classes A to F; the vertical one increases logarithmically for the most stable weather condition (F), linearly for the neutral stability class (D) and as the third power (D^3) for the most unstable stability class (A).

Approximations: It is interesting to note that the Gaussian model is similar to our rule-of-thumb model; if we set $y = 0$ and expand the first term of the exponent we obtain an approximation for the maximum concentration downwind. Assuming that $H^2/2\,\sigma_z^2 > 1$, we get

$$e^{-\frac{H^2}{2\sigma_z^2}} = \frac{1}{e^{\frac{H^2}{2\sigma_z^2}}} \approx \frac{1}{1+\frac{H^2}{2\sigma_z^2}} \approx \frac{1}{\frac{H^2}{2\sigma_z^2}} = \frac{2\sigma_z^2}{H^2} \qquad (21\text{-}11)$$

$$C_{\max} = \frac{Q}{\pi\sigma_y\sigma_z U}\frac{2\sigma_z^{\ 2}}{H^2} \qquad (21\text{-}12)$$

Now, assuming that the horizontal and vertical dispersion coefficients are equal, $\sigma_y = \sigma_z$, we get:

$$C_{\max} = \frac{2Q}{\pi UH^2} = \frac{2Q}{3UH^2} \qquad (21\text{-}13)$$

This is twice the concentration suggested by Equation 21-8.

5. CORROSION

There are several corrosive agents, and among them is sulfur dioxide (SO_2), which is particularly aggressive. Concentrations even below WHO health guidelines of 5 μg m^{-3} may corrode materials if the humidity is high enough (>80%), and the temperature is above freezing.[19] When it is possible to rehabilitate buildings, monuments or other artifacts from the damage of airborne pollutants, the cost of damage may be assumed equal to the annual monetary cost of rehabilitation. If rehabilitation is impossible, it is necessary to use decision panels or WTP studies to assess the damages.

5.1 City Level Relations

Aggregated corrosion damage in towns, measured in terms of maintenance costs, is correlated with population and average SO_2 concentration (μg m^{-3}). Regression analysis of 17 cities in Norway gave this relationship:

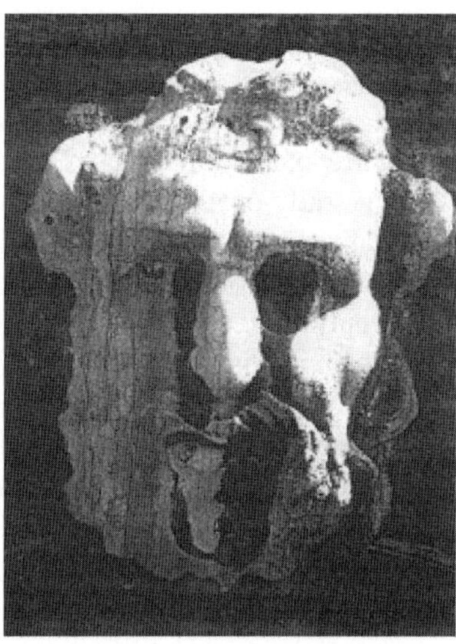

Figure 21-7. Effect of air pollutants. With splashing rain droplets, the pollutants deposited on the surface react with the underlying rock and form black crusts. Where there is no run-off or wash out, the dry deposit is not removed. On masks a tear shaped black crust is typical below the eyebrows. (Venice)[20] Camuffo and Bernardi (1990) p. 12.

$$\log_{10} D = 0.5 \log_{10} P + 0.07\, SO_2 - 0.2,\ r^2 = 0.9,\ p < 0.0001 \qquad (21\text{-}14)$$

D (€ yr^{-1}) is aggregate damage cost. P is number of inhabitants (1000).[21]

5.2 Materials and Corrosion

For some materials, there are clear dose-response relationships between exposure and corrosion. The most significant variables are SO_2 concentration and "humidity-hours" defined as the number of hours per year with humidity >80% and temperature >0°C.

In temperate climates the following relation holds for unpainted galvanized steel:

$$v_{corr} = 0.29 + 0.039\, SO_2 \qquad (12\text{-}15)$$

Here, v_{corr} = corrosion rate (μm yr^{-1}), SO_2 concentration (μg m^{-3}). That is, 260 μg m^{-3} will give 0.01 mm erosion on unpainted galvanized steel per year.

6. VISIBILITY

Visibility is part of the esthetic experience at recreational sites, and an important contribution to life quality in cities. But there are two sorts of problems in any assessment study of visibility. One is to present changes in visibility that are both meaningful to the respondent and that can be related to pollution control policies. The other is separating valuation of health effects from the value of visibility change. Many people assume that health effects from pollution diminish as visibility improves, but wet fog cleans the air more than dry haze.[22]

Most studies define visibility as visual range – the distance at which a large black object disappears from view.[23] A study has shown that households would pay approximately $101 annually for a 10% improvement in visibility in San Francisco.

7. BIRDS

The sky and the air is the common compartment for all birds, although some birds are almost exclusively confined to the waters and some to terrestrial environment. Birds perform a series of ecological services, like eating insects that are potentially detrimental to crops and they distribute seeds.

Estimates of the average household's annual willingness-to-pay to preserve endangered bird species is US$31.80 for the whooping crane, $15.40 for the bald eagle, $11.66 for the red-cockaded woodpecker.[24]

Background information: Birds are abundant over the whole globe. The number of species per area unit is largest around the equator (1,400 species on about 611,000 km^2) and declines to about 200 species in the same area at 60° North or South. The belt between ±20° appears to be fairly uniform with respect to species diversity.[25] About 2600 species (27%) of the world's bird population have breeding ranges of less than 50,000 km^2. Of these species, 760 (27%) are threatened in some way or other. The main habitats used by endemic species are forests (69%) and shrubs (12%).[26] Although the arctic tundra is one of the harshest areas on Earth, it is the summer home of millions of birds. Waders, ducks, geese, swans, divers, skuas, gulls, birds of prey and passerines migrate to breed in the thawing soils and wetlands on top of the permafrost.[27] The density of geese in the arctic may exceed 16 geese km^{-2}.[28] In the USA, the 15 ha large Wintergreen Lake in Michigan had 6,518 visits days per year of Geese (weight 2.56 kg), 3,845 of Dabblers (weight 1.18 kg) and 397 of Divers (weight 1.01 kg). Most visits were during the months of October and November.

Dose-response relationships: As far as we know, there are few models or index systems that predict bird species' vulnerability to human interference. A problem with such models for birds is that the birds have to be present at the site where the interfering activity takes place, like spraying of pesticide or an oil pollution accident. Some bird species are distributed evenly in the landscape, whereas others are strongly clustered; puffins may occur in colonies counting millions of individuals on one rock formation. Some models and index systems to predict seabird vulnerability to oil pollution, however, have been developed.[29]

Table 21-13 shows parameters that were used to assess species' vulnerability in two of the studies, and the range of the resulting vulnerability indexes. The index by King and Sanger (1979) was an additive index, whereas the results by Seip et al. (1991) were based on simulation models. Both indexes and models distinguish between the individuals' susceptibility to oil pollution, and the population's ability to recover from an oil incident. The potential for recovery is a function of migration lengths compared to the distance to nearby seed areas, such as burrows. Thus the individual vulnerability for Kittiwake was 41 and the population vulnerability 58 on a scale from 0 to 100 (highest vulnerability) in the study by Anker- Nielsen (1987).

Windmill effects: We do not know of any models of the effects of windmills on birds, but studies are presently being conducted in connection with the construction of new windmill parks.

Table 21-13. Sea bird vulnerability to oil pollution. Data Seip et al. (1991), King and Sanger (1979).

Parameter	Black-legged kittiwake	Common guillemot	Common eider	Comment
Marine orientation (1)	5	5	5	
Migration (2)	3	5	5	
Vulnerability to oil pollution (3)	1	5	5	
At sea for foraging	0.5	0.5	0.2	0.0 = max
Foraging radius (km)	20	20	20	
Migration length in one week (km)	10	20	20	
Species avoidance of oil	0.8	1.0	1.0	1.0 Least avoidance
Overall vulnerability index	41–58	69–99	39–96	100 Highest vulnerability

(1) 1 = Coastal zone; 2 = intertidal zone; 3 = open water.
(2) 1 = long; 3 = medium; 5 = short.
(3) 1 = small; 3 = medium, 5 = high.

8. APPLICATION: REPOPULATION AFTER A NUCLEAR DISASTER – RETURNING HOME, BUT TO WHAT?

The Chernobyl nuclear meltdown on 26 April 1986 is estimated to have released 50–200 million curies of radiation plus a variety of chemical and metal pollutants. A **curie** is a unit of radioactivity, equal to the amount of a radioactive isotope that decays at the rate of 3.7×10^{10} disintegrations per second. 2500 people are estimated to have died in the accident, and about 116,000 people were evacuated from the inner 30 km radius area around the accident site. The effects were closely related to how pollutants spread through the atmosphere. Animals such as rodents closer to the reactor than one km, were subject to high levels of genetic change 10 years after the disaster, whereas none have been recorded at less exposed sites, such as 32 km southeast of Chernobyl. On the other hand, enhanced levels of radioactivity could be detected already on May 2 and 3 as far away as Great Britain and the effect lasted for several years. For example, 150 hill sheep farmers in Lake District could still in 1989 not sell their sheep freely.

8.1 Decision Analysis

When a disaster like Chernobyl occurs, a time comes when the question arises of when and how to repopulate the afflicted area. Let us have a look at important issues in the decision-making process.

Decision makers: The decision maker is the local government, and most probably the government will take advice from environmental authorities.

Stakeholders: The most important stakeholders are the people that had to move from the site, nuclear protest groups and government authorities. How to identify decision makers and stakeholders is discussed in Chapter 3, section 2.3.

Decision objectives: The main objective is to treat the dislocated people decently, and maximize the utility of the afflicted area. This translates to the following *sub-goals*:

- Maximize the number of happy people, returning them to their old homes.
- Minimize health hazards.
- Minimize opposition (protests) to nuclear power.
- Maximize the utility of the land area that has been abandoned.
- Minimize cost of supporting dislocated people.

Alternatives: The following is a list of rather general alternatives; more detailed alternatives must be specified in an actual analysis:

- Recommend that people return to their previous homes.
- Do not recommend a return.
- Prohibit return.
- Recommend a restricted return.

Consequence analysis: Areas that were exposed to nuclear contamination are slowly becoming habitable again, with a time span of 10 to 50 years, but safety depends on susceptibility to radiation, and pregnant women and children are more vulnerable than the average person. See Chapter 14 on toxicity. Safety also depends on what you eat and where the food is grown. One needs to build models of cumulative concentrations over time, and expected health effects. The dispersion models included in this chapter may therefore be helpful; see section 4.

Utility of consequence: Some of the primary utility functions will most probably be risk-averse, like the one that represents health hazards. Risk-averse utility functions are discussed in Chapter 2, section 2.4, and Chapter 3, section 6. The analysis should result in a consequence- and a utility table like the Tables 3-5 and 3-6, and primary utility graphs, like those in Figure 3-3.

Preferences: People probably have a strong emotional preference to return home. Some people want to return home even if the government advises not to. The expected life expectancy is rather short in Russia, 60 years for men (Chapter 8 section 5.4), and the standard of living is for many people rather poor. Should these factors be taken into account in a relocation decision, for instance with help of the purchasing power parity concept. In any case, preferences among the people that may move home should probably be solicited in an unbiased way, but under realistic scenarios, see Chapter 7 on preferences. The results should be importance weights for each criterion as in Table 3-8.

8.2 What Happens?

Approximately 150,000 people live in areas where radio cesium levels exceed 555 kbq m^{-2}, and protection measures are still required. In 2002, people had not been allowed to return within a 30 km radius. Medical observations have shown an increase in thyroid cancer of those living in contaminated regions of the former Soviet Union (about 1800 cases), but no increase in other cancers that can be attributed to the Chernobyl accident. Neither has any adverse pregnancy consequences been recorded. It is a paradox that in spite of enhanced genetic changes in rodents close to the reactor site, the rodents thrive and reproduce in the radioactive regions around the Chernobyl reactor.[30] However, animal responses only partially reflect responses that can occur in humans and safety factors probably imply that there are many years left before people can use the site.[31]

---

[1] Finkelman (1996)
[2] Rennings and Wiggering (1997)
[3] Scorer (1990)
[4] Pyne (2002)
[5] Data: Schultz (1986) for W.Germany and Opschoor (1986) for The Netherlands
[6] collected by Iversen et al. (1991)
[7] The science of the Total Environment, 1995, special issue 167: The deterioration of monuments
[8] Gauri et al. (1981)
[9] Camuffo and Bernardi (1996)
[10] Data Camuffo and Bernardi (1996) p. 23, 38
[11] Klumpp et al. (1996), and others
[12] Wotawa and Trainer (2000) p. 324
[13] Rodan et al. (1999)
[14] Grennfelt et al. (1994)
[15] Landsberg (1970)
[16] Bøhler (1987) p. 10:10
[17] developed by Scorer (1990) p. 63
[18] Scorer (1990)
[19] Glomsrød et al. (1997)
[20] Camuffo and Bernardi (1990) p. 12
[21] Glomsrød et al. (1997), the cost of renovating the buildings
[22] Scorer (1990)
[23] Cropper and Oates (1992) p. 719
[24] Bulte et al. (1999)
[25] Gaston (2000)
[26] Jarvis (2000)
[27] Alerstam and Jønsson (1999) p. 212-223
[28] Alerstam and Jønsson (1999) p. 214
[29] King and Sanger (1979), Anker-Nilsen (1987), and Seip et al. (1991)
[30] Baker et al. (1996)
[31] http://www.nea.fr/html/rp/chernobyl/chernobyl-update.pdf, UNSCEAR (2000)

Chapter 22

CLIMATE

Climate, ecosystems and anthropogenic emissions

We first discuss weather and climate in general and then show how the effect of anthropogenic emissions may affect the climate and ecosystems. Climate warming is predicted to increase the probability of severe weather conditions, elevate the sea level, and impact soil, bio-geochemistry, vegetation and wildlife. Since a major source of the emissions is burning of fossil fuels, we include information on energy use. Models that may be used to calculate climate effects are described in the separate chapters on the particular environments.

●

"But I, that was born to be my own destroyer, could no more resist the offer than I could restrain my first rambling designs, when my father's good counsel was lost upon me" Defoe, Daniel 1987 Edition. Robinson Crusoe (p. 51)

1. INTRODUCTION

Climate variations are natural, occurring on time scales of many thousands of years, as can be seen from the many glacial periods that interchange with warm periods.[1] Climate affects the environment profoundly thorough temperature, precipitation patterns, and wind. The climate was until some decades ago assumed to be independent of human activity. However, there is now general consensus – although there still are some dissidents – that man-made emission of gases into the atmosphere contributes to an increase of the global temperature. This is called the **greenhouse effect**. Release of CO_2 into the atmosphere by burning fossil fuels is the major cause, and deforestation contributes by reducing CO_2 storage on earth. Another important greenhouse gas is methane CH_4, which is released from waste depots, rice paddies, and through animal husbandry. The first cause is closely related to our need for energy, on the second to our need for food of different varieties. Other human activities like discharge of dust, ignition of forest fires and airplane emissions directly into the atmosphere adds to the greenhouse effect. Among the most important natural sources for methane are swamps, marshes and floodplains.

In this chapter we address, but in no way solve, questions such as:

- What measures can arrest or reduce global warming?
- What are the most cost-effective ways to achieve reductions?
- Which biological mechanisms are working to change ecosystems during global warming?
- Since energy consumption is related to affluence, like use of cars, heating and air conditioning, how can the developing world achieve the developed world's living standard without increasing global warming?

2. BACKGROUND INFORMATION

We shall in this context restrict the term "Climate change" to change brought about by emissions to the atmosphere related to human activity. These inputs are predicted with a 90% probability to cause an overall global warming between 1.7°C and 4.9°C within the year 2100, but the effects will be uneven, with temperature rises here and cooling there.[2]

Figure 22-1 shows the development of CO_2 concentration, solar radiation and temperature from 1850 to 2000.

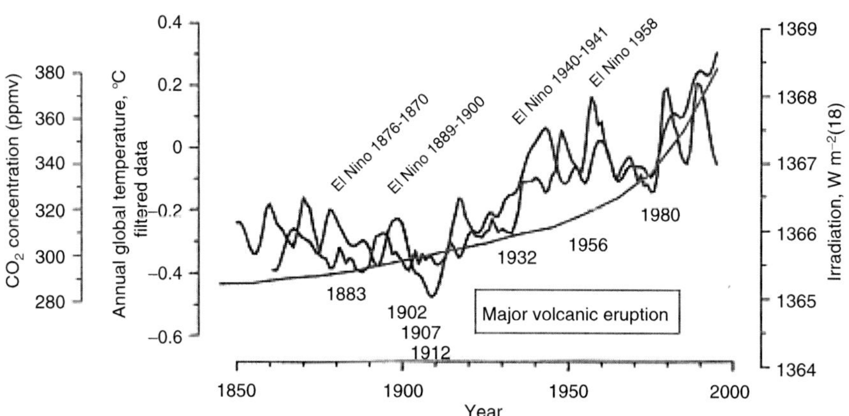

Figure 22-1. Global temperature (the oscillating curve starting second), solar irradiation (the oscillating curve starting first) and atmospheric concentration of greenhouse gases (smooth curve) from 1850 to 2000. An 11 year binominal filter is applied to temperature observations. Major volcanic eruptions and El-Niño events are marked. Data Karlén (2001) p. 349.

The Krakatau eruption in Indonesia in 1883, Mt. Pelée in Martinique in 1902, and Mt. Pinatubo on the Philippines in 1991, have been suggested as antecedents of major large-scale climate changes because of the cooling effects of volcanic aerosols.[3] Mt. Pinatubo ejected 20 megatons of SO_2 into the atmosphere and cooled the Northern Hemisphere by about 2°C in the summer of 1992 and lowered winter temperatures 1991–1992 by up to 3°C.[4] It probably retarded global warming for several years because of the cooling effects of volcanic aerosols.[5] Even the extinction of dinosaurs has been attributed to volcanic eruption; the theory being that dust made the earth dark long enough to prevent plant growth, thus no fodder was available for these large animals.[6] Although most **global climate changes** have occurred over geological times, the current trend is our concern, and it is so fast that natural adaptive evolutionary processes may not keep pace.

2.1 Climate Services and Damage Cost Estimates

The atmosphere contributes to regulation of global temperature, and distribution of temperatures over the earth's surface. Swamps and flood-plains give the highest contribution to climate regulation per unit area, Table 22-1.[7]

Forest and woodlands contribute to climate regulation by sequestering carbon, and by changing evaporation rates. Many studies on the cost of global warming are dispersed in scientific literature worldwide. The cost of damages caused by global warming has been estimated per ton of carbon by at least three methods, direct damage cost estimates, prevention costs, and market prices for carbon emission permissions.

Unit costs based on *damage cost methods* have been estimated by two independent groups (1992) to be in the range US$5–125 and in the range US$2–23 per ton of carbon emission respectively.[8]

EU-funded projects like ExterneE have studied global warming externalities. Marginal damage depends upon the discount rate used: With 1% and 3% discount rates the EU ExterneE suggests 170 Euro per ton and 70 Euro per ton respectively.[9] The corresponding numbers for methane were €530 per ton of CH_4 and €350 per ton of CH_4^{-1} and for nitrous oxide €17,000 per ton of N_2O and €6400 per ton of N_2O^{-1}.

Table 22-1. Estimated values of ecosystems as climate regulators. Data: Constanza et al. (1997).

Biome	Open ocean	Grass / rangelands	Wetlands	Swamps / floodplains
Area (ha 10^6)	32,200	3898	320	165
Climate regulation value (1994 USD ha^{-1} yr^{-1})	38	7	133	265

The prevention cost method is based on the assumption that estimated costs of preventing CO_2 from being emitted actually reflect the cost of future damages.

A third method is based on the actual prices obtained for transferable carbon emission permits and were found to be between US$ 25 and US$ 50 per ton of carbon. The market price for avoided emissions that can be credited under the Kyoto protocol is in the range US$5–35 per ton of carbon. CH_4 cost estimates are often calculated as four times the CO_2 estimates, although a factor of 20 has also been used.

Considering all the uncertainties involved, there is every reason to regard current damage estimates with uttermost caution. First, one has to predict to what extent greenhouse gases will influence the climate and the sea level, and then how these changes will affect ecosystems, human health, economy and other end impacts. Finally, one has to take into consideration willingness-to-pay to avoid unwanted external effects.

2.2 Normal and Extreme Weather

Life on earth is profoundly affected by average climatic conditions as well as extreme events. When we describe a climate, we therefore need to take into account the temporal and spatial distribution of important climatic features like temperature, humidity, precipitation, wind and waves. Natural disasters may serve as a backdrop for understanding some of the potentially negative effects of global warming.

Wind: The global average wind speed is 3–5 ms^{-1}. Strong winds may reach 213 km hr^{-1}, which was the speed of Allen, the strongest Caribbean hurricane in the last century. Ecosystems are adapted to normal and strong winds, but extreme winds that occur infrequently may have devastating effects.[10] Some land areas are particularly prone to high winds, such as southeastern USA, the Caribbean, western Pacific islands and coastal regions in the Indian Ocean. Strong winds are given names according to a set of criteria that reflects local climate conditions. The distinguishing feature of **a tropical cyclone** includes a central core or "eye" with very low barometric pressure. A storm with wind rotating around the eye that exceeds sustained speed of 33 ms^{-1} is classified as a **hurricane** in the Atlantic Ocean, a **typhoon** in western Pacific and as a **cyclone** in the Indian Ocean. Hurricane wind speeds are in the range 33 to 70 ms^{-1}. Hurricanes travel at speeds between 3 and 8 ms^{-1} and may impact millions of square km, last for a week or longer and travel several thousand kilometers. **Tornados** have stronger winds –90 to 235 ms^{-1} – and travel faster than hurricanes, 10–20 ms^{-1}, but cover transects less than 2 km wide and travel less than 300 km. The wind speed decreases inversely with the square root of the distance from the center. About 80 tropical cyclones develop over the world oceans each year, and 50 develop into hurricanes. Some areas in the Caribbean may experience up to five storms stronger than 33 ms^{-1} per year.[11]

Waves: Waves are set up by wind. The height of the waves increases with the wind's **fetch**, but only up to a distance that is a function of the wind speed. With normal trade winds of 6–8 ms^{-1}, the waves saturate when the fetch is 50 to 100 km. Hurricane winds, however, need nearly three days to raise waves to maximum, and require a fetch of more than 2250 km. The power of the waves is proportional to the square of their height. Landslides or volcanic eruptions under the ocean floor cause **tsunamis**. They are broad waves with surfs up to 40 m high that wash shores and cause great damages 10 to 20 m above normal sea level.

Ecosystem damages: Storms may cause damage to crop, forest, and built land. There is evidence that damages to trees and buildings are proportional to the fifth power of the wind speed.[12] In spite of the exposure of coral reefs and rainforests to storms and hurricanes, little damage to these two ecosystems is reported. When a coral reef is damaged, it recovers quickly because of high growth rate and its ability to regenerate from fragments that reattach to the bottom.

2.3 Climate Change

The two main concerns about the climate are the greenhouse effect and the ozone depletion effect. Currently, the greenhouse effect is probably of greatest concern. In this section we start our discussion with what has been named the Global Warming Potential, GWP. This term summarizes all natural and anthropogenic factors that may contribute to global warming. Thereafter we discuss the causes and effects of ozone depletion in the stratosphere, which reduces most life-forms' protection from ultraviolet radiation.

Global Warming Potential: The gases contributing to the climatic impacts are called **greenhouse gases**, GHG. Carbon dioxide CO_2 is the main culprit, but other GHGs like methane CH_4, nitrous oxide N_2O, nitrogen oxides NO_x, carbon monoxide CO, and several volatile organic compounds contribute directly or indirectly to global warming. The concept of the GWP has been developed for policy-making purposes. The GWP of a greenhouse gas is its warming potential relative to that of CO_2.[13] Methane is approximately 50 times more effective in absorbing infrared radiation than CO_2, but there are two complicating factors. The first is that some compounds – like methane – have indirect effects by affecting concentrations of other GHGs.[14] The second is that the residence time in the atmosphere varies; for methane it is 15 years versus 125 years for CO_2. The

GWP of methane is therefore somewhere in the range 4–50, with 20 as a working compromise. Because of the huge amount of CO_2 present, the carbon cycle is of great importance for global warming.

The global carbon cycle: Carbon is at the core of climate change. It is released from reservoirs and enters the atmosphere, from where it again precipitates and is trapped in reservoirs. Figure 22-2 illustrates different pathways of the carbon cycle. It is stored in fossil fuel reservoirs, and released through burning of coal, natural gas, and oil.

Emission from waste depots and some natural sources: Organic material that is deposited in waste depots or landfills emits methane under anaerobic decomposition.

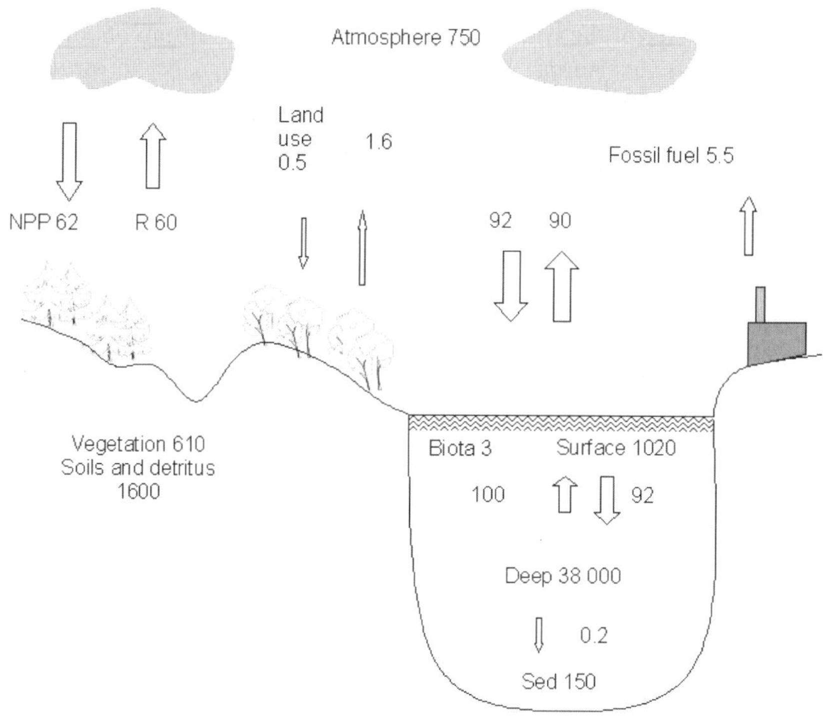

Figure 22-2. The global carbon cycle. Reservoirs are measured in gigatons of carbon, GtC, and transfers in GtC yr^{-1}. 11 Gt = 10^9 ton. NPP is Net Primary productivity, R is respiration. Man-made discharges are indicated with 5,5 GtC yr^{-1} from burning of fossil fuels, and 1,6 GtC yr^{-1} from land use. These are average figures for the period 1980–89. The amount of CO_2 in the atmosphere increased with 3,3 GtC yr^{-1} in this period. The cycles are simplified considerably and the figures are uncertain, and may change considerably from year to year.

Under completely anaerobic conditions, one ton of paper can theoretically generate 400 kg of methane along with 540 kg of carbondioxide.[15] Since methane has a GWP four to twenty times higher than CO_2, it would, with the last ratio, correspond to 8540 kg of CO_2 equivalents per ton of waste. How much methane is actually generated, and how fast it is released into the atmosphere, varies from site to site. Methane is also released from wetlands, husbandry, and hydropower dams where the water level varies with tapping and with the season. It has been estimated that hydropower dams in Brazil release 10,000 ton CO_2 equivalents per megawatt and year from decaying biomass. A coal fired plant releases about 7000 tons per MWe. A significant amount of gasses is also released when methane rich water passes through turbines.[16]

Emissions from industry: Emissions from the paper mill industry contribute 13–590 kg carbon per ton of paper products to the GWP depending upon the technology used. There are no significant emissions of CH_4 from the factories, but considerable emissions from anaerobic decay of paper products. Lifetime emissions of CO_2 from producing and driving a car have been estimated at 83 tons of carbon per kg of car-weight.[17]

Emissions per capita: The most important sources of GHG emission are burning of fossil fuels and forest fires.[18] More than 85% of the world's energy is provided by fossil fuel.[19] In 1995 emissions were 6.5 gigatons (10^9) of carbon per year. Emissions per capita in six continents are: Europe 2.68 tons, Africa 0.39 ton, North America 4.26 tons, South America 0.55 ton, Asia 1.05 tons, and Oceania 3.85 tons.[20] Cumulative emissions in the world from burning of conventional fossil fuels are about 714G tC. This is sufficient to sustain a level of 450 ppmv (parts per million by volume), which is approximately 30% above the current CO_2 level.

2.4 End Impacts of Global Warming

Greenhouse effects are the major climate concerns. The assumption – and one may say the general agreement – is that human discharge of so-called greenhouse gases contributes to an increase in average global temperature. Global warming has several expected end impacts – negative as well as positive – that cause serious concern. Among them are:

- Increase in the sea level with consequent submerging of land.
- Change in fresh water availability.
- Increase in ambient ground level air temperatures.
- Change of extreme weather patterns.
- Change in the prevalence of forest and bush fires.
- Change in the extent of deserts.

- Change in agricultural crops.
- Change in flora and fauna.
- Change in human health.

Indicators: It is important to monitor effects along the impact pathways, so that we get warning signals as early as possible about what is happing. We therefore try to monitor:

- CO_2 emissions.
- CO_2 concentrations in the atmosphere.
- Increase in average global air temperature.
- Increase in average sea temperature.
- Melting of glaciers.
- Changes in flora and fauna.

To observe changes in glaciers is particularly interesting, since drilling of ice cores that go back several thousand years can tell us about changing climatic patterns before we entered the scene, thus making it easier to distinguish between what is man-made and what is natural variation. In addition, while air temperatures are volatile and changes with time and place, glaciers act like calculators of moving average, giving us a good picture of recent history.

2.5 Ozone Depletion

Ozone, O_3, in the stratosphere protects the earth from ultraviolet radiation, which is harmful to most kinds of life and a cause of skin cancer in humans. Ozone concentrations vary considerably and are particularly low when there is high pressure. Sometimes, low ozone events or **ozone holes** arise spontaneously, offering almost no protection against ultraviolet radiation. A well-known episode occurred in central Europe during the hot summer of 2003, when the radiation increased by about 15%. It is especially serious when this happens during the summer, when many people are out enjoying the sun.

Although variations are natural, humans contribute to the hazard by causing emission of gases that deplete ozone. Chemicals with high ozone depletion potentials are chlorofluorocarbons, carbon tetrachloride, 1,1, 1-triclorate, hydrochlorofluorcarbones and methylbromides. It has turned out to be much easier to reduce emissions of such chemicals than gases with global warming potential.

3. CLIMATE STANDARDS

GWP concentration in the atmosphere is now used as an indicator of climate state. The state is seen relative to the Climate Standard, which is a new term, and by using it one implicitly assumes that the climate around 1990 was an optimum. More important is probably that one does not want large and fast deviations from the present climate conditions, because the ecosystems would not be able to adapt to rapid changes, and therefore be in a worse condition after a climate change.

The annual average CO_2 concentration has risen from 316 ppmv in 1959 to 364 ppmv in 1997.[21] CO_2 measurements display seasonal variations, caused by seasonal uptake and release by terrestrial organisms. Atmospheric CO_2 increased at a rate of 3.4 ± 0.2 gigaton carbon per year during the 1980s. The current goal is to remain at the 1990-level, that is, constant emissions of 6GtC yr^{-1}.

To reduce the growth of CO_2 concentration, the Kyoto meeting in 1997[22] set a limit on future annual CO_2 emissions equal to the emissions in 1990, and countries were assigned goals relative to this. The combined commitment by all countries is a 5% reduction from the 1990 rate. Table 22-2 shows emissions in selected countries and their goals.

Table 22-2. Emission of CO_2 in selected countries, and emission reduction goals Data: Romm et al. (1998), Cooney (1997) Bach (1998).

Country	Emission 1990 10^9 tons	Emission per capita, ton $capita^{-1}$	Goal 2008-2012
United States	1.6	6.4	reduce 7%
China	0.8	0,7	–
Japan	0.3	2,7	Reduce 6%
Norway [(1)]	0.004	1.0	Increase 1%

[(1)] Most of Norway's energy comes from hydropower, but Norway extracts oil.

4. DOSE-RESPONSE

The impact pathways of emission of GHGs to the atmosphere are many and complicated, and large models such as the Terrestrial Ecosystem Model are being developed to improve our understanding of what is happening.[23] In this section, we will therefore restrict ourselves to a qualitative discussion of some of the more important cause-effect relations. According to the Intergovernmental Panel on Climate Change, IPCC, a doubling of CO_2 from 1990 is likely to cause an increase in the global temperature of 1.5–4.5°C.[24] With the present emissions of CO_2 and other greenhouse gases, this may occur before 2100, but with strong regional variations. Thus, considerable impacts on sea level, water supply, weather, plant growth, health, and ecosystems seem rather imminent.

Carbon emission from energy production: An identity that makes it easier to estimate the rate at which carbon, C as CO_2 is emitted by energy production is shown in Equation 22-1.[25]

$$C = N \times (GDP/N) \times (E/GDP) \times (C/E) \tag{22-1}$$

The first term, *N*, on the left side is the population of a region, or the world (the world population is 6.0×10^9 persons). The next term is gross domestic product, *GDP*, per capita which is available for most regions (USA: 4100 US$ capita^{-1} yr^{-1}). The third term represents primary energy use, *E*, per GDP (USA: 0.49W yr^{-1} US$$^{-1}$). *C* is carbon use per energy unit (USA: 0.56 kgC W^{-1}). With a high proportion of for example nuclear energy, carbon use per energy unit is low. The equation is similar to Equation 21-1 for emissions of SO_2, which also depends upon the population, *N*. *GDP* cancels out, but it has a dummy term for former eastern European countries (because of low energy efficiency) as opposed to western European countries.

Effect of sea and air temperatures on the sea level: Two processes contribute to an expected rise in the sea level: melting of polar ice, and thermal expansion of water. The West Antarctic ice sheet seems particularly unstable, and may retreat rapidly. Were it to collapse completely – an unlikely event –3.2 million km^3 of ice would contribute dramatically to the global sea level with a five meter rise. At the present rate of accumulation, the time needed to rebuild the ice sheet is more than ten thousand years. Estimates of sea level rise from ice melting range from 25 cm per century to 120 cm per century.[26] A doubling of CO_2 is likely to cause the sea level to rise 15–95 cm due only to thermal expansion. The current scenario is that we expect a sea level rise of 20 cm by 2100, where 13.2 cm is due to melting glaciers and 7.6 cm comes from the melting of Greenland's ice sheet.[27] A higher sea level will inundate shores and lowlands. Coastal zones and small islands are especially vulnerable. Such areas often have highly diverse ecosystems. Delta areas like Bangladesh will be inundated, and 23 million people will be exposed to coastal flooding by the year 2050.[28] Since the increase in sea level is relatively rapid (0.0025–0.0012 m yr^{-1}), coral reefs will probably not be able to rebuild themselves fast enough to higher elevations.

Effect on atmospheric temperature on water availability: Increasing atmospheric temperatures allow the atmosphere to hold more vapors. Increased precipitation is therefore one of the predicted effects of global warming. Some regions may benefit from a temperature rise with regard to fresh water, but water supply may still be reduced considerably in other places – especially arid regions. Even if precipitation increases, the added amount of water may be offset by increased evaporation. We have no reliable estimates of areas that will experience loss of water supply, or deserts that will expand, but unmitigated global warming is predicted to expose an additional 1000 million people to water shortages.

Effect of global warming on extreme weather: In August 1992, the hurricane Andrew ripped through southern Florida and Louisiana, causing total damages of more than US$ 30×10^9. Subsequent hurricanes, like Katrina in September 2005 broke down levees and flooded much of New Orleans. It was followed by several less strong hurricanes. If the frequency and patterns of hurricanes change and they start to hit densely populated areas, there will be dire consequences. Analyses suggest that a tropical cyclone hitting a large city can generate losses up to US$ 100×10^9. This exceeds all reinsurance capital in the world and would deplete much of the total pool of capital.[29] The main reason for an increase in extreme weather frequency is increasing amounts of water vapor in the atmosphere. The atmospheric vapor content increases 6% for every extra centigrade, and a moisture-laden atmosphere generates more tropical-like downpours.[30] There are not yet reliable dose-response models that relate CO_2 increase to frequency and location of natural catastrophes, but the observed increase in annual forest fires from 6000 in the period 1930–1960 to 10,000 in the 1980s may be due to global warming.[31]

Effect of CO_2 concentration on plant growth: Increased CO_2 levels stimulate plant growth. There is evidence that a higher CO_2 concentration has increased oceanic production in the nutrient-poor central Atlantic Ocean by 15–19%.[32] Hopefully, it would also store more carbon in the vegetation, thus counteracting global warming. However, experiments over several years now suggest that elevated CO_2 concentrations rather increase the rate of carbon cycling within ecosystems than carbon storage.

Effects of increase in temperature on flora and fauna: Higher temperature generally stimulates plant growth. There are at least three causes: Higher average temperature over the growth period, longer growth seasons, and higher minimum temperatures. There is evidence that the growth season for plants has increased by roughly eight days at northern latitudes, and that birds in the United Kingdom are laying eggs on the average nine days earlier.[33] In marginal areas such as Australia, crop loss because of freezing is avoided.[34]

Effects of higher temperatures on health: People in tropical and semi-tropical regions are already experiencing life-threatening episodes because of severe heat waves, and some are dying because of hyperthermia or dehydration. In the summer of 1980, a severe hot spell hit the United States, and almost 1700 people lost their lives from heat-related illness. Also, higher temperature increases the moistness of the atmosphere, generating tropical weather conditions that create breeding advantages for mosquitoes, and drive rodents from burrows.[35] No firm data exist however, for effects on health.

Effects of climatic changes on ecosystems: Three basic mechanisms are involved in the impacts on ecosystems, *geographic displacement, ecosystem mismatch and switch among alternative stable states.* Increased temperatures will *displace* species relative to their current *niche.* If there are no cold mountain peaks, a species preferring a cold niche may be driven to extinction because temperatures increase above the species' tolerance limits, assuming that it cannot cross the lowlands.[36] Species *mismatch* means that many organisms depend on other organisms to provide certain services, for example to supply food at critical periods in their life stages. If such symbiotic relations are upset, ecosystems will suffer. An example is when fish larva have used up their yolk, it is then important that other food items are available. If one of these processes is triggered by temperature, the chances of mismatch increase, but we have so far seen no evidence of this effect. A stable state, like the arctic ice caps or the arctic tundra, may switch into alternative states, and the process may be reinforcing. Melting tundra can release methane from the soil, causing rising temperatures, which again increases the melting process. This is called a *runaway process*, known from nuclear power plant disasters.

5. MITIGATION MEASURES

There are basically two mitigation strategies, a proactive and a retroactive one: One may proactively try to reduce global warming, or one may retroactively try to abate the detrimental effects. The proactive strategy – to slow down global warming – has basically two options: one must limit emissions of CO_2, or reduce concentrations in the atmosphere by CO_2 trapping. This leaves us with three main strategies, which by no means are mutually exclusive:

1. Limitation of CO_2 emission.
2. CO_2 trapping.
3. Mitigating consequences of global warming.

5.1 Limitation of CO_2 Emission

Energy production by burning of fossil fuels is the singularly most important cause of CO_2 emissions. But energy can be generated in two basically different ways, by using renewable or non-renewable sources.

Current efforts are therefore directed towards relying less on exhaustible resources like coal, oil and gas, and more on renewable resources like sun power, wood, wind, hydropower, thermal energy, and even nuclear energy. But all this is done in the face of an ever-increasing demand for energy, since energy consumption increases all over the world. And renewable resources also have environmental impacts, some of which were discussed in Chapter 20 on agriculture and land use.

5.1.1 Production Costs and Environmental Costs of Renewable Resources

To calculate end impacts of energy production from renewable resources, we have to know the amount of wood, peat, pellets and hydropower used for electricity, heating and other purposes. A total of 17,147 MW of installed electric power generating capacity was available from US non-utility generating facilities in 1996, Table 22-3. The national average cost for renewable power was 8.78 ¢ kWh^{-1}. This compares with 1.8–5.5 ¢ kWh^{-1} for nonrenewable sources, or 3 ¢ kWh^{-1} from new high-efficiency gas turbines.[37]

Table 22-3. Energy costs in the United States. Energy generating devices. Data Cooney,[38] Jacobson (J);[39] and EC.[40] Some technologies are presently on an experimental stage and few reliable data are available.

Technology	Capacity MW	Device	Cost ¢ kWh^{-1}	Environmental costs ¢ kWh^{-1}
Municipal solid waste and landfills	3063		7	none
Wood and wood waste, pellets	7053	Incineration	10	Source dependent[(1)]
Other biomass	267		13	Source dependent[(1)]
Conventional hydro	3419	Turbines	7	Site dependent
Solar heating	–	Water filled panels		Aesthetics
Solar electricity	354	Solar panels	16	Aesthetics
Solar Power satellites	–	–	–	–
Wind	1670	Windmills	4 (J)–12	
Geothermal	1346		12	
Wave energy	–	Floating elements		
Coal		Coal fired plant	3.5–4	2–4.3 (J); 3–15 (EC)
Oil	–	Oil fired plant	–	3–11 (EC)
Gas	–	Gas fired plant	–	1–4 (EC)
Nuclear fusion	–	Nuclear fusion reactors	–	–
Fossil fuel from which carbon has been sequestered			20	–

(1) Fuel wood forests may function as suitable ecosystems

Energy costs from a new coal power plant are also low, 3.5–4 ¢ kWh^{-1}, but including externalities, like coal mine dust, acid depositions, smog, visibility degradation, and global warming, the cost may be as much as 5.5 to 8.3 ¢ kWh^{-1}.

5.1.2 Alternatives to Fossil Fuels

Biomass: When wood is used as fuel, parts of it are usually natural undergrowth forest, and parts of it come from tree farming. Tree farming has a negative and a positive side. Tree farms are only partially natural ecosystems. They have for instance logging roads, which disturb wildlife and the function of the forest as a natural ecosystem. On the other hand, tree farming increases the stock of trees and therefore the amount of wood that traps CO_2.

Wind: Wind power could provide 10% of the words electricity requirements – 1.2 million MW by 2020, cutting global CO_2 emissions by more than 10×10^9 tons. Denmark already generates more than 12% of its electricity from wind.[41] A windmill may extract kinetic energy from moving air that is proportional to the cube of the wind speed. To be commercially viable, average wind speeds should be above about 7 to 8 ms^{-1}, which in the United States occurs all across North Dakota, in 70% of South Dakota, and over large tracts of the West, Great Plains, and in the Northeast. Windmill areas are typically developed with six turbines per square kilometer. To reduce greenhouse gasses by 1% in the US and thus replacing 8% of the coal fired plants, wind turbines could be spread over a 73×73 km^2 large patch of farmland or ocean.[42] There is, however, considerable opposition to windmills from an environmental point of view. They are by some considered unsightly and noisy, and require new roads and power transmission lines. It is therefore not obvious that the benefits are worth the costs.

Waves: Several attempts to extract energy from waves have been made. Some contraptions use moving objects floating in the water. A second category is land-based installations with devices extending out into the water to harvest energy. A third category funnels waves up into elevated reservoirs on land, where the energy is tapped from the water as it runs out of the reservoir, like an ordinary hydroelectric plant.

Waterfall: Hydroelectric plants with water reservoirs that store water for continuous use account for about 6% of global energy production. The external costs of hydropower consist of damages to recreation, cultural heritage objects, and ecosystems. Large tracts of land may be inundated; local climate conditions may change, and decaying organic matter may emit methane. The ecosystem of reservoirs may be less desirable than the ecosystem of natural water bodies.

Nuclear energy: Nuclear energy accounts for 5% of global energy production and 16% of electricity production. In the European union, 31% of the commercial electricity in 2003 was provided by nuclear reactors. About 150 of the 438 nuclear reactors in the world are in Europe, and 58 of them are in France. When the development of nuclear energy started in the middle of the 20th century, there were high hopes that this would solve the world's future energy problems. Nuclear power plants in normal operation are rather clean from an environmental point of view. There are almost no emissions to the atmosphere; no greenhouse gases, only small amounts of radioactive krypton gas with a half-life of 19 days. However, there is a significant problem how to find safe storage for nuclear waste products, which release radioactive radiation that can last for thousands of years. The largest obstacle to general embracement of nuclear power, however, is fear – fear of accidents like Chernobyl or worse, and fear of nuclear proliferation where nuclear technology is used to create nuclear weapons. Nuclear energy has therefore had a bad reputation for quite a while, but it now seems that reconsideration is taking place – especially in light of the problems with fossil fuels. One example is the Organization for Economic Cooperation and Development – OECD, which now apparently wants to evaluate nuclear power as a means to reverse global warming.[43]

5.2 Trapping CO_2

Current projects consider tree farming, incorporation of CO_2 in soils and deposition of CO_2 below the ocean floor as means to reduce the CO_2 concentration in the atmosphere.

There are approximately 610 Gt carbon (1 Gt = 10^9 tons) trapped in the land biomass, and an additional 1600 Gt carbon in the form of soil and detritus. This is about three times as much carbon as in the CO_2 in the Earth's atmosphere. Carbon is stored in forests, soils and wetlands. Tropical forests can sequester as much as 200 metric tons of carbon per hectare, about two times the storage capacity of forests at mid- and high latitudes.[44] However, the annual net uptake of CO_2 by a deciduous forest in New England (1991–1995) was only 1.4 to 2.8 tons carbon ha^{-1}.[45] Clearing forests in Amazon released 194 t of carbon per hectare as net emission after deduction of re-growth of secondary forest and other components of the replacement landscape.[46] A new interesting storage is forest re-growth on abandoned farmland and formerly logged forests.

Carbon may also be stored in soil, peat, and coral reefs. There is 150,000 km^2 of peat land in the Amazon, often more than a meter thick. Peat land may sequester 1.2 ton of carbon per hectare during summer, but on the other hand release up to 0.83 ton of carbon per hectare when the water table is low. The net sequestering capacity is not known, but probably positive.[47]

Over long time spans there will be a balance between wood growth and wood decay so that the net contribution from forest standing stock will be nil. However, if we assume that industrial use of wood contributes to a larger standing stock of forest than if the wood was not used for industrial purposes, this fraction can be ascribed as a net gain for storage of CO_2 and credited to the industry.

Some researchers claim that carbon storage in wood used for building materials, and thus intended for permanent use, should be counted as a net sink. A storage time of 10 years has been suggested as a minimum requirement for giving wood storage "permanent sink" status.

5.3 Mitigating Consequences of Global Warming

Given that global warming with consequential elevation of the sea level is a reality, detrimental effects may be abated in different ways. The notion of "the dumb farmer" has been devised to characterize a farmer that continues her farming practice in spite of increased temperatures or decreased water availability. A wise farmer would probably plant new varieties that fit the new temperature and precipitation regimes, build new irrigation systems if required, and so on.

To mitigate the effects of a higher sea level, coastal protection structures can be built, like bulkheads and levees, pumping sand and raising barrier islands. Associated with the global warming benefits of mitigating greenhouse gases is also air pollution and human health benefits, because decreases in CO_2 emissions also decrease emissions of particulates, SO_2 and NO_x.[48]

6. APPLICATION: WINDMILLS[49]

Noisy energy with a wiev

Windmills are favored as a renewable and emission-free energy source, but they do have impacts that depend on the location. As windmill parks spread, it becomes more and more important to perform rational analyses of alternative sites. Here is an outline of a framework.

Figure 22-3. Windmill park in Navarre, Spain. Photo EHN, Navarre, Spain.

6.1 Decision Analysis

Decision problem: Identify the most suitable site to locate a windmill park, judged according to environment and economy.

Decision makers: Who are the decision makers depends on the political system, but there are usually two levels – a political and a commercial, for example the local administration and the CEO of the commercial electricity company in charge of the development. (See Chapter 3, section 2.3)

Stakeholders: Stakeholders are people living near the candidate sites, environmentalist groups, the general public that wants to use "renewable" energy, and the government that wants to increase the share of energy from renewable sources, as well as other governmental bodies. In the USA, this includes the Army Corps of Engineers, the Coast Guard, the Federal Aviation Administration, the Energy Facilities Siting Board, the State

Executive Office of Environmental Affairs, and in coastal areas the State office of Coastal Zone Management.

Objectives: Many countries have Renewable Portfolio Standards that set goals for introduction of non-renewable energy. In this case, the overall goal is to maximize the amount of energy at the least economic and social costs. Natural sub-goals include:

- Maximize annual production of energy.
- Minimize periods with little or no production.
- Minimize construction and operating costs.
- Minimize impacts on the local fauna.
- Minimize the number of people exposed to noise.
- Minimize landscape deterioration (windmills, roads and electric cables).
- Minimize effects on commercial navigation and leisure boating.

Alternatives: The *alternatives* are the candidate sites

Consequence analysis: To maximize energy production, the windmill park has to be sited at a windy place, with wind speeds at least $8ms^{-1}$ (This chapter, section 2). The effect of windmills on birds is uncertain, but probably small. Chapter 21, section 5 deals with conditions that are detrimental to birds, in contrast to the models in Chapter 19, section 3.2 which deals with conditions that are optimal for a species. Sea mammals are observed at about the same density close to windmills as at similar habitats. Windmills generate noise, but about 40 m away the noise is at the level of normal conversation. A population weighted sound impact estimate can be made as shown in Chapter 10, section 3.4. People differ with regard to appreciation of the aesthetic value of windmills; some find them fascinating, many think they are unsightly; a questionnaire type survey among neighbors to windmills may be useful as described in Chapter 11. If offshore windmills are spaced at least 700 m apart, they do not interfere with navigation.

Utility functions: Some of the utility functions of this analysis will probably be judged non-linear, for example for landscape deterioration. From the analysis should result a consequence and a utility table like Tables 3-5 and 3-6, and primary utility graphs, like those in Figure 3-3.

Preferences: Members of the board of the windmill park company will express preferences among the end-points. However, these preferences will be influenced by those of the general public, and by the people living in the neighborhood of the alternative sites. Pictures, such as Figure 22- 3, with and without windmills, could be used in a WTP study (Chapter. 8). The results should be importance weights for each criterion as in Table 3-8.

6.2 What Happens?

Windmill parks are now abundant. They are usually sited in sparsely populated areas on hillcrests or open land with high average wind velocity. Windmill parks are increasingly also built offshore. Windmill farms have to some degree become tourist destinations, but this may change when they become more common. A willingness-to-pay survey in Norway has showed that people would be willing to pay about 70 USD yr^{-1} for not having a windmill park in their neighborhood.[50]

---

[1] May et al. (1995)
[2] Wigley and Raper (2001); Allen et al. (2001)
[3] Robock (2002)
[4] Robock (2002), Newhall et al. (2002)
[5] Robock (2002)
[6] Palmer (2000)
[7] Constanza et al. (1997)
[8] first estimate by Pearce (1996), second by Matthews and Lave (2000)
[9] European Commission (1999)
[10] Lugo et al. (2000)
[11] Lugo et al. (2000) Figure 2
[12] Lugo et al. (2000) p. 107, 111
[13] Houghton et al. (1995)
[14] Houghton et al. (1995)
[15] Hocking (1991)
[16] Fearnside (1996) p. 28, Fearnside (1999)
[17] MacLean et al. (1998) p. 327A
[18] Levine et al. (1995)
[19] Wuebbles et al. (1999) p. 60
[20] Folke et al. (1997)
[21] Wuebbles et al. (1999) p. 63
[22] Kyoto (1997)
[23] The terrestrial ecosystem model by (1998), others are listed in Kickert et al. (1999)
[24] Houghton et al. (2001)
[25] Kaya (1989) and discussed by Hoffert et al. (1998) p. 861
[26] Oppenheimer (1998) p. 329
[27] Gregory et al. (1998)
[28] Parry et al. (1998)
[29] Michaelis et al. (1997) p. 225
[30] Epstein (1999) p. 347
[31] Wotawa and Trainer (2000:327)
[32] Carbon cycling: Hungate et al. (1997); oceanic production: Hein et al. (1997) p. 527
[33] Crick et al. (1997)
[34] Australia: Nicholls (1997); UK and Spain: Porter et al. (1999)
[35] Epstein (1999)
[36] Kareiva (1998) p. 21, Scott (1999)

[37] Cooney (1999) p. 495A
[38] Cooney (1999)
[39] Jacobson and Masters (2001)
[40] EC (1995)
[41] Betts (2000)
[42] Jacobson and Masters (2001)
[43] Gran (2005)
[44] Moffat (1997) p. 316
[45] Goulden et al. (1996)
[46] Fearnside (1999) p. 312
[47] Schulman et al. (1999) p. 523
[48] Cifuentes et al. (2001)
[49] http://www.capewind.org/index.htm
[50] Nordahl (2000)

REFERENCES

(1997). The international dolphin conservation program act.

(NERC), (1975). Flood studies report 1975. London, NERC, Institute of hydrology.

Aanderaa, R., J. Rolstad and S. M. Søgnen, Eds. (1996). *Biological diversity in forest*. Oslo, The Norwegian Forest Owners Federation.

Ahlborg, U. A. (1996). "Methods of risk assessment." *The Science of the total environment* **188 Supply**: 75-77.

Alerstam, T. and P. E. Jönsson (1999). "Ecology of tundra birds: patterns of distribution, breeding and migration along the Northeast passage." *Ambio* **28**: 212-224.

Allan, J. D. (1995). *Stream ecology. Structure and function of running waters*. London, Chapman and Hall.

Allen, A. W. (1982). Habitat suitability index models. Fort Collins, Colorado, U.S. Fish and Wildlife service.

Allen, M., S. Raper and J. Mitchell (2001). "Uncertainty in the IPCC's assessment report." *Science* **293**: 430-433.

Allen, M. and K. Tugend, I. (2004). "Effects of large-scale habitat enhancement project on habitat quality for age -0 largemouth bass at Lake Kissimmee, Florida." *Journal of lakes and reservoir management* **20**(1): 54-64.

Alston, L., J., G. Libecap, D. and B. Mueller (2000). "Land reform politics, the sources of violent conflict, and implications for deforestation in the Brazilian Amazon." *Journal of environmental economics and management* **39**: 162-188.

Ambrosius, E. (1923). *World Atlas*. Copenhagen, Heinrich Koppel.

Andersen, T. (1997). *Pelagic nutrient cycles, herbivores as sources and sinks*. Heidelberg, Springer Verlag.

Ando, A., J. Camm, S. Polasky and A. Solow. (1998). "Species distribution, land values, and efficient conservation." *Science* **279**: 2126-2127.

Andreoni, J. and A. Levinson (2001). "The simple analytics of the environmental Kuznets curve." *Journal of public economics* **80**: 269-286.

Andrews, M. (1989). *The search for the picturesque: landscape aesthetics*. Aldershot, Scholar press.

Anker-Nilsen, T. (1987). Methods for consequence analysis oil/seabirds. Trondheim, Directorate for nature management.

Anonymous (1998). Standard for sustainable forest in region in Western parts of Eastern Norway. Oslo, Living forest: 1-12.

Anonymous (2005). A survey of corporate social responsibility. *The Economist*.

Appleton, J. (1975). *The experience of landscape*. London, Wiley.

Ashton, P. J. (2002). "Avoiding conflicts over Africa's water resources." *Ambio* **31**(3): 236-242.

Atkinson, S. F. (1985). "Habitat-based methods for biological impact assessment." *The environmental professional* **7**: 265-282.

Aunan, K., J. H. Fang, H. Vennemo, K. Oye and H. M. Seip (2004). "Co-benefits of climate policy - lessons learned from a study in Shanxi, China." *Energy Policy* **32**(4): 567-581.

Bach, W. (1998). "The climate protection strategy revisited." *Ambio* **27**: 498- 505.

Baker, R., J., R. Van der Bussche, A., A. J. Wright, L. Wiggins, E., M. Hamilton, J., E. P. Reat, M. Smith, H., M. Lomakin, D. and R. Chesser, K. (1996). "High levels of genetic change in rodents of Chernobyl." *Nature* **380**: 707-709.

Balmford, A., J. E. Moore, T. Brooks, N. Burgess, L. A. Hansen, J. C. Lovett, S. Tokumne, P. Williams, F. I. Woodward and C. Rahbek (2001). "People and diversity in Africa." *Science* **293**: 1591-1592.

Balmford, A., J. E. Moore, T. Brooks, N. Burgess, P. Williams and C. Rahbek (2001). "Conservation conflicts across Africa." *Science* **291**: 2616-2619.

Barbier, E. B., J. C. Burgess and C. Folke (1994). *Paradise Lost? The ecological economics of diversity* London, Earthscan publications.

Barlaz, M. A., A. Rooker, P., P. Kjeldsen and R. C. Borden (2002). "Critical evaluation of factors required to terminate the post closure monitoring period at solid waste landfills." *Environmental Sience and Technology* **36**(16): 3457-3464.

Basu, B. K. and F. R. Pick (1996). "Factors regulating phytoplankton and zooplankton biomass in temperate rivers." *Limnol. Oceanogr.* **41**: 1572-1577.

Bays, J. S. and T. L. Chrisman (1983). "Zooplankton and trophic state relationships in Florida Lakes." *Can. J. Fish. Aquat. Sci.* **40**: 1813-1819.

Begon, M., J. L. Harper and C. Townsend, R. (1995). *Ecology. Individuals, populations and communities*. Cambridge, MA, Blackwell science.

Belovsky, G. E. (1987). Extinction models and mammalian persistence. *Viable populations for conservation*. Soulé. Cambridge, Cambridge university press: 35-57.

Belovsky, G. E., C. Mellison, C. Larson and P. A. V. Zandt. (1999). "Experimental studies of extinction dynamics." *Science* **286**: 1175-1177.

Belton, V. and T. J. Stewart (2002). *Multiple Criteria Decision Analysis*. Dordrecht, Kluwer academic publishers.

Betts, K., S. (2000). "The wind at the end of the tunnel." *Environmental Science and Technology* **July 1**: 306A-312A.

Bhattacharya, A. and S. K. Sarkar (2003). "Impact of overexploitation of shellfish: Notheastern coast of India." *Ambio* **32**(1): 70-75.

Biggs, B. J. F. (2000). "Eutrophication of streams and rivers: dissolved nutrients - chlorophyll relationships for benthic algae." *Journal of the North American benthologcal society* **19**: 17-31.

Blackburn, S. (1994). *The Oxford Dictionary of Philosophy*. Oxford:, Oxford University Press.

Blackburn, S. (1998). *Ruling passions. A theory of practical reasoning*. Oxford, Clarendon.

Blackburn, T. M. and R. P. Duncan (2001). "Determinants of establishment success in introduced birds." *Nature* **414**: 195-197.

Blacksell, M. and A. Gilg (1981). *The countryside: Planning and change*. London, Allen and Unwin.

Bolt, S., P. Dyson, L. K., T. J., S. Lumley and B. Lane (2001). Options for the Productive Use of Salinity. Canberra, (OPUS), PPK, Adelaide, and the National Dryland Salinity Program.

Bonvechio, T. F. and M. S. Allen (2005). "Relations between hydrological variables and year class- strength of sports fish in eight Florida water bodies." *Hydrobiologia* **to appear**.

Boyce, J. K. (1994). "Inequality as a cause of environmental degradation." *Ecological economics* **11**: 169-179.

Boyle, K. J. (1994). "An investigation of part-whole biases in contingent-valuation studies." *Journal of Environmental Economics and Management*, **27**: 64-83.

Brabrand, Å., B. Faafeng and J. P. M. Nilssen (1990). "Relative importance of phosphorus supply to phytoplankton production: fish excretion versus external loading." *Can. J. Fish Aquat. Sci.* **47**(364-372).

Bradley, M. M., M. Codispoti, B. N. Cuthbert and P. J. Lang (2001). "Emotion and motivation I: Defensive and appetitive reactions in picture processing." *Emotions* **1**(3): 276-298.

Brembs, B. (1996). "Chaos, cheating and cooperation: potential solutions to the Prisoner's dilemma." *OIKOS* **76**: 14-24.

Brennan, A. (2002). *Environmental ethics*, Stanford University.

Briggs, D. and F. Courtney (1994). *Agriculture and environment. The physical geography of temperate agricultural systems*. Harlow, Longman Scientific and Technical.

Briggs, D. J. and J. France (1981). "Assessing landscape attraction." *Landscape Research* **6**: 2-5.

Briggs, D. J. and J. France (1981). "Landscape evaluation: a comparative study." *J. Environmental Management* **10**: 159-168.

Briggs, G. A. (1973). Diffusion estimates for small emissions, Atmospheric turbulence and diffusion laboratory.

Brooks, T. (2000). "Living on the edge." *Nature* **403**: 26-27.

Brooks, T. and A. Balmford (1996). "Atlantic forest extinction." *Nature* **780**(115 (only)).

Bruna, E. M. (1999). "Seed germination in rainforest fragments." *Science* **402**: 139 (only).

Brussaard, L., V. M. Behan-Pelletier, B. D.E., V. K. Brown, W. Didden, P. Folgarait, C. Fragoso, D. W. Freckman, V. V. S. R. Gupta, T. Hattori, H. D.L., C. Klopatek, P. Lavelle, D. W. Malloch, J. Rusek, B. Søderstrøm, J. M. Tiedje and R. A. Virginia (1997). "Biodiversity and ecosystem functioning in soil." *Ambio* **26**: 563-570.

Bruvoll, A. and H. Medin (2003). "Factors behind the environmental Kuznets curve. A decomposition of the changes in air pollution." *Environmental and resource economics* **24**: 27-48.

Buerge, I. J., T. Poiger, M. D. Müller and H.-R. Busher (2003). "Caffeine, as an anthropogenic marker for wastwater contamination of surface waters." *Environmental Science and Technology* **37**(4): 691-700.

Bult, T., S.C. Riley, R.L. Haedrich, R. J. Gibbson and J. Heggenes. (1999). "Density - dependent habitat selection by juvenile Atlantic salmon (Salmo salar) in experimental river habitats." *Can. J. fish. Aquat. Sci.* **56**: 1298-1306.

Bulte, E. H. and G. C. v. Kooten (1999). "Marginal valuation of charismatic species: implications for conservation." *Environmental and resource economics* **14**: 119-130.

Burton, I., R. W. Kates and G. F. White. (1993). *The environment as hazard.* New York., The Guilford press,.

Burton, P. J., A. C. Balisky, L. P. Coward, S. G. Cumming and D. D. Kneeshaw (1992). "The value of managing diversity." *The forestry chronicle* **68**(2): 225-237.

Bøhler, T. (1987). Users guide for the Gaussian type dispersion model CONEX and CONDEI. Lillestrøm, Norwegian Institute for Air Research**:** 30.

Callan, S., J. and J. Thomas, M. (1996). *Environmental economics and management. Theory, policy, and application.* Chicago, IRWIN.

Camuffo, D. and A. Bernardi (1990). *Atmospheric pollution and deterioration of monuments. Advanced from workshop.* Analytical methods for the investigation of damaged stones, Pavia, Consiglio nazionale delle ricerche, ICTR, Padova.

Camuffo, D. B., A. (1996). "Deposition of urban pollution on the Ara Pacis, Rome." *The Science of the total environment* **189/ 190**: 235- 245.

Canfield, D., E., J. Shireman, V., D. E. Colle, W. Haller, T., C. Watkins, E. and M. Maceina, J. (1984). "Prediction of chlorophyll a concentrations in Florida lakes: importance of aquatic macrophytes." *Can. J. fish. Aquat. Sci.* **41**: 497-501.

Canter, L. W. (1996). *Environmental impact assessment.* New York, McGraw-Hill.

Carlsen, A. J., J. Strand and F. Wenstøp (1993). "Implicit environmental costs in hydroelectric development: An analysis of the norwegian plan for water resources." *Journal of environmental economics and management* **25**: 201-211.

Carnegie-Mellon (2003). Economic input-output life cycle assessment (EIO-LCA) Model, Carnegie Mellon University, http://www.eiolca.net.

Carpenter, S. R., D. Bolgrien, C. Lathrop, C. A. Stow, T. Red and M. Wilson (1998). "Ecological and economic analysis of lake eutrophication by nonpoint pollution." *Australian journal of ecology* **23**: 68-79.

Carson, R. (1962). *Silent Spring.* Boston, Houghton Miffin.

Carson, T. N. and L. Gangadharan (2002). "Environmental labeling and incomplete consumer information in laboratory markets." *J Environ. Economics and Management* **43**: 113-134.

Cederlund, G., J. Bergquist, P. Kjellander, R. Gill, J. M. Gaillard, B. Boisaubert, P. Ballon and P. Duncan (1998). *Managing roe deer and their impact on the environment: Maximising the net benefits to society.* Oslo, Scandinavian University Press.

Chameides, W. L., R. D. Saylor and E. B. Cowling. (1999). "Ozone pollution in the rural United States and the new ANNQS." *Science* **276**: 916 (only).

Channell, R. and M. V. Lomolino (2000). "Dynamic biogeography and conservation of endangered species." *Nature* **403**: 84-86.

Chow, V. T., D. R. Maidment and L. Mays (1988). *Applied Hydrology*. New York, McGraw Hill.

Cifuentes, L. (2000). Generación de Instrumentos de Gestión Ambiental para la Actualización del Plan de Descontaminación Atmosférica para la Región Metropolitana de al Año 2000. Parte I. Estimación de los Beneficios Sociales de la Reducción de Emisiones y oncentraciones de Contaminantes Atmosféricos en la Región Metropolitana. Santiago, Chile,. Santiago, Universidad Católica de Chile.

Cifuentes, L., V. Borja-Aburto, H., N. Gouveia, G. Thurston and D. L. Davis (2001). "Hidden health benefits of greenhouse gas mitigation." *Science* **293**: 12561259.

Cochrane, M., A., A. Alencar, M. Schulze, D, C. Souza, Jr, D. Nepstad, C., P. Lefebvre and E. A. Davidson (1999). "Postive feedbacks in the fire dynamic of closed canopy tropical forests." *Science* **284**: 1832-1835.

Cochrane, M., A. and M. Schulze, D. (1999). "Fires as a recurrent event in tropical forests on the Eastern Amazon: Effects on forest structure, biomass, and species composition." *Biotropica* **31**(1): 2-16.

Colborn, T., F. S. V. Saal and A. M. Soto "Developmental effects of endocrine disrupting chemicals in wildlife and humans." *Environ Health Perspect* **101**: 378- 384.

Cold, B., A. Kolstad and S. Larssæther (1998). *Aesthetics, well-being and health*. Oslo, Norsk Form, Center for design, archirecture and the built environment.

Cole, S. (1998). "The emergence of treatment wetlands." *Environmenta Science and Technology* **May 1.**: 218A- 223A.

Coleman, A. (2003). "Depth of strategic reasoning in games." *Trends in cognitive science* **7**(1): 2-4.

Colman, A. M. (1990). *Game theory and experimental games. The study of strategic interaction*. Oxford, Pergamon press.

Constanza, R., R. d´Arge, R. d. Groot, S. Farber, M. Grasso, B. Hannon, K. Limburg, S. Naeem, R. V. O'Neill, J. Paruelo, R. G. Raskin, P. Sutton and M. v. d. B. . (1997). "The value of the world's ecosystem services and natural capital." *Nature* **387**: 253- 260.

Conway, G. and G. Toenniessen (2003). "Science for African food security." *Science* **299**(1187-1188).

Cooney, C., M. (1997). "Nations seek "fair" greenhouse gas treaty in Kyoto." *Environmental Science and Technology* **31**(11): 516A-518.

Cooney, C., M. (1999). "Can renewable energy survive?" *Environmental Science and Technology* **December 1**: 495A-499A.

Cooper, C., M. and W. Lipe, M. (1992). "Water quality and agriculture:Mississippi experiences." *Journal of Soil and Water Conservation* **47**(3): 220-223.

Cowen, T. (1993). "The scope and limits of preference sovereignty." *Economics and philosophy* **9**: 253-269.

Crick, H. Q. P., C. Dudley, D. E. Glue and D. L. Thomson (1997). "UK birds are laying eggs earlier." *Nature* **388**: 526 (only).

Crisp, D. T. (1996). "Environmental requirements of common European species in freshwater with particular reference to physical and chemical aspects." *Hydrobiologia* **323**: 201-221.

Crooks, K. R. and M. E. Soulé (1999). "Mesopredator release and avifaunal extinctions in a fragmented system." *Nature* **400**: 563-566.

Cropper, M. L. and W. E. Oates (1992). " Environmental economics: A survey." *J. Economic literature.* **30**: 675-740.

Currie, D. J., P. Dilworth- Christie and F. Chapleau. (1999). "Assessing the strength of top-down influences of plankton abundance in unmanipulated lakes." *Can. J. Fish. Aquat. Sci.* **56**: 427-436.

Cushing, C. E., K. W. Cummins and G. W. Minshall. (1995). *River and stream ecosystems. Ecosystems of the word*, Elsevier.

Cyr, H. and M. L. Pace (1993). "Magnitude and patterns of herbivory in aquatic and terrestrial ecosystems." *Nature* **361**: 148-150.

Czeh, B. and P. R. Krausman (1997). "Distribution and causation of species endangerment in the United States." *Science* **277**: 1116. (only).

Dalton, R. (2005). "Fish futures." *Nature* **435**: 473-474.

Damasio, A. R. (1994). *Descartes' error. Emotion, reason and the human Brain.* New York, G P Putnam's sons.

Davis, S. H. and A. Wali (1994). "Indigenous land tenure and tropical forest management in Latin America." *Ambio* **23**: 485- 490.

De Lucio, J. V., M. Mohamadian and e. al. (1994). "Visual landscape exploration as revealed by eye movement tracking." *Landscape and urban planning* **29**: 135-142.

Dernbach, L. S. (2000). "The complicated challenge of MTBE cleanups." *Environmental science and technology* **Dec 1.**: 516A-521A.

Dietz, T., E. Ostrom and P. C. Stern (2003). "The struggle to govern commons." *Science* **302**: 1907-1912.

Dijk, H. v., N. A. Onguene and T. Kuyper (2003). "Knowledge and utilization of edible mushrooms by local populations on the rain forest of South Cameroon." *Ambio* **32**(1): 19-23.

Dobson, A. P., J.P. Rodrigues, W.M. Roberts and D. S. Wilcove. (1997). "Geographic distribution of endangered species in the United States." *Science* **275**: 550- 553.

Dobson, A. P. and A. Lyles (2000). "Black-footet ferret recovery." *Science* **288**: 985-988.

dos Santos, B. L. and M. L. Bariff (1988). "A study of user interface aids for model oriented decision support systems." *Management science* **34**(4): 461-468.

Downing, J. A. and E. McCauley (1992). "The nitrogen: phosphorus relationship in lakes." *Limnol. Oceanogr.* **37**: 936-945.

Doyle, R. (2005). "Measuring beauty." *Scientific American*: 18.

Duarte, C. M. and J. Kalff (1986). "Littoral slope as predictor of the maximum biomass of submerged macrophyte communities." *Limniol. Oceanogr.*: 1072-1080.

Duarte, C. M. and J. Kalff (1987). "Latitudinal influences on the depths of maximum colonization and maximum biomass of submerged anginosperms in lakes." *Can J. Fish: Aquat. Sci.* **44**: 1759-1764.

EC (1995). ExterneE externalities of energy. Luxembourg, European Commission DG XII. **6: Wind and hydro**.

Edmondson, W. T. (1991). *The uses of ecology*, University of Washington press,.

Edvardsen, B., F. Moy and E. Paasche (1990). Hemolyte activity in extracts of chrysochromulina polyepis grown at different bethic levels of selenite and phosphate. *Toxic marine phytoplankton.* E. Graneli and e. al. Amsterdam, Elsevier science publishers**:** 284- 289.

Effler, S. W., S. R. Boone, C. Siegfrid and S. L. Ashby (1998). "Dynamics of zebra mussel oxygen demand in Seneca river, New York." *Environmental Science and Technology* **32**(807-812).

Ehrlich, P. and A. Ehrlich (1990). *The population explosion.* New York, Simon and Schuster.

Ellerman, A. D. and A. Decaux (1998). Analysis of Post-Kyoto CO2 Emissions Trading Using Marginal Abatement Curves. Cambridge, USA, Massachusetts Institute of Technology, Joint Program on the Science and Policy of Global Change, Cambridge, MA 02139 , USA.

Enserink, M. (1999). "Biological invaders sweep in." *Science* **285**: 1834-1836.

Epstein, P. R. (1999). "Climate and health." *Science* **285**: 347-348.

Erikson, R. (1993). *Description of inequalities: The Swedish approach to welfare research.* Oxford, Claredon Press.

EuropeanCommission (1999). ExterneE Externalities of Energy. Luxembourg, European Commission DG XII.

Ezcurra, E. and C. Montana (1990). Los rrecursos naturelles renovables en el norte arido de Mexico. *Medio ambiente y desarrollo en México.* E. Leff and M. A. Porrína. Mexico. **1:** 279-327 (In Spanish).

Ezzati, M., B. Singer, H. and D. Kammen, M. (2001). "Towards an integrated framework for development and environmental policy: The dynamics of environmental Kuznets curves." *World development* **29**: 1421-1434.

FAO (1994). *FAO yearbook.* Rome, Food and Agriculture organization of the United Nations.

FAO (1994). Soil map of the world. Revised legend with corrections. Wageningen, International soil reference and information centre.

Fearnside, P. M. (1993). "Deforestation in Brazil Amazon: the effect of population and land tenure." *Ambio* **22**: 537-545.

Fearnside, P. M. (1996). "Amazonian deforestation and global warming: carbon stocks in vegetation replacing Brazil's Amazon forest." *Forest ecology and management.* **80**: 21-34.

Fearnside, P. M. (1996). Socio-economic factors in the management of tropical forest for carbon. *Forest ecosystems, forest management and the global carbon cycle.* M. Apps and D. Price. Berlin, Springer verlag. **40.:** 349-361.

Fearnside, P. M. (1999). "Biodiversity as an environmental service in Brazil's Amazonian forests: Risks, value and conservation." *Environmental conservation* **26**(4): 305-321.

Fedorff, N. (2005). "The difficulties of defining the term "GM"." *Science* **303**: 1765-1767.

Ferng, J.-J. (2001). "Using composition of land multiplier to estimate ecological footprints associated with production activity." *Ecological Economics* **37**: 159-172.

Finkel, E. (1999). “Australian biocontrol beats rabbits, but not rules.” *Science* **285**: 1842 (only).

Finkelman, J. (1996). “Chemical safety and health in Latin America. An overview.” *The Science of the Total Environment.* **188 Suppl.**: 3-29.

Fisher, A. C. and J. V. Krutilla (1975). “Resource conservation, environmental preservation, and the rate of discount.” *Quart. J. Econ.* **89**: 358-370.

Fitter, C. (1995). *Poetry, space, landscape : toward a new theory.* Cambridge, Cambridge University Press.

Folke, C., Aa. Janson, J. Larson and R. Constanza (1997). “Ecosystem appropriation by cities.” *AMBIO* **26**: 167-172.

Freckman, D. W., T. H. Blackburn, L. Brussard, P. Hutchings, M. A. Palmer and P. V. R. Snelgrove. (1997). “Linking biodiversity and ecosystem functioning of soil and sediments.” *Ambio* **26**: 556-562.

Freeman, R. E. (1984). *Strategic management: A stakeholder approach*, Pitman.

Froch, R., A. and N. Gallopoulos (1989). “Strategies for Manufacturing.” *Scientific American* **Sept.**: 94-102.

Føllesdal, D. (1982). “The status of rational assumptions in interpretation and in the explanation of action.” *Dialectica* **36**: 301-316.

Faafeng, B., A., D. Hessen, O., Å. Brabrand and J. P. Nilssen (1990). “Biomanipulation and food-web dynamics- the importance of seasonal stability.” *Hydrobiologia* **200/201**: 119-128.

Galbraith, J. K. (1973). *The Age of uncertainty.* Oslo, Dreyer (Norwegian translation).

Gangadharan, L. and M. Valenzula, Rebecca (2001). “Interrelationships between income, health and the environment, extending the environmental Kuznets curve hypothesis.” *Ecological Economics* **36**: 513-531.

Gascon, C., G. B. Williamson and A. B. d. Fonseca (2000). “Receding forest edges and vanishing reserves.” *Science* **288**: 1356-1358.

Gaston, K. J. (2000). “Global patterns in biodiversity.” *Nature* **405**: 220-227.

Gauri, K. L. and G. C. Holdren (1981). *Environmental Science and Technology* **15**: 386-390.

Gerking, S. D., Ed. (1978). *Ecology of freshwater fish production.* Oxford, Blackwell scientific publications.

Gibbons, R. (1992). *A primer in game theory.* Hertfordshire, Harvester Wheatsheaf.

Gibbs, W. W. (2001). “On the termination of species.” *Scientific American* **November**: 28-37.

Giller, P. S. and B. Malmqvist (1998). *The biology of streams and rivers.* Oxford, Oxford University Press.

Giller, P. S., N. Sangpradub and H. Twomey (1991). “Catastrophic flooding and maroinvertebrate community structure.” *Inter. Verh. limnol.*

Gittenberger, E., D. S. J. Groenenberg, B. Kokshoorn and R. C. Preece (2006). “Molecular trails from hitch-hiking snails.” *Nature* **439**(7075): 409-409.

Glomsrød, S., O.Godal, J. F. Henriksen, S. E. Haagenrud and T. Skancke (1997). Air pollution - impacts and values. Corrosion costs of building materials and cars in Norway. Oslo, State pollution Control Authority.

Gordon, N., D., T. A. McMahon and B. L. Finlayson (1992). *Stream Hydrology. An introduction for ecologists.* Chichester, John Wiley and Sons.

Goulden, M. L., J. W. Munger, S.-M. Fan, B. Daube, c. and S. C. Wofsy (1996). "Exchange of carbon dioxide by deciduous forest: Response to interannual climate variability." *Science* **271**(1576-1578).

Gram, S., L. P. Kvist and A. Cáseres (2001). "The economic importance of products extracted from the Amazonian flood plain forests." *Ambio* **30**: 365-368.

Gran, J. (2005). Editorial. *Cicerone.*

Graves, J. and D. Reavey (1996). *Global environmental change: plants animals and communities*, Harlow: Longman.

Greer, L. and C. v. L. Sels. (1997). "When pollution prevention meets the bottom line." *Environ. Science and Technol.* **31**: 418A - 422A.

Gregory, J. M. and J. Oerlmans (1998). "Simulated future sea-level rise due to glacier melt based on regionally and seasonally resolved temperature changes." *Nature* **391**(474-476).

Grennfelt, P., Ø. Hov and D. Derwnt (1994). "Second generation abatement strategies for NOx, NH3, SO2 and VOCs." *Ambio* **23**: 425-433.

Grenouillet, G., D. Pont and K. L. Seip (2002). "Abundance and species richness as a function of food resources and vegetation structure: juvenile fish assemblages in rivers." *Ecography* **25**: 641-650.

Grigalunas, T. A., T. J. Tyrell, J. B. Dirlam and R. Congar (1983). The tourist industry. *Assessing the social costs of oil spills:The Amoco Cadiz case story*. C. Ehler. Washington, D.C., U.S. department of commerce. NOAA.

Grønn, E. (1990). *Samfunnsøkonomiske emner*. Oslo, Bedriftsøkonomen forlag.

Guilford, J. P. (1954). *Psychometric methods.* New York., McGraw-Hill Book Company.

Guinée, J., B., M. Gorrée, R. Heijungs, G. Huppes, A. d. Koning, L. van Oers, A. Sleeswijk, Wegener, S. Suh, H. A. U. de Haes, H. d. Bruijn, R. van Duin and M. Huijbregts, A. J. (2002). *Handbook on Life Cycle assessment. Operational guide to the ISO standard.* Dordrecht, Kluwer Academic Publisher.

Haberl, H., K.-H. Erb and F. Krausmann (2001). "How to calculate and interpret ecological footprints for long periods of time: the case of Austria 1926-1995." *Ecological Economics* **38**(25-45).

Haftorn, S. (1971). *Norges fugler*. Oslo, Oslo University Press.

Haksver, C., R. Chaganti and R. Cook, G. (2004). "A model of value creation: strategic view." *Journal of business ethics* **49**(291-305).

Hale, M. L., P. Lurz, W.W., M. Shirley, D.F., S. Rushton, R. M. Fuller and K. Wolff (2001). "Impact of landscape management on the genetic structure of red squirrel populations." *Science* **293**: 2246-2248.

Hall, C., J. Pontius, A.S., R. Gil, L. Coleman and J.-Y. Ko (1994). "The environmental consequences of having a baby in the United States." *Population and environment* **15**(6): 505-524.

Hammerstein, P. (1995). "A twofold tragedy unfolds." *Nature* **377**: 478 (only).

Hanna, M. (1990). "Evaluation of models predicting mixing depth." *Can J. Fish: Aquat. Sci.* **47**: 940-947.

Hanna, S., R., G. Briggs, A. and R. Hosker, P. (1982). *Handbook of atmospheric diffusion.* Washington, DC, Dept. of energy. Office of energy research. US Dept. of energy.

Hardin, G. (1968). "The tragedy of the commons." *Science* **162**: 1243-1248.

Harestad, A. S. and F. L. Brunell (1979). "Home range and body weight: A reevaluation." *Ecology* **60**: 389-402.

Hastings, H. M. and G. Sugihara (1993). *Fractals. A users guide for the natural sciences*, Oxford science publications.

Hayes, R. and A. Harestad (2000). "Demography of a recovering wolf population in the Yukon." *Canadian Journal of Zoology* **78**(1): 36-48.

Hayes, R. D. and A. S. Harestad (2000). "Demography of recovering wolf population in the Yukon." *Can.J.Zool.* **78**: 36-48.

Hector, A. e. a. (1999). "Plant diversity and productivity experiments in European Grasslands." *Science* **286**: 1123- 1127.

Hedin, L. O. (2006). "Physiology - Plants on a different scale." *Nature* **439**(7075): 399-400.

Heerwagen, J. H. and G. H. Orians (1993). Humans, habitats, and aesthetics. *The biophilia hypothesis*. S. R. Kellert and O. E. Wilson. Washington D.C., Island Press: 138-172.

Heiberg, A. and K. L. Seip (1994). Water quality Simulations in the river Moselle. Environmental modelling Seminar, Trondheim, SINTEF Group.

Henrich, J., R. Boyd, S. Bowles, C. Camerer, E. Fehr, H. Gintis and R. McElreath (2001). "In search of Homo economicus: Behavioral experiments in 15 small-scale societies." *American Economic Review* **91**(2): 73-78.

Henriques, P. and R (1987). Macrophytes. *The ecological effects of Hydro power development in New Zealand*. P. R. Henriques. Oxford, Oxford University Press.

Herbert, C., E., K. Hobson, A. and J. L. Shutt (2000). "Changes in food web structure affect rates of PCB decline in Herring gulls (Larus argentatus) eggs." *Environmental Science and Technology* **34**(9): 1609-1614.

Heyman, U., S.-O. Ryding and C. Forsberg (1984). "Frequency distribution of water quality variables. Relationships between mean and maximum values." *Water. Res.* **18**(7): 787-794.

Hocking, M. B. (1991). "Paper versus polystyrene: A complex choice." *Science* **251**: 504-505.

Hofbauer, J. and K. Sigmund (2003). "Evolutionary game dynamics." *Bulletin of the American mathematical society* **40**(4): 479-519.

Hoffert, M. I., K. Caldeira, A. K. Jain, E. F. Haites, L. D. D. Harvey, S. D. Potter, M. E. Schlesinger, S. H. Schneider, R. G. Watts, T. M. Wigley and D. J. Wuebbles (1998). "Energy implications of future stabilization of atmospheric CO2 content." *Nature* **395**(881- 884).

Holland, Ø., A. Mysterud, A. Wannag and J. Linnell, D.C. (1998). *Roe deer in northern environments: Physiology and behaviour*. Oslo, Scandinavian University Press.

Holme, J. A. and E. Dybing (1997). "Environmental chemicals with hormone like properties: a human health problem?" *Tidskrift for den norske lægeforening* **117**: 0-73.

Holmgren, M., M. Scheffer, E. Ezcurra, J. R. Gutierrez and M. G.M.J. (2001). "El Nino effects on the dynamics of terrestrial systems." *Trends in ecology and evolution* **16**(2): 89-94.

Hornberger, G., M., J. Reaffensperger, P., P. Wiberg, L. and K. Eshlman, N. (1998). *Elements of physical hydrology*. Baltimore, The Johns Hopkins University Press.

Horton, R., E. (1945). "Erosional development of streams and their drainage basins: Hydrophysical approaches to quantitative morphology." *Bull. Geol. Soc. Am.* **56**: 275-370.

Horton, R. E. (1933). "The role of infiltration in the hydrological cycle." *Trans. Am. Geophs. Union* **14**: 446-460.

Horvath, A., C. T. Hendrickson, L. B. Lave, F. C. McMichael and T.-S. Wu (1995). "Toxic emissions indices for green design and inventory." *Environ. Science & Techn* **29**: 86A - 90A.

Hosper, H. and M.-L. Meijer (1993). "Biomanipulation, will it work for your lake? A simple test for the assessment of chances for clear water, following drastic fish-stock reduction in shallow, eutrophic lakes." *Ecological engineering* **2**: 63-72.

Houghton, J. J. and e. al., Eds. (2001). *Climate change 2001: The scientific basis.* Cambridge, Cambridge University Press.

Houghton, J. J., T. M. Filho, H. Lee, B. A. Callander, E. Haites, N. Harris and K.Maskell, Eds. (1995). *Intergovernemental panel on climate change 1995. Climate change 1994. Radiative forcing of climate change and evaluation of the IPCC IS92 emission scenarios.* Cambridge., Cambridge University Press,.

Howe, C., W. (2002). "Policy issues and international impediments in the management of groundwater: Lessons from case studies." *Environment & Development Economics* **7**: 625-641.

Howe, C. W. (2005). "The functions, impacts and effectiveness of water pricing: Evidence from the United States and Canada." *Water Resources Development* **21**(1): 43-53.

Hsieh, C., S. Glaser, A. Lucas and G. Sugihara (2005). "Distinguishing random environmental fluctuations from ecological catastrophes for the North Pacific Ocean." *NATURE* **435**: 336-340.

Hubin, D. C. (1994). "The moral Justification of benefit/cost analysis." *Economics and philosophy*(10): 169-194.

Hughes, J. B., G. C. Daily and P. R. Ehrlich (1997). "Population diversity, Its extent and extinction." *Science* **278**: 689-692.

Hui, X., Q. Yi, P. Bu-zhuo, J. Xiliu and H. Xiao-mei (2003). "Environmental pesticide pollution and its counter -measures in China." *Ambio* **32**(1): 78-80.

Hull, R. B. and I. D. Bishop (1988). "Scenic impacts of electric transmission towers: The influence of landscape types and observers distance." *J. environmental Management* **27**: 98-108.

Hume, D. (1988). *An enquiry concerning human understanding (1748)*. Chicago, William Benton.

Huston, M. (1993). "Biological diversity, soils, and economics." *Science* **262**: 1676-1680.

Imboden, D. M., U. Lemmin, T. Joller and M. Scurter (1983). "Mixing processes ion lakes: mechanisms and ecological relevance." *Schweiz. Z. Hydrol.* **45**: 11-44.

Inamdar, A., H. de Jode, K. Lindsay and S. Cobb (1999). "Capitalizing on nature: protected area management." *Science* **283**(1856-1857).

Iversen, T., N.E. Halvorsen, S. Mylona and H. Sandnes. (1991). *Calculating budgets for airborne components in Europe 1985, 1987, 1988, 1989 and 1990. EMEP Co-operative programme for monitoring and evaluation of the long range transmission of air pollutants in Europe*. Oslo, The Norwegian meteorological institute.

Jacobson, M., Z. and G. Masters, M. (2001). "Exploiting wind versus coal." *Science* **293**: 1438 (only).

James, A. N., K. J. Gaston and A. Balmford. (1999). "Balancing the earth's accounts." *Nature* **401**: 323-325.

James, D. (1994). Economic impact analysis. *The application of economic techniques in environmental impact assessment*. D. James. Dordrecht, Kluwer Academic Publishers: 97-109.

Janzen, D. (1998). "Gardenification of wildland nature and the human footprint." *Science* **279**: 1312-1313.

Jarvis, P., J. (2000). *Ecological principles and environmental issues*. Harlow, Prentice Hall.

Jensen, M. C. (2001). "Value maximization, stakeholder theory, and the corporate objective function." *J. Applied Corporate Finance* **14**(3).

Jetz, W., C. Carbone, J. Fulfurd and J. H. Brown (2004). "The scaling of animal space use." *Science* **306**: 266-268.

Johnson, C. J., D. Seip, R. and M. Boyce, S. (2004). "A quantitative approach to conservation planning: using resource selection functions to map distribution of mountain caribou at multiple scales." *Journal of applied ecology* **41**: 238-251.

Johnson, N., C. Revenga and J. Echeverria (2001). "Managing water for people and nature." *Science* **292**: 1071-1072.

Jones, C. G., J. H. Lawton and M. Shacha (1997). "Positive and negative effects of organisms as physical ecosystem engineers." *Ecology* **78**: :1946-1957.

Jones, J. A. A. (1997). *Global hydrology*, Longman.

Jowett, I., G. (1993). "A method for objectively identifying pool, run, and riffle habitats from physical measurements." *New Zealand Journal of marine and freshwater research* **27**: 241-248.

Kahneman, D. and J. L. Knetsch (1992). "Valuing Public Goods: The Purchase of Moral Satisfaction." *Journal of Environmental Economics and Management,* **22**: 57-70.

Kahneman, D. and A. Tversky (1979). "Prospect theory: An analysis of decisions under risk." *Econometrica* **47**: 263-291.

Kahneman, D. and A. Tversky (1996). "On the reality of cognitive illusions." *Psychological review* **103**(3): 582-591.

Kaiser, J. (2001). "Bold corridor project confronts political reality." *Science* **293**(2196-2199).

Kaly, U. L. and G. P. Jones (1998). "Mangrove restoration. A potential tool for coastal management in tropical developing countries." *Ambio* **27**: 656- 661.

Kant, I. (1983). *Grundlegene zur Metaphysic der Sitten*. Indianapolis, Hackett Publisher, Co.

Kaplan, S. and R. Kaplan (1983). "A model of person-environment compatibility." *Environment and behaviour* **15**: 311-332.

Kareiva, P. (1998). *Application: Tomorrow's extinction of mammals*, Prentice Hall.

Karlén, W. (2001). "Global temperature forced by solar irradiation and greenhouse gases?" *Ambio* **30**: 349-350.

Karr, J. R., K. D. Fausch and I. J. Schlosser (1987). "Spatial and temporal variability of the index of biotic integrity in Midwestern streams." *Trans. Am.Fish. Soc.* **116**: 1-11.

Kaya, Y. (1989). "Impact of carbon dioxide emission on GNP growth: Interpretation of proposed scenario." *Strategic working group memorandum*, IPCC-report.

Keeney, R. L. (1980). *Siting of energy facilities*. New York, Wiley.

Keeney, R. L. (1992). *Value focused thinking.* Cambridge, MA, Harward University Press.

Keller, A. E. and T. L. Crisman (1990). "Factors influencing fish assemblages and species richness in subtropical Florida lakes and a comparison with temperate lakes." *Can J. Fish: Aquat. Sci.* **47**: 2137-2146.

Kerr, J. T. and L. Parker (1997). "Habitat heterogeneity as a determinant of mammal species richness in high-energy regions." *Nature* **385**: 252- 254.

Kickert, R. N., G.Tonella, A. Simovov and S. Krupa (1999). "Predictive modelling of effects under global change." *Environmental pollution* **100**: 87-132.

Kiely, G. (1997). *Environmental Engineering*, McGraw Hill.

King, J. R. and G. A. Sanger (1979). Oil vulnerability index for marine oriented birds. *Conservation of marine birds of Northern America*. J. C. Bartonek and D. N. Nettleship., US dept. Interior. Fish Wildl. Serv. Wild. Res. **Rep. No 11:** 227-239.

King, S. E. and J. Lester, N (1995). "The value of salt march as a sea defence." *Marine Pollution Bulletin* **30**(3): 180-189.

Kirchmann, H., A. E. J. Johnston and L. F. Bergstrøm (2002). "Possibilities for reducing nitrate leaching from agricultural land." *Ambio* **31**(5): 404-408.

Klieve, H. and T. G. MacAuley (1993). "A game theory analysis of management strategies for the Southern bluefin tuna industry." *Australian J. agricultural Economics* **37**: 17-32.

Knapp, B. J. (1978). Infiltration and storage of soil water. *Hillslope hydrology*. M. J. Kirby. Chichester, John Wiley & Sons.: 44- 72.

Knapp, R., A. and H. K. Preisler (1999). "Is it possible to predict habitat use by spawning salmonids. A test using California golden trout (Oncorhynchus mykiss aquabonita)." *Can.J.Fish.Aquat. Sci.* **56**: 1576-1584.

Knisel, W. G. (1980). CREAMS: A field scale model for chemicals, runoff and erosion from agricultural management systems. Washington, D.C., USDA.

Knowlton, M. F., M. V. Hoyer and J. R. Jones (1984). "Sources of variability in phosphorus and chlorophyll and their effects on use of lake survey data." *Water resources bulletin* **20**(3): 397-407.

Knutzen, J. (1999). "The significance of reference values for monitoring of toxic chemicals in water." *Vann* **3**: 569-581 (In Norwegian).

Kohnke, H. (1970). *Soil physics*. Bombay, Tata - McGraw Hill.

Kotak, B. G., A. K.-Y. Lam, E. E. Prepas and S. Hrudey (2000). "Role of chemical and physical variables in regulating microcystin-LR concentrations in phytoplankton of eutropic lakes." *Can J. Fish. Aquat. Sci.* **57**: 1584-1593.

Krajick, K. (2001). "Defending deadwood." *Science* **293**: 1579-1581.

Krebs, C. (1995). *Ecology. The experimental analysis of distribution and abundance*, Harper & Row.

Krebs, C. (2001). *Ecology. The experimental analysis of distribution and abundance.* San Francisco, Benjamin Cummings.

Krewitt, W., T. Heck, A. Trukenmüller and R. Friedrich (1998). "Environmental damage from fossil electricity generation in German and Europe." *Energy Policy* **27**: 173-183.

Kula, E. (1994). *Economics of natural resources, the environment and policies.*, Chapman & Hall.

Kyoto, P. (1997). The Third conference of the parties (COP-3) to the United Nations Framework on Climate Change (UNFCCC), December 1997. Kyoto, United Nations.

Lampert, W. and U. Sommer (1997). *Limnology: The ecology of lakes and streams.* Oxford, Oxford university press.

Lande, R., S. Engen and B. E. Saether (1994). "Optimal Harvesting, Economic Discounting and Extinction Risk in Fluctuating Populations." *Nature* **372**(6501): 88-90.

Landsberg, H. E. (1970). "Man-made climate changes." *Science* **170**(1265-1274).

Laurance, W. F., P. Delamônica, S. G. Laurance, H. Vasconcelos, L. and T. E. Lovejoy (2000). "Rainforest fragmentation kills big trees." *Science* **404**: 836 (only).

Laurence, W., F., S. G. Laurance, L. V. Ferreira, J. Rankin-deMerona, C. Gascon and T. E. Lovejoy (1997). "Biomass collapse in Amazonian fragments." *Science* **278**(7): 1117.

Lee, K. (1996). "The Source and Locus of Intrinsic Value: A Reexamination." *Environmental Ethics* **18**: 297-309.

Levine, J., S, W. R. Cofer III, j. Cahoon, Donald, R. and E. Winstead, L. (1995). "Biomass burning. A driver for global change." *Environmental Science and Technology* **29**(3): 120A-125A.

Li, W. and Q. Yang (1995). "Wetland utilization in Lake Taihu for fish farming and improvement of lake-water quality." *Ecological engineering* **5**(1): 107-121.

Lid, J. and D. T. Lid (1994). *Norsk Flora.* Oslo, Det Norske Samlaget.

Linkov, I., D. Burmistrov, J. Cura and T. S. Bridges (2002). "Risk-based management of contaminated sediments: Consideration of spatial and temporal patterns in exposure modelling." Environmental science and technology **36**(2): 238-246.

Lugo, A. E., C. S. Rogers and S. W. Nixon (2000). "Hurricanes, coral reefs and rainforests: Resistance, ruin and recovery in the Caribbean." *Ambio* **29**(2): 106-116.

Lumley, S. (1997). "The environment and the ethics of discounting: An empirical analysis." *Ecological Economics* **20**: 71-82.

Lumley, S. (2002). *Sustainability and degradation in less developed countries. Immolatng the future?* Burlington, Ashgate.

Lægreid, M., O.C. Bøckman and O. Kaarstad (1999). *Agriculture fertilizers and the environment.*, CABI Publishing.

Låg, J. (1976). *Soil types, soil profiles, and landscapes in color.* Ås, Landbruksforlaget.

Mace, R. (2000). "Fair game." *Nature* **406**: 248-249.

MacGuire, A. S. and M. J. Childs (1998). "Wastepaper management and protection of forest biodiversity." *J. Environmental planning and management* **41**: 403-410.

MacLean, H. L. and L. B. Lave (1998). "A life-cycle model of an automobile." *Environmental Science and Technology.* **July 1**: 322A-330A.

Madhusudan, M. D. and K. U. Karanth (2002). "Local hunting and the conservation of large mammals in India." *Ambio* **31**(1): 49-54.

Malavoi, J. R. and Y. Souchon (1989). "Methodologie de description et quantification des variables morphodynamiaues d'un cours d'eau à fond caillouteux. Exemple de la Filiére (Haute Savoie)." *Revue geographique de Lyon* **64**: 252- 259.

Mann, C. C. (1999). "Genetic Engineers aim to soup up crop photosynthesis." *Science* **283**: 314-316.

Mann, L. and V. Tolbert (2000). "Soil sustainability in renewable plantings." *Ambio* **29**(8): 492-498.

Mann, R. H. K. (1996). "Environmental requirements of European non-salmonid fish in rivers." *Hydrobiologia* **323**: 223-235.

Manny, B. A., W. C. Jonson and R. G. Wetzel (1994). "Nutrient additions by waterfowl to lakes and reservoirs: predicting their effects on productivity and water quality." *Hydrobiologia* **279/280**: 121-132.

Marlatt, R. M., T. A. Hale and R. G. Sullivan (1993). "Video simulations as part of army environmental decision-making:observations from Camp Shelby, Mississippi." *Environ Impact assessment rev* **13**: 75-88.

Marsh, G. P. (1857). Report made under authority of the legislature of Vermont on the artificial propagation of fish, Burlington, VA, Free press print.

Marshall, C. T. and R. H. Peters (1989). "General patterns in the seasonal development of chlorophyll a for temperate lakes." *Limnio. Oceanogr.* **34**: 856-867.

Mather, A. S. and K. Chapman (1995). *Environmental resources*, Longman Scientific Technical.

Matthews, H. S. and L. Lave, B (2000). "Application of environmental valuation for determining externality costs." *Environmental Science and Technology* **34**(8): 1390-1395.

Matthews, H. S., L. B. Lave and H. L. MacLean (2002). "Life cycle impact assessment: A challenge for risk analysis." *Risk Analysis* **22**(853-860).

Matuszek, J. E. and G. L. Beggs (1988). "Fish species richness in relation to lake area, pH, and other abiotic factors in Ontario lakes." *Can J. Fish: Aquat. Sci.* **45**(1931-1941).

May, R. M. (1976). Models for single populations. *Theoretical ecology*. R. M. May. Oxford, Blackwell scientific publications: 4-25.

May, R. M. (1994). "The economics of extinction." *Nature. London.* **372**: 42-43.

May, R. M., J. H. Lawton and N. E. Stork (1995). Assessing extinction rates. *Extinction rates*. J. H. Lawton and R. M. May. Oxford, Oxford University Press: 1- 21.

Maynard Smith, J. (1993). *Did Darwin get it right? Essays on games, sex and evolution*, Penguin books.

Mazumder, A. and K. E. Havens (1998). "Nutrient chlorophyll-Secchi relationships under contrasting grazer communities of temperate versus subtropical lakes." *Can J. Fish. Aquat. Sci.* **55**: 1652-1662.

McCabe, T. (2004). *Cattle bring us to our enemies: Turkana ecology, politics and raiding in a disequilibrium system*. Ann Arbor, Michigan University Press.

McCann, K., A. Hastings and G. R. Huxel (1998). "Weak trophic interactions and the balance of nature." *Nature* **395**: 794-797.

McCauley, E., J. A. Downing and S. Watson (1989). "Sigmoid relationships between nutrients and chlorophyll among lakes." *Can. J. Fish. Aquat. Sci.* **46**: 1171- 1175.

McEachern, P., E. E. Prepas, J. J. Gibson and W. P. Dinsmore (2000). "Forest fire induced impacts on phosphorus, nitrogen, and chlorophyll a concentrations in subarctic lakes of Northern Alberta." *Can J. Fish. Aquat. Sci.* **57**((suppl 2)): 73-81.

McGranahan, D. A. (1999). Natural resources drive rural population change, Economic research service, U.S. Department of agriculture.

McIntosh, R., W., C. R. Goeldner and J. R. B. Ritchie (1995). *Tourism, Principles, Practices, Philosophies*, John Wiley and Sons.

McNaughton, S., J. M. Oesterheld, D. A. Frank and K. J. Williams. (1989). "Ecosystem-level patterns of primary productivity and herbivory in terrestrial habitats." *Nature* **341**: 142- 144.

McRae, M. (1997). "Is "Good wood" bad for forests?" *Science* **275**: 1868- 1869.

Mercer, D. (1994). Native peoples and tourism: Conflict and compromise. *Global tourism. The next decade.* W. Theobald, F. Oxford, Butterworth-Heineman: 124-145.

Mertz, G. and R. A. Myers (1998). "A simplified formulation for fish production." *Can.J.Fish.Aquat. Sci.* **55**: 478-484.

Michaelis, A., D. Malmquist, A. Knap and A. Close (1997). "Climate science and insurance risk." *Nature* **389**: 225-227.

Mill, J. S. (1848 (1965)). *Principles of political economy.*

Moen, J. and L. Oksanen (1991). "Ecosystem trends." *Nature* **353**: 510 (only).

Moffat, A. S. (1997). "Resurgent forests can be greenhouse gas sponges." *Science* **277**: 315-316.

Montagnini, F. and R. Mendelsohn (1997). "Managing forest fallows: improving the economics of swidden agriculture." *AMBIO* **26**: 118- 123.

Montgomery, D. R., E. M. Beamer, G. R. Pess and T. P. Quinn (1999). "Channel type and salmonid spawning distribution and abundance." *Can. J. Fish. Aquat. Sci.* **56**: 377-387.

Morell, V. (1997). "Counting creatures of the Serengeti, great and small." *Science* **278**: 2058-2060.

Morgan, M. G. (2000). "Risk management should be about efficiency and equity." *Environmental Science and Technology* **January 1**: 32A- 34A.

Morgan, R. P. C. (1995). *Soil erosion and conservation.*, Longman group.

Moss, B. (1990). "Engineering and biological approaches to the restoration from eutrophication of shallow lakes in which plant communities are important components." *Hydrobiologia* **200/201**: 367-377.

Mourato, S., E. Ozdemiroglu and V. Foster (2000). "Evaluating health and environmental impacts of pesticide use: implications for design of ecolabels and pesticide taxes." *Environmental Science and Technology* **34**: 1456-1461.

Mourato, S., E. Ozdemiroglu and V. Foster (2000). "Evaluating health and environmental impacts of pesticide use: implications for the design of ecolabels and pesticide taxes." *8* **34**(8): 1456-1461.

Musters, C. J., H. J. deGraaf and W. J. ter Keurs (2000). "Can protected areas be expanded in Africa?" *Science* **287**(1759-1760).

Myers, N. (2000). "Sustainable consumption." *Science* **287**: 2419 (only).

Myers, N., R. A. Mittelmeier, C. G. Mittelmeier, G. A. B. d. Fonseca and J. Kent (2000). "Biodiversity hotspots for conservation priorities." *Nature* **403**: 853-858.

Navrud , S. (1995). "What is the value of nature experiences? Economical Valuation of natural land resources." *LØF* **4**: 39-53. (In Norwegian).

Navrud, S. (2001). "Environmental costs of hydro compared with other energy options." *Hydropower and dams* **2**: 44-46.

Naylor, R. (1996). "Invasion in Agriculture: assessing the cost of the Golden Apple Snail in Asia." *Ambio* **25**: 443 - 448.

Naylor, R., L., S. Williams, L. and D. R. Strong (2001). "Aquaculture - a gateway for exotic species." *Science* **294**: 1655-1656.

Nee, S. and R. M. May (1997). "Extinction and loss of evolutionary history." *Science* **278**: 692-694.

Neumayer, E. (2002). "Do democracies exhibit stronger international environmental commitment?" *Journal of peace research* **39**(2).

Newhall, C., G. (2000). "Mount St. Helens, master teatcher." *Science* **288**: 1181-1182.

Newhall, C., G., J. A. Power and R. S. Punongbayan (2002). ""To make grow"." *Science* **295**: 1241-1242.

Nicholls, K. H. and G. J. Hopkins. (1993). "Recent changes in Lake Erie (North Shore) phytoplankton: Cumulative impacts of phosphorus loading reductions and the zebra mussel introduction." *J. Great Lakes Res.* **19**: 637-647.

Nicholls, N. (1997). "Increased Australian wheat yield due to recent climate trends." *Nature* **387**: 484-485.

Nieuwenhuyse, E. E. v. and J. R. Jones (1996). "Phosphorus - chlorophyll relationships in temperate streams and its variation with stream catchments area." *Can. J. Fish. Aquat. Sci.* **53**: 99-105.

Nilsson, C., R. Jansson and U. Zinko. (1997). " Long-term responses of river-margin vegetation to water level regulation." *Science* **276**: 798-800.

Nilsson, C. and P. Keddy (1988). "Predictability of change in shoreline vegetation in a hydroelectric reservoir, Northern Sweden." *Can J. Fish: Aquat. Sci.* **45**: 1896-1904.

Ninan, K. N. and S. Lakshmikanthamma (2001). "Social cost-benefit analysis of a watershed development project in Karnataka, India." *Ambio* **30**(3): 157-161.

Nordahl, E. (2000). Environmental costs of wind mill parks on Smøla, Norway (In Norwegian). *Institute for economy and social sciences*. Ås, Norways Agricultural University.

Novak, M. and K. Sigmund (1993). "A strategy of win-stay, loose-shift that outperforms tit-for-tat in the Prisoners dilemma game." *Nature* **364**: 56-58.

Nowak, M., A., R. M. May and K. Sigmund (1995). "The arithmetics of mutual help." *Scientific American* **June**: 50-55.

Nowak, M. A. and K. Sigmund (1992). "Tit for tat in heterogeneous populations." *Nature* **364**: 56-58.

Nowak, M. A. and K. Sigmund (2004). "Evolutionary dynamics of biological games." *Science* **303**: 793-799.

NS (1999). Norways Official Statistics. Oslo, Statistics Norway.

O'Neill (1997). "Managing without prices.The monetary valuation of biodiversity." *Ambio* **26**(8): 546-550.

O'Riordan, T., Ed. (1995). *Environmental science for environmental management.* Harlow, Longman scientific and technical.

Oberdorff, T., J.-F. Guégan and B. Hugueny (1995). "Global scale patterns of fish species richness in rivers." *WEcography* **18**: 345-352.

Oberdorff, T., D. Pont, B. Hugueny and D. Chessel (2000). "A probabilistic model characterizing fish assemblages of French rivers: A framework for environmental assessment." *Freshwater Ecology* **46**: 399-415.

OECD (1982). *Eutrophication of waters. Monitoring, assessment and control.* Paris., OECD.

Oppenheimer, M. (1998). "Global warming and the stability of the West Antarctic Ice Sheet." *Nature* **393**: 325-318.

Opschoor, J. B. (1986). *The benefits of environmental policy and decision-making,.* OECD Workshop on the benefits of environmental policy and decision-making., Avignon, France.

Østbye, E. and Mysterud (1980). Reeindeer ecology. Oslo, University of Oslo.

Ostrom, E. (2000). "Crowding out citizenship." *Scandinavian Political Studies* **23**(1): 3-16.

Ostrom, E., J. Burger, C. B. Field, R. B. Norgaard and D. Policansky (1999). "Revisiting the commons: local lessons, global challenges." *Science* **284**: 278-282.

Padmore, C. L., M. D. Newson and E. Charlton (1998). Instream habitat in gravel-bed rivers: Identification and characterization of biotopes. *Gravel- bed rivers in the environment.* P. L. Klingeman, R. L. Beschta, P. D. Komar and J. B. Bradly. Englewood, Colorado, Water Resources Publications: 345-364.

Palmer, M. A., C. Swan, M., K. Nelson, P. Silver and R. Alvestad (2000). "Streambed landscapes: evidence that stream invertebrates respond to the type and spatial arrangement of patches." *Landscape ecology* **15**: 563-576.

Palmini, D. (1999). "Uncertainty, risk aversion, and the game theoretical foundations of the safe minimum standard: A reassessment." *Ecological economics* **29**: 463-472.

Parry, M., N. Arnell, M. Hulme, R. Nicholls and M. Livermore (1998). "Adapting to the inevitable." *Nature* **395**: 741 (only).

Pauly, D. and V. Christensen (1995). "Primary production required to sustain global fisheries." *Nature* **374**(6519): 255-257.

Pearce, D. W., W. R. Cline, A. N. Achanta, S. Frankhauser, R. K. Pachauri, R. S. J. Tol and P. Vellinga (1996). The social costs of climate change: Greenhouse damage and the benefits of control. *Environmental and social dimensions of climate change. IPPC Second assessment report.* IPPC. Cambridge, Cambridge University Press: 183-224.

Pearce, D. W. and Turner (1990). *Economics of natural resources and the environment.* New York., Harvester Wheatsheaf.

Pelley, J. (2000). "US report strengthens scientists' call for a critical load approach to controlling acid rain." *Environmental science and technology* **June 1**: 248A (only).

Perman, R., Y. Y. Ma and J. McGilveray (1996). *Natural resources & environmental economics*, Longman.

Peters, R. H. and J. V. Raelson (1984). "Relationship between individual size and mammalian population density." *Am. Nat* **104**: 498-517.

Peters, R. H. (1991). *A critique for ecology*. Cambridge, Cambridge University Press.

Peterson, D. L. and K. C. Ryan (1986). "Modeling postfire mortality for long-range planning." *Environ. Management* **10**: 797-808.

Pilon, P. J., R. Condie and K. D. Harvey (1985). Consolidated frequency package CFA User manual for version 1 - Dec pro series. Ottawa, Ontario, K1A 0E7., Water Resources branch, Inland Water Directorate, Environment Canada.

Pimentel, D. (1996). "Green revolution and use of chemicals." *The Science of the Total Environment*. **188 Suppl.**: 86-100.

Pimentel, D. and E. Skidmore (1999). "Rates of soil erosion." *Science* **286**(5444): 1477.

Pimentel, D., C. Wilson, C. McCullum, R. Huang, P. Dwen, J. Flack, Q. Tran, T. Saltman and B. Cliff (1997). "Economic and environmental benefits of biodiversity." *BIOSCIENCE* **47**(11): 747-757.

Pimentel, D. C., P. Harvey, K. Resosudarmo, D. Sinclair, M. Kurz, S. McNair, L. Crist, R. Shpritz, Saffouri and R. Blair. (1995). "Environmental and economic costs of soil erosion and conservation benefits." *Science* **267**: 1117- 1123.

Pimm, S. L. and P. Raven (2000). "Extinction by numbers." *Nature* **403**: 843-845.

Pont, D., O. T., H. B. and P. J.P. (2000). *Modelling ecological requirement of European fishes: the importance of thermal conditions and large-scale effects*. Freshwater Biological Association Annual Scientific Meeting., Birmingham, United Kingdom.

Porter, J., R. and M. Semenov, A. (1999). "Climate variability and crop yields in Europe." *Nature* **400**: 724.

Porteus, J. D. (1996). *Environmental aesthetics: Ideas, politics and planning*. London, Rutledge.

Possingham, H. (2000). "Preserving diversity the U.S. way." *Science* **288**: 983 (only).

Postel, S. L., G.C. Daily and P. R. Ehrlic. (1996). "Human appropriation of renewable fresh water." *Science* **271**: 785-788.

Poulos, C. and D. Whittington (2000). "Time preferences for life-saving programs: Evidence from six less developed countries." *Environmental Science and Technology* **34**(8): 1445-1455.

Power, M. and L. McCarty (1997). "Fallacies in ecological risk assessment practices." *Environmental Science and Technology* **31**(8): A370-A37.

Praire, Y. T., C. M. Duarte and J. Kalff. (1989). "Unifying nutrient- chlorophyll relationships in lakes." *Can. J. Fish. Aquat. Sci.* **46**: 1176-1182.

Preiser, W., F.E. and K. P. Rohane (1988). A survey of aesthetic controls in English-speaking countries. *Environmental aesthetics*. J. L. Nasar. Cambridge, Cambridge University Press: 422-433.

Prendergast, J. (1997). Applying concepts to cases: four African studies. *Building peace. Sustainable reconciliation in divided societies*. J. P. Lederach. Washington D.C., United states institute of peace press: 153-180.

Prepas, E. E., B. Pinel-Alloul, P. A. Chambers, T. Murphy, P., S. Reedyk, G. Sandland and M. Serediak (2001). "Lime treatment and its effect on the chemistry and biota of hardwater, eutrophic lakes." *Freshwater biology* ***To be**.

Purvis, A., P.-M. Agapow, J. L. Gittleman and G. M. Mace (2000). "Nonrandom extinction and the loss of evolutionary history." *Science* **288**: 328-330.

Pyne, S. (2002). "Small particles add up to big disease risk." *Science* **295**: 1994 (only).

Rajpurohit, K. S. (1999). "Child lifting: Wolfes in Hazaribagh, India." *Ambio* **28**: 162-166.

Rawls, J. (1993). *Political liberalism*. New York, Columbia university press.

Regan, H., M., R. Lupia, A. N. Drinnan and M. Burgman, A. (2001). "The currency and tempo of extinction." *Am. Nat.* **157**(1): 1-10.

Reich, P. B., M. G. Tjoelker, J. L. Machado and J. Oleksyn (2006). "Universal scaling of respiratory metabolism, size and nitrogen in plants." *Nature* **439**(7075): 457-461.

Rennings, K. and H. Wiggering (1997). "Steps towards indicators of sustainable development: Linking economic and ecological concepts." *ECOLOGICAL ECONOMICS* **20**(1): 25-36.

Renskaug (1998). Unit pollution values. *Environmental technology*. Porsgrunn, Høgskolen i Telemark.

Reynolds, C. J. (1984). "Phytoplankton periodicity: the interactions of form, function and environmental variability." *Freshwater biology 1* **14**: 111-142.

Rice, R. E., R. E. Gullison and J. W. Reid. (1997). "Can sustainable management save tropical forests?" *Scientific American* **April**: 34-39.

Richardson, C. J. and S. S. Quian. (1999). " Long-term phosphorus assimilative capacity in freshwater wetlands: a new paradigm for sustaining ecosystem structure and function." *Environ. Science & Tchnol.* **33**: 1546-1551.

Ricklefs, R. E. (1990). *Ecology, 3'rd ed.* New York., Freeman and company.

Robinson, J. G. and E. L. Bennett (1999). Carrying capacity limits to sustainable hunting in tropical forests. *Hunting sustainability in tropical forests.* J. G. Robinson and E. L. Bennett. Columbia, Columbia university press.

Robinson, J. G., K. H. Redford and E. L. Bennett (1999). "Wildlife harvest in logged tropical forests." *Science* **284**: 595- 596.

Robock, A. (2002). "The climatic aftermath." *Science* **295**(1242-1244).

Rodan, B. D., D.W. Pennington, N. Eckley and R. S. Boethling (1999). "Screening for persistent organic pollutants: techniques to provide a scientific basis for POP´s criteria in international negotiations." *Environmental Science and Technology* **33**: 3482-3488.

Rodrígues, J. P., G. Ashenfelter, F. Rojas-Suárez, J. J. Garcia, F.L. Súarez and A. P. Dobson (2000). "Local data are vital to worldwide conservation." *Nature* **403**: 241 (only).

Romm, J., M. Levine, M. Brown and E. Petersen (1998). "A road map for U.S. carbon reductions." *Science* **279**: 669-670.

Rosenblum, J., A. Horvath and C. Hendrickson (2000). "Environmental implications of service industries." *Environmental Science and Technology* **34**: 4669-4676.

Rosenzweig, M., L. (2001). "Loss of speciation rate will impoverish future diversity." *PNAS* **98**(10): 5404-5410.

Rosenzweig, M. L. (1999). "Heeding the warning in biodiversity's basic law." *Science* **284**: 276-277.

Roughgarden, J. (1997). *Theory of population genetics and evolutionary ecology: an introduction*. New York, Macmillan publishing Co.

Ruthenberg, H. (1980). *Farming systems in the trophics (third edition)*. Oxford, Claredon Press.

Rutherford, S., S. D´Hondt, D. and W. Prell (1999). "Environmental controls on the geographic distribution of zooplankton diversity." *Nature* **409**(749-753).

Saitoh, T., N. C. Stenseth and O. Bjørnstad, N. (1998). "The population dynamics of the vole Clethrionomys rufocanus in Hokkaido, Japan." *Res. Popul. Ecol.* **40**(1): 61-76.

Sakamoto, M. (1966). "Primary production by phytoplankton community in some Japanese lakes and its dependence on depth." *Arch. Hydrobiol.* **62**: 1-28.

Saleh, M. A. E. (2000). "Value assessment of cultural landscapes in Alckas settlement, Soutwestern Saudi Arabia." Ambio **29**: 60-66.

Sanchez, P. A. and M. S. Swaminathan (2005). "Cutting world hunger in half." *Science* **307**: 357-359.

Sandvik, G., C. Jessup, K. L. Seip and B. J. M. Bohannan (2004). "Using the angle frequency method to detect signals of competition and predation in experimental time series." *ECOLOGY LETTERS* **7**(8): 640-652.

Scheffer, M., S. H. Hosper, M.-L. Meijer, B. Moss and E. Jeppesen (1993). "Alternative equilibria in shallow lakes." *Trends in ecology and evolution* **8**(8): 275-279.

Scholin, C. A., F. Gulland, G. J. Doucette, S. Benson, M. Busman, F. P. Chavez, J. Cordaro, R. DeLong, A. De Vogelaere, J. Harvey, M. Haulena, K. Lefebvre, T. Lipscomb, S. Loscutoff, L. J. Lowenstine, R. Marin, P. E. Miller, W. A. McLellan, P. D. R. Moeller, C. L. Powell, T. Rowles, P. Silvagni. M. Silver, T. Spraker, V. Trainer and F. M. Van Dolah (2000). "Mortality to sea lions along the central California coast linked to toxic diatom bloom." *Nature* **403**: 80-84.

Schreve, R.-L. (1966). "Statistical law of stream numbers." *Journal of Geology* **74**: 17-37.

Schulman, L., K. Ruokolainen and H. Tuomisto (1999). "Parameters for global ecosystem models." *Nature* **399**: 535-536.

Schultz, W. (1986). *The benefits of environmental policy and decision-making,*. Workshop on the benefits of environmental policy and decision-making, Avignon, France.

Scorer, R. S. (1990). *Metereology of air pollution*. New York, Ellis Horwood.

Scott, J. M. (1999). "Lynx reintroduction." *Science* **286**: 49 (only).

Seip, H. K. (1996). *Forestry for human development*. Oslo, Scandinavian University Press.

Seip, K. L. (1980). "A computational model for the growth and harvesting of the marine algae Ascophyllum Nodosum." *Ecol. Modelling* **8**: 189-199.

Seip, K. L. (1984). "The Amoco Cadiz Oil Spill - At a glance." *Mar. poll. Bull.* **15**: 218-220.

Seip, K. L. (1991). Decisions with multiple environmental objectives. The siting of oil drilling wells in Norway. *Water resources engineering risk assessment*. J. Ganoulis. Berlin., Springer-Verlag.

Seip, K. L. (1994). "Phosphorus and nitrogen limitation of algal biomass across trophic gradients." *Aquatic Sciences* **56**: 16-28.

Seip, K. L., H. Betele and K. Johnsen (2000). "Siting of paper mills: is a pristine environment an industrial resource?" *Environmental Science and Technology* **34**: 546-551.

Seip, K. L., E. Sandersen, F. Mehlum and J. Ryssdal (1991). "Damages to seabirds from oil spills: comparing simulation results and vulnerability indexes." *Ecol. Modelling* **53**: 39-59.

Seip, K. L. and H. Ibrekk (1987). "Regression equations for lake management-how far do they go?" *Verh. Internat. Verein. limnol.* **23**: 778-785.

Seip, K. L. and P.Botterweg (1988). *Sediment yield/Surface runoff Models (modellers perspective)*. International Symposium on Water Quality Modeling of Agricultural Non Point Sources., Utah State University, Utah.

Seip, K. L. and C. S. Reynolds (1995). "Phytoplankton functional attributes along trophic gradient and season." *Limnology and Oceanography*, **40**: 589-597.

Seip, K. L., H. Sas and S. Vermij (1991). "The ecosystem of a mesotrophic lake-I. Simulating plankton biomass and the timing of phytoplankton blooms." *Aquatic Sciences* **53**: 239-262.

Seip, K. L., H. Sas and S. Vermij (1992). "Changes in Secchi disk depth with eutrophication." *Archiv für Hydrobiologie* **124**: 149-165.

Selden, T. M. and D. Song (1995). "Neoclassical growth, the J curve for abatement, and the inverted U curve for pollution." *J Environ. Economics and management* **29**: 162-168.

Sen, A. (1995). "Rationality and social choice." *The American economic Review* **85**(1): 1-24.

Shih-táo (1997). *Quotes on painting*. Oxford, Blackwell.

Short, C. (2000). "Common land and ELMS: a need for policy innovation in England and Wales." *Land use policy* **17**: 121-133.

Short, C. and M. Winter (1999). "The problem of common land: Towards stakeholder governance." *J. Environmental Planning and Management* **42**(5): 613-630.

Shugart, H., H. (1998). *Terrestrial ecosystems in changing environments*. Cambridge, Cambridge university press.

Shuter, B. J., M. L. Jones, R. M. Korver and N. P. Lester (1998). "A general, life history based model for regional management of fish stocks: the inland lake trout (Salvelinus namaycush) fisheries of Ontario." *Can J. Fish: Aquat. Sci.* **55**: 2161-2177.

Siegel, S. and J. Castellan, N. John (1988). *Nonparametric systems for behavioural sciences*, McGraw Hill.

Sigmund, K. and M. A. Nowak (2000). "Playing for keeps." *Science* **290**(5490): 281.

SimaPro (2003). LCA program, SimaPro, http://www.pre.nl/.

Simon, H. (1982). Theories of bounded rationality. *Models of Bounded Rationality: Behavior Economics and Business Organization*. H. A. Simon. Cambridge, Mass, MIT Press. **2**: 408-423.

Simon, H. (1997). *Administrative behavior*. New York, The free press.

Simons, P. and L. Wallace (1995). "A comment on Klieve- MacAuley's southern bluefinn tuna game." *Austr. J. Agric. Econom.* **39**: 289-291.

Skonhoft, A. and H. Solem (2001). "Economic growth and land-use changes: the declining amount of wilderness land in Norway." *Ecol. Econ* **37**: 289-301.

Skulberg, O. M., C. A. Codd and W. W. Carmichal. (1984). "Toxic blue-green algae blooms in Europe: a growing problem." *Ambio* **13**: 244-247.

Smaling, E. M., A. L. O. Fresco and A. de Jager (1996). "Classifying, monitoring and improving soil nutrient stocks and flows in African Agriculture." *Ambio* **25**: 492-496.

Smith, D. W. (1997). "A critical review of the benefits analysis for the Great Lakes initiative." *Env. Sci. Technol.* **31**: 34A-38A.

Smith, V. and J. Shapiro (1981). "Chlorophyll-phosphorus relations in individual lakes. Their importance to lake restoration strategies." *Environmental Science and Technology* **15**: 444-451.

Smith, V. A., G., D. Tilman and J. C. Nekola (1999). "Eutrophication: impacts of excess nutrient inputs on freshwater, marine, and terrestrial ecosystems." *Env. Poll.* **100**: 179-196.

Smith, V. H. (2003). "Eutrophication of freshwater and coastal marine ecosystems - A global problem." *Environmental Science and Pollution Research* **10**(2): 126-139.

Smith, V. K. and J.-C. Huang (1995). "Can Markets Value Air Quality? A Meta-Analysis of Hedonic Property Value Models." *Journal of Political Economy* **103**(1): 210-227.

Soulé, M. E. (1982). Genetic aspects of ecosystem conservation. *Coservation of ecosystems: Theory and practice.* W. R. Siegfrid and B. R. Davies. Pretoria, South African Natl. Sci. Prog. Rep. 61 CSIR: 34-45.

Soulé, M. E. and M. A. Sanjayan (1998). "Conservation targets. Do they help?" *Science* **279**: 2060-2061.

Sparling, G., L. Lilburne and M. Vojvodi´c-Vukovi´c (2003). *Provisional targets for soil quality indicators in New Zealand.* Palmerstone North, Landcare research New Zealand.

Stamps III, A., E. (2004). "Mystery, complexity, legibility and coherence: A meta-analysis." *Journal of environmental psyhology* **24**: 1-16.

Stark, C. P. and M. Stieglitz (2000). "The sting in a fractal tail." *Nature* **403**: 493-495.

Statzner, B. and F. Sperling (1993). "Potential contribution of system-specific knowledge (SSK) to stream-management decisions: ecological and economic aspects." *Freshwater biology* **29**: 313-342.

Stenseth, N. C., O. N. Bjørnstad and T. Saitoh (1998). "Seasonal forcings on the dynamics of Clethrionomysrufocanus: Modelling geographical gradients in population dynamics." *Res. Popul. Ecol.* **40**(1): 85-95.

Sterner, T. (2003). Policy Instruments for Environmental and Natural Resource Management. Washington, Resources for the Future.

Stevens, C., J., N. B. Dise, J. O. Mountford and D. Gowing, J. (2004). "Impact of nirtrogen deposition on the species richness of grasslands." *Science* **303**: 1876-1879.

Stigum, B. and F. Wenstøp, Eds. (1987). *Foundation of utility and risk theory with applications.* Dordrecht, Reidel publishing Company.

Streeter, H. W. and E. B. Phelps (1925). "A study of the pollution and natural purification of the Ohio river." *US Public Health Bulletin* **146**.

Stum, W. and J. J. Morgan (1996). *Aquatic chemistry: Chemical equilibria and rates in natural waters.* New York, John Wiley.

Staats, H., A. Kieviet and T. Hartig (2003). "Where to recover from attentional fatigue: An expectancy-value analysis of environmental preference." *Journal of environmental psychology* **23**: 147-157.

Sugihara, G. and R. M. May (1990). "Nonlinear forecasting as a way of distinguishing chaos from measurement errors in time series." *Nature* **344**: 731-741.

Swithart, M. M. and C. Petrich (1988). "Assessing the aesthetic importance of small hydropower development." *The environmental professional* **10**: 198-210.

Syers, J., J. Lingard, C. Pieri, E. Ezcurra and G. Faure (1996). "Sustainable land management for the semiarid and sub-humid tropics." *Ambio* **25**(8): 484-491.

Saaty, T. L. and L. Vargas, G. (1994). *Decision making in economic, political, social and technological environments*. Pittsburg, Universtity of Pittsburg.

Teisl, M., F., B. Roe and R. L. Hicks (2002). "Can Eco-labels tune a market? Evidence from dolphin safe labeling." *J Environ. Economics and management* **43**(339-359).

Tellus (1991). Disposal cost fee study. 89 Broad street, MA 02110, USA, Tellus institute.

Tian, H. and e. al. (1998). "Terrestrial Ecosystem Model." *Nature* **396**: 664-667.

Tientenberg, T. H. (1992). *Environmental and natural resource economics. Third edition*. New York, Harper Collin publishers.

Tilley, D. R. and M. T. Brown (1998). "Wetland networks for stormwater management in subtropical urban watersheds." *Ecological Engineering* **10**: 131-158.

Tomkins, S. S. (1991). *Affect, Imagery, Conciousness. Volume III. Anger and fear*. New York, Springer Publishing Company.

Tomter, S. M. (1994). Statistics of forest conditions and resources in Norway. Ås, Norwegian institute of land inventory: 103.

Trimbee, A. M. and E. E. Prepas (1987). "Evaluation of total phosphorus as a predictor of the relative biomass of blue-green algae with emphasis on Alberta lakes." *Can. J. Fish. Aquat. Sci.* **44**: 1337-1342.

Trimble, S. W. (1999). "Response to Pimentel and Skidmore." *Science* **286**: 1477-1478.

Troeh, F. R., J. A. Hobbs and R. L. Donahue (1991). *Soil and water conservation*. Englewood Cliffs, N.J., Prentice Hall.

Tupper, M., H. (2002). "Marine reserves and fisheries management." *Science* **295**: 1233 (only).

Turner, R. K., S. georgiou, I.-M. Gren, F. Wulff, S. Barrett, T. Söderqvist, I. J. Bateman, C. Folke, S. Langaas, T. Zylicz, K.-G. Mäler and A. Markowska (1999). "Managing nutrient fluxes and pollutuion in the Baltic: an interdisiplinary simulation study." *Ecological economics* **30**: 333-352.

Tutin, T. G., V. H. Heywood, N. A. Burges, D. M. Moore, D. H. Valentine, S. M. Walters and D. A. Webb (1976). *Flora Europaea*. Cambridge, Cambridge University Press.

Tømte, O., K. L. Seip and N. Christophersen (1998). "Evidence that loss in predictability (and possibly dynamic chaos) increase with increasing trophic level in aquatic ecosystems." *Oikos* **82**: 325-332.

U.S. Army Corps of Engineers (1980). A habitat evaluation system for water resource planning. Vicksburg, Miss, U.S. Army Corps of Engineers.

UK. (2000). "A better quality of life." 2000, from http://www.defra.gov.uk/environment/sustainable/quality99/pdf/chap3.pdf.

UN (2000). Human development report 2000. New York, United Nations.

UNESCO (1972). Convention concerning the protection of the world cultural and natural heritage. Adopted by the general conference at its seventeenth session, 16 November 1972. Paris, UNESCO.

UNI/ISPRI (1991). "Global assessment of Soil degradation."

UNSCEAR (2000). Sources and effects of inonizing radiation. New York, The Genera Assembly, United Nations.

Van Jaarsveld, A., S. Freitag, S. Chown, C. Muller, S. Koch, H. Hull, C. Bellamy, M. Kruger, Endrody, S. Younga, M. Mansell and C. Scholtz (1998). "Biodiversity assessment and conservation strategies." *Science* **279**: 2106-2108.

van Praag, B. M. (1993). *The relativity of the welfare concept*. Oxford, Claredon press.

Vannote, R. L., G. W. Minshall, K. W. Cummins, J. R. Schell and C. E. Cushing (1980). "The river continuum concept." *Can J. Fish. Aquat. Sci.* **37**: 130-137.

Vollenweider, R. A. (1976). "Advances in defining critical loading levels for phosphorus in lake eutrophication." *Mem Ist. Ital.Idrobiol.* **33**: 53-83.

von Gierke, H. E. (1977). Guidelines for preparing environmental impact statements on noise. Washington, D.C., National research council.

von Winterfeldt, D. and W. Edwards (1986). *Decision analysis and behavioural research*. Cambridge, Cambridge university press.

Vought, L. B.-M., J. Dahl, L. Pedersen and J. O. Lacourière. (1994). " Nutrient retention in riperian ecotones." *Ambio* **23**: 342 - 347.

Vrana, S. R. and D. Gross (2004). "Reactions to facial expressions: effects of social context and speech anxiety on responses to neutral, anger, and joy expressions." *Biological Psychology* **66**(1): 63-78.

Wackernagel, M., M. Lewan, L. and C. B. Hansson (1999). "Evaluating the use of natural capital with ecological footprint. Applications in Sweden and subregions." *Ambio* **28**: 604-611.

Wackernagel, M., L. Onisto, P. Bello, A. C. Linares, I. S. L. Falfan, J. M. Garcia, A. I. S. Guerrero and M. G. S. Guerrero (1999). "National natural capital accounting with the ecological footprint concept." *Ecolo. economics* **29**: 375-390.

Wackernagel, M. and W. Rees (1995). *Our Ecological Footprints: Reducing human impacts on earth*. Philadelphia, New Society publisher.

Wada, Y. (1993). The appropriated carrying capacity of tomato production: The ecological footprint of hydroponic greenhouse versus mechanized open field operations. *School of Community and Regional Planning*. Vancouver, University of British Columbia.

Waite, S. (2000). *Statistical ecology in practice*. Harlow, Prentice Hall.

Wang, X. and K. Smith, R. (1998). Near-term health benefits of greenhouse gas reductions: A proposed method assessment and application to China. Geneva, Office of integrated and global environmental health. WHO.

Warren-Rhodes, K. and A. Koenig (2001). "Escalating trends in the urban metabolism of Hong Kong: 1971-1997." *Ambio* **30**(7): 429-438.

Watkiss, P., S. Baggot, T. Bush, S. Cross, J. Goodwin, M. Holland, F. Hurley, A. Hunt, G. Jones, S. Kollamthodi, T. Murrells, J. Stedman and K. Vincent (2004). "An evaluation of the air quality strategy."

Weber, M., F. Eisenfuhr and D. von Winterfeldt (1988). "The effects of splitting attributes on weights in multiattribute utility measurement." *Management Science* **34**: 431-445.

Weber, M. G. and B. J. Stocks (1998). "Forest fires and sustainability in the boreal forests of Canada" *Ambio* **27**(7): 545-550.

Welch, E. (1980). *Ecological effects of waste water*. Cambridge, Cambridge University Press.

Wellman, J. D. and G. J. Buhyoff (1980). "Effects of regional familiarity on landscape preferences." *Journal of environmental management* **11**: 105-110.

Wenstøp, F. and A. J. Carlsen (1998). Using decision panels to evaluate hydropower development projects. *Multi criteria evaluation in land use management*. E. Beinat and P. Nijkamp. Dordrecht, Kluwer academic publishers.

Wenstøp, F. and K. Seip (2001). "Legitimacy and quality of multi-criteria environmental policy analysis: A meta analysis of five MCE studies in Norway." *Journal of multi-criteria decision analysis* **10**: 53-64.

Weterings, R. A. P. M. and J. B. Opschoor (1992). The ecocapacity as a challenge to technological development. Rijswijk, Advisory council for research and nature environment.

Wieriks, K. and Schulte-Wülwer-Leidig (1997). "Integrated water management for the Rhone river basin, from pollution prevention to ecosystem improvement." *Natural Resources Forum* **21**(2): 147-156.

Wigley, T. M. and S. Raper (2001). "Interpretation of high projections for Global-Mean Warming." *Science* **293**: 451-452.

Wilson, E. M. (1983). *Engineering hydrology*. London, The Macmillan press, Ltd.

Windolf, J. (1997). Retention of nutrients in river basins. A. Grimwal. Linkøping, University of Linkøping.

Winiwarter, W., H. Haberl and D. Simpson (1999). "On the boundary between man-made and natural emissions:Problems in defining European ecosystems." *Journal of geophysical research* **104**(D7): 8153-8159.

Wischmeier, W. H. and D. D. Smith (1978). Predicting rainfall erosion losses. Washington, D.C., USDA Agricultural research services handbook.

Wotawa, G. and M. Trainer (2000). "The influence of canadian forest fires on pollutant concentrations in theUnited States." *Science* **288**: 324-328.

WRI (1994). World resources 1994-1995. A report by the World Resource Institute in collaoboration with The United Nations Environmental programme and the United Nations development programme. Oxford, Oxford University Press.

Wu, J., J. Huang, X. Han, Z. Xie and X. Gao (2003). "Three-Gorges Dam-experiment in habitat fagmentation?" *Science* **300**: 12391240.

Wuebbles, D. J., A. Jain, J. Edmonds, D. Harvey and K. Hayhoe (1999). "Global hange: State of the sience." *Environmental pollution* **100**: 57-86.

Yin, C. and B. Shan (2001). "Multipond systems: a sustainable way to control diffuse phosphorus pollution." *Ambio* **30**(6): 369-375.

INDEX

R

S